64-94 BK Bud July 94

NORTH CAROLINA
STATE BOARD OF COMMUNITY COLLEGES
LIBRARIES
SOUTHEASTERN COMMUNITY COLLEGE

For Reference
Not to be taken from this room

SOUTHEASTERN COMMUNITY
COLLEGE LIBRARY
WHITEVILLE, NC 28472

PEOPLES AND CULTURES

Ref
G
128
.P46
1992

PEOPLES AND CULTURES

GENERAL EDITOR

Alisdair Rogers

SOUTHEASTERN COMMUNITY
COLLEGE LIBRARY
WHITEVILLE, NC 28472

For Reference
Not to be taken from this room

New York
OXFORD UNIVERSITY PRESS
1992

CONSULTANT EDITOR
Professor Peter Haggett, University of Bristol

Dr Sheena Asthana, Exeter University, UK
The Indian Subcontinent

Noel Castree, University of British Columbia, Vancouver, Canada
Southern Africa

Dr David Cleary, University of Edinburgh, UK
South America

Professor T.H. Elkins, Oxford, UK
Central Europe, France and its neighbors

Dr Kevin Frawley, Canberra, Australia
Australasia, Oceania and Antarctica

Dr Roger Goodman, University of Essex, UK
Japan and Korea

St. John B. Gould, London, UK
The Middle East

Dr Sarah Harper, Royal Holloway and Bedford New College, University of London, UK
China and its neighbors

George Joffé, London, UK
Northern Africa

Professor David Ley, University of British Columbia, Vancouver, Canada
Canada and the Arctic

Finn Rindom Madsen, Institute of Ethnography and Social Anthropology, Aarhus, Denmark
The Nordic Countries

Dr Nick Middleton, University of Oxford, UK
Central Africa

Dr Michael Pincombe, University of Newcastle Upon Tyne, UK
The Low Countries

Dr Jonathan Rigg, School of Oriental and African Studies, London, UK
Southeast Asia

Dr Alisdair Rogers, University of Oxford, UK
Peoples in the World, USA, The British Isles

Dr Andrew Ryder, Portsmouth Polytechnic, UK
Eastern Europe, USSR

Dr Alwyn Scarth, University of Dundee, UK
Italy and Greece

Dr Steven Vertovec, University of Oxford, UK
Central America and the Caribbean

Dr Mary Vincent, University of Sheffield, UK
Spain and Portugal

AN EQUINOX BOOK

Copyright © Andromeda Oxford Limited 1992

Planned and produced by
Andromeda Oxford Limited
11-15 The Vineyard, Abingdon
Oxfordshire, England OX14 3PX

Published in the United States of America by
Oxford University Press, Inc.,
200 Madison Avenue,
New York, N.Y. 10016

Oxford is a registered trademark of
Oxford University Press

Library of Congress
Cataloging-in-Publication Data

Peoples and cultures / edited by Alisdair Rogers.
p. cm.
"An Equinox book"--Verso t.p.
Includes bibliographical references and index.
ISBN 0-19-520928-1
1. Geography. I. Rogers, Alisdair.
G128.P46 1992
304.2--dc20 91-37909
CIP

Editor Victoria Egan
Designer Jerry Goldie
Cartographic manager Olive Pearson
Picture research manager Alison Renney
Picture researcher David Pratt

Project and volume editor Susan Kennedy
Art editor Steve McCurdy

All rights reserved. No part of this publication may be reproduced, stored in a retrieval system or transmitted in any form or by any means, electronic, mechanical, photocopying, recording or otherwise, without the permission of the publishers and copyright holder.

ISBN 0-19-520928-1

Printing (last digit): 9 8 7 6 5 4 3 2 1

Printed in Spain by Heraclio Fournier SA, Vitoria

INTRODUCTORY PHOTOGRAPHS
Half title: *Albanians in Macedonia, Yugoslavia (Magnum Photos, Steve McCurry)*
Half title verso: *Carnival dancers, Trinidad (Explorer/M. Moisnard)*
Title page: *Hunza woman, Pakistan (Christine Osborne)*
This page: *Punk youths (Robert Harding Picture Library)*

Contents

PREFACE

WE LIVE IN A WORLD OF RAPID SOCIAL AND CULTURAL CHANGE, IN which traditional peoples are fighting to preserve their ways of life while the inhabitants of the industrial world have access to an unprecedented variety of cultural experiences via their television screens and tourist holidays. Ideas and commodities are swept into an international maelstrom of circulation and exchange, while millions of individuals have to choose between staying in the places of their childhood or throwing themselves into the world's migration streams to secure their livelihoods.

The aim of this volume is to explore the diversity of humankind's social and cultural organization and account for the ways that people make sense of their world and locate themselves within it. It examines the perceptions of the natural world that underlie the use of resources and the transformation of the natural environment. How do people view themselves in relation to their physical landscapes and how have they adapted their ways of life to them? What cultural landscapes have they produced and how have they filled them with meanings and associations?

The historical processes of migration, conquest, and the mixing of peoples that has created today's cultural mosaics are identified. Few societies are culturally uniform, and in each the forces of social differentiation compete with the attempts by governments and dominant groups to forge common national identities. How do the passions of religion, language and ethnicity become translated into the political conflicts of the modern world? How does faith provide a sense of identity in a world increasingly dominated by the values of money? Can community still provide a sense of belonging in a world increasingly dominated by global processes?

Finally, the tensions between modernization and tradition are considered. What impact does urbanization have on family life and the changing relationships between women and men, for example, and what effect does state control have on the lives of pastoral nomads, hunter–gatherers and other traditional cultures?

The world is viewed not as a mosaic of isolated and self-enclosed peoples, but as a place of openness, movement and fascinating juxtapositions and connections. The notion of culture involved in this volume is one that embraces a diversity of meanings and practices, styles and forms, from art to soccer, local to global, and tradition to the newest forms of personal identity and expression. Each of us has to face the possibilities and challenges of placing ourselves in this world, in our families and settlements, our religious and linguistic communities, and our countries. This volume describes the rich variety of such cultural positions and identities.

Dr Alisdair Rogers
SCHOOL OF GEOGRAPHY, UNIVERSITY OF OXFORD

Village elders wearing red for mourning, Ghana

Hindu faithful on the banks of the Ganges at Varanasi, India (*overleaf*)

PEOPLES IN THE WORLD

People in their Environment

EXCEPT FOR THE EXTREME POLAR REGIONS, some deserts and high mountain areas, people inhabit the whole of the Earth's land surface. They do so in greater or lesser numbers and occupy an immense variety of physical environments. It is from the physical environment that humans obtain their means of subsistence: food, water, and material for food and clothing, and consequently it is not surprising that environmental diversity is matched by a range of human culture – the nexus of beliefs, customs and social relations that influence the way groups of people organize themselves. Culture has both mental and material aspects: in changing nature to provide the elements of material culture as such, people also change themselves. This is always a social process. These changes are enabled by and embodied in the systems for communicating the ideas and meanings, in words, rituals, myths and stories, that serve to tie individuals to one another and establish distinct peoples.

Yet variations in human culture are not simple reflections of the physical world. People are neither the products nor the prisoners of nature, not even those nonurbanized, nonindustrialized peoples whose way of life is variously termed indigenous, primitive or tribal. Similar physical environments have not necessarily given rise to the same cultural adaptations. For example, the methods of obtaining subsistence, forms of social organization, systems of belief and cultural practices of the reindeer-herding Sami of the Nordic Countries are very different from those of the Inuit of North America, who survive by hunting, though they both inhabit one of the world's harsher environments, the northern tundra. Furthermore, few if any peoples exist in complete isolation from others. Over the centuries, patterns of movement, of borrowing crops and tools as well as social practices, and of commercial exchange meant that cultures did not develop as insular pockets. Even in the scattered islands of Oceania obsidian blades were being traded great distances across the Pacific Ocean at least 2,500 years ago.

A Tuareg man, one of the Berber pastoral peoples whose way of life is adapted to the arid conditions of the central Sahara of Africa. They are true nomads, perpetually on the move with their herds from one pasture to another. Traditionally they are also caravan traders, and maintain a strictly hierarchical society.

Forms of social organization

There are many different forms of social organization. These are often able to tell us how peoples have adapted to their environments. Hunter–gatherers, once a majority of humankind, now form only a tiny fraction of the world's population. Typically, their social organization consists of small, loose-knit bands, with a marked division of labor between female plant-gatherers and male hunters. This suits a lifestyle based on occasional hunting, and gathering any food plants that are naturally available. Examples of these peoples are the Aborigines of Australia, the San of Southern Africa and the Pygmies of Central Africa.

Pastoral nomads possess more diverse forms of organization. They include the Africa's and Asia's dry regions and the reindeer-herding Sami of the far north of Europe. Some are egalitarian, others strongly stratified, and their social activity commonly adjusts to the seasonal requirements of their animals. Some range over wide areas in search of pastures while others, known as transhumants, follow a regular annual trek making use of long-established rights to lowland and upland pastures.

Settled farmers practice either shifting

SOMALIS AND THEIR HERDS

Somali pastoral nomads live a life that is in some ways typical of all groups that subsist by herding in arid environments. The availability of water and pasture is unpredictable, and both resources are regarded as common property. There is no strong sense of territorial ownership among Somalis who are patrilineal, meaning that property is owned and inherited on the male side of families, and clan-based, meaning that family units identify themselves with distinct groups of descendants. But they adjust their social organization to the seasons.

In the dry season groups of unmarried male kin cover huge distances with their camel herds, while smaller family units, including unmarried daughters, stay with the less mobile herds of sheep and goats. Somali social life therefore increases dramatically in the spring and summer when families come together. At such times the conflicts that erupt over scarce resources in the dry period may be resolved.

Life in the frozen tundra (*above*) offers little in the way of vegetation. The Inuit people of North America live entirely by hunting – both for food and for furs that they trade. They inhabit the coast during the winter months and move inland for the brief summers looking for game and fish.

The Pygmies of Central Africa (*below*) are a typical hunter–gatherer people. Traditionally they live in bands, where the men have total responsibility for hunting for meat while the women collect fruit, seeds and nuts. In some places Pygmies have become farmers and herders in the surrounding savannah.

or permanent cultivation. The former include peoples in Central Africa, Southeast Asia and the rainforests of Central and South America. Their forms of social organization, which often include the periodic breaking up of village units into smaller bands and the creation of new settlements, are flexible enough to prevent the local overexhaustion of soils.

Where the land is more fertile and water from rain or irrigation more plentiful, permanent societies are found. This characterizes not only the specialized agriculturalists of the developed world but also the great rice-based societies of Asia. Among these, the need to organize supplies of water is accompanied by larger forms of social and political organization. In Bali, for example, the island-wide coordination of water supplies is conducted by the network of Hindu temples. Where such societies were productive enough to produce large agricultural surpluses, the world's first cities were formed. The fixed location of such cultivation means that these farmers generally have a stronger sense of territory and territorial boundaries than is found among other ways of life.

Migration, Diffusion and Exchange

THE LONG HISTORY OF WORLDWIDE PATTERNS of migration, diffusion and exchange means that peoples cannot be viewed in isolation within their local environments. The human species, which is at most 250,000 years old, spread from Africa across Eurasia, crossing into the Americas and Australia on land bridges left by the last glacial advance some 75,000 years ago. After rising sea levels cut off these continents, the peoples of the world began to develop specialized ways of life to match their natural environments. The islands of the South Pacific were probably the last places settled, some of them perhaps only 1,000 years ago.

Once the major geological changes were over, movement took the form of exchanges between peoples. New foods and technologies spread across the globe; rice from Southeast Asia to Mediterranean Europe, yams and bananas from Africa to Asia during the first millennium BC. Major migrations displaced existing peoples, as newcomers often possessed new technologies that enabled them to exploit the land more effectively. For example, the Bantu-speaking peoples from central Cameroon carried their iron-working culture across sub-Saharan Africa from 2000 BC onward.

An interlocking world

The world's peoples became linked to each other by systems of local, regional and, eventually, of global trading. New ideas, religions, technologies, luxury commodities and diseases crisscrossed the continents. In Northern and Central Africa for example, the kingdoms of the tropical south swapped gold, slaves and minerals with Arabs from the north in exchange for horses, textiles and salt. The Roman and Chinese empires were linked by a long sea route via India by 100 BC, and by an overland passage known as the Silk Road. The process of global intercommunication was made complete by the European arrival in the Americas in 1492, and in Australasia a little more than a century later.

The rise of Europe from a relative backwater to a position of worldwide mercantile influence brought major changes. Trade links, which had once centered on China, began to find a new focus in Europe. As other peoples were brought into this system of exchange it created an intensely specialized world economic pattern. This generated a new geography of labor supply and demand, which was met by a worldwide redistribution of peoples. Although slaves had

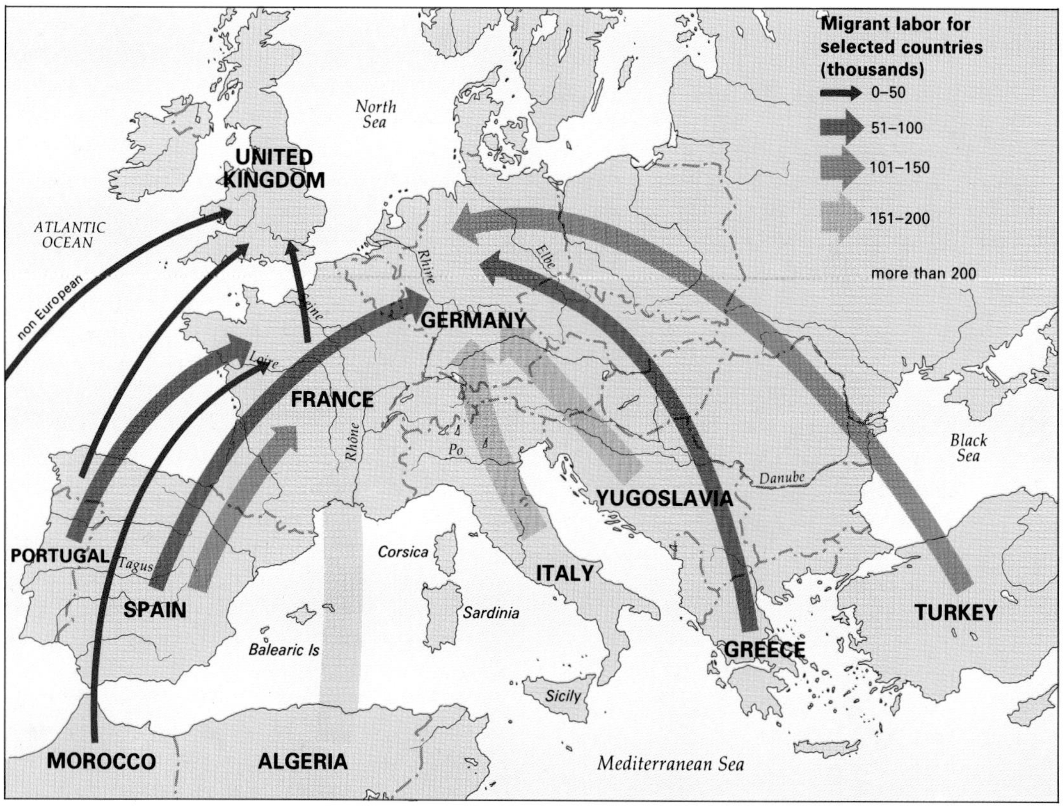

A Hindu temple (*above*) on the island of Mauritius in the Indian Ocean. With the abolition of slavery, large numbers of laborers from India were imported by the ruling British to work on the island's sugar plantations. Over two-thirds of the inhabitants now are Indo-Mauritians and 50 percent of the population practice the Hindu religion, making this temple a monument to a migrant culture.

Migrant labor for selected countries (thousands)
0–50
51–100
101–150
151–200
more than 200

UNITED KINGDOM
FRANCE
GERMANY
PORTUGAL
SPAIN
ITALY
YUGOSLAVIA
GREECE
TURKEY
MOROCCO
ALGERIA
non European
ATLANTIC OCEAN
North Sea
Black Sea
Mediterranean Sea
Corsica
Sardinia
Sicily
Balearic Is
Rhine
Elbe
Loire
Seine
Rhône
Po
Danube
Tagus

Guest workers (*left*) During the boom economic years of the 1960s and 1970s nearly all the industrialized countries of Western Europe received large flows of immigrants to fill labor shortages. In the early years most came from southern Europe and Turkey; later they were drawn increasingly from North Africa. Britain's labor requirements were met by immigrants from its former colonial possessions. This map shows the flow in 1970.

A bill of sale (*right*) announcing the arrival of negro slaves in the American colonies. The slave trade from Africa to North America reached its peak in the late 18th century. The numbers of Africans taken by the trade are estimated at 12.5 million. Many others died resisting seizure, or from disease on board ship, and the total drain of people meant that Africa's population remained static for the next 200 years.

INVOLUNTARY MIGRANTS

With numbers estimated at 14 million or more, and rising all the time, refugees – individuals displaced from their homelands by fear of persecution, war, or natural disaster – far outnumber legal immigrants in the world today. Most are found in Africa, Asia and the Middle East, sheltering in countries adjacent to their own. The more fortunate ones may escape to the safety of stable and wealthy countries – for example 875,000 Vietnamese have been admitted to the United States since the early 1970s. But few countries are keen to admit them any more, and many make it difficult for persons to qualify for asylum.

The United Nations defines a refugee as someone fleeing his or her country for reasons of persecution, or a "well-founded" fear of persecution. But since poverty and persecution so often go together, reluctant governments can easily justify their exclusion by claiming that refugees are simply seeking a richer life, and not necessarily a safer one. From the downtowns of American cities to the banks of the Jordan river and the camps along the Thai-Cambodian border, refugees are people denied one of the most basic of human needs, home.

TO BE SOLD, on board the Ship *Bance-Yland*, on tueſday the 6th of *May* next, at *Aſhley-Ferry*; a choice cargo of about 250 fine healthy

NEGROES,

juſt arrived from the Windward & Rice Coaſt. —The utmoſt care has already been taken, and ſhall be continued, to keep them free from the leaſt danger of being infected with the SMALL-POX, no boat having been on board, and all other communication with people from *Charles-Town* prevented.

Auſtin, Laurens, & Appleby.

N. B. Full one Half of the above Negroes have had the SMALL-POX in their own Country.

previously been taken from many peoples, from 1600 onwards Africa became the major source as first the Portuguese, and later the Dutch and English, transported them to their new colonies in North and South America: during the 17th and 18th centuries over 6 million people were enslaved.

In the 19th century Indians left to labor in the Caribbean, eastern Africa and Southeast Asia, while many Chinese migrated to the United States and Southeast Asia. Rapid economic change, combined with high land rents and low food prices, drove 50 million Europeans abroad in the 19th century. Two-thirds of them were destined for the United States, others for Siberia, South America, Australia and New Zealand.

Migration today

Today, only the United States accepts large numbers of permanent settlers, though Australia, Canada, Israel and New Zealand all receive some. The other affluent countries have generally closed their borders to new citizens, and those who do come are accepted principally as temporary or migrant workers, most of them from the Indian subcontinent, Africa, the Caribbean and Mexico. The major destinations for these migrant workers are the oil-rich Gulf states, some developing African countries such as South Africa, Nigeria and the Ivory Coast, the United States and Western Europe. It is estimated that there are now more than 3 million North Africans working in Europe, a third of them illegally.

Although this international exchange of labor separates families, some migrant workers are able to send money home. These remittances can contribute significantly to the economies of their countries of origin. Once they have earned enough, many immigrants return home and set themselves up as small business owners. Others, however, fail to achieve this, and have to face the difficult decision of whether to settle permanently in their new country and send for their families to live with them.

Conquest and Colonialism

A MAJOR CAUSE OF CULTURAL CHANGE HAS been the conquest of some peoples by others, and the creation of empires. Conquest is an age-old event. Much of the history of Eurasia can be described as a struggle between the settled cultivators of water-rich regions and the mobile pastoralists who followed the natural routeways of desert and steppe to spread throughout the continent. In the 4th and 5th centuries AD, nomadic pastoralists from Central Asia invaded China, Persia, India and parts of the Mediterranean. In the 13th century new waves again invaded Russia, Central Europe and China. During the 15th century the Incas overran almost the entire Pacific coast of South America before falling to conquerors from Spain and Portugal. Only China withstood the subsequent European global onslaught, though it too succumbed following the Opium Wars (1839–42), the result of British attempts to control the lucrative trade in this drug.

The effects of conquest have not always been the same. At the worst they have led to the total eradication of a people; for example, the Arawaks and Caribs of the Caribbean were wiped out by foreign diseases brought by European invaders in the 15th and 16th centuries. Elsewhere entire peoples were removed from the lands occupied by the colonizers or were otherwise completely separated from them. This happened in China, North America and Australia. The British in sub-Saharan Africa segregated the local peoples into specific areas within cities in order to keep their daily labor; South Africa's apartheid policy is in part a legacy of this. Others fought and resisted: the Yaqui in Mexico, the Zulus in southern Africa and the Maori in New Zealand, for example. Lastly, particularly in Central and South America, European and local people intermarried, giving rise to mestizo and creole cultures. It was often from among these groups that the leaders of the subsequent struggle for independence emerged.

The colonial imprint

The impact of European colonialism had many and diverse consequences on particular cultures and ways of life. New languages and religions were introduced, often by coercion. Old practices were forbidden. Existing systems of land rights were swept away by new ones based on

Colonial propaganda Spaniards meet a horrible death in this manuscript drawing depicting Inca brutality. By portraying indigenous peoples as primitive savages, the colonizers used the imposition of their own culture as a moral justification for conquest.

Enduring influences The Gothic-style cathedral in Lomé is a reminder of Togo's period as a French colony. Even after independence, the presence of French goods and automobiles in this Central African state denotes lasting economic and cultural links.

the idea that land could be divided up and sold rather than held communally. In so doing, the close relationship between peoples and nature was often severed. New taxes forced people into a cash economy and into towns; money replaced not only indigenous forms of exchange but became a new measure of value for all things and people. Different opportunities for men and women in the new cash economy altered traditional gender roles and family structures.

It was in such encounters that ideas of race as an indelible characteristic of peoples were formed, creating new categories such as Red Indians and Negroes that were, at least to begin with, of much greater significance to the people who imposed them than to the people they described. The collapse of empires in the second half of the 20th century brought many such peoples to the colonizing countries in search of work, forcing the inhabitants to adjust their views of their country's past.

Plural societies

In a number of instances colonialism did more than leave behind a small class of European origin. By importing labor from around the world it created wholly new societies. Where cultural differences between incomers in such things as language, family and marriage assumed greater significance, for example in economic or political spheres, plural societies were formed. Such societies were held together by forms of colonial coercion or European domination. In the post-colonial period they may not have found the means to integrate.

In many countries this has led to political instability. In Trinidad, Guyana and Fiji, for example, the post-colonial political divisions followed cultural lines of race and ethnicity, and national identity has been hard to achieve. In Fiji, colonialism created a population in which imported Indian laborers eventually outnumbered the indigenous Melanesian population. In the late 1980s tension between these groups culminated in violence and a coup. Many Indians subsequently left the islands and indigenous Fijians are once again the majority.

WORLD PATTERNS/4

Race and Racism

ALTHOUGH THE PEOPLES OF THE WORLD have clear systematic differences in their physical appearance, the behaviors and practices associated with these differences are social or cultural rather than biological in origin. Racism is the belief that humans can be classified into distinct races, that such differences explain or determine social behavior and that races can be ranked hierarchically.

Racism is not restricted to the modern era or to a small number of societies or individuals, nor is it fading away as some would like to believe. All peoples develop sets of ideas to categorize both themselves and others, and many of these carry negative evaluations. The ancient Greeks and Romans termed other peoples "barbarians" (from the Greek word for "stammer") because they thought they lacked the ability to speak articulately and, therefore, the capacity to reason. Throughout the world various peoples have claimed that their neighbors are cannibals, though there is almost no evidence for it as a systematic practice. Each society develops its own marks of acceptability, such as skin color, hair texture or even diet.

European racism is an historically changing set of ideas and associated practices. The Greek historian Herodotus (c.484-420 BC) traveled widely and encountered a variety of races. Early biblical interpretations of the differences in humankind were given substance in the 16th century by travelers' accounts of foreign lands. By the 17th and 18th centuries environmental or climatic explanations of race were developed. Slavery could be justified by arguing that Africans were uniquely suited to laboring in tropical conditions, but incapable of governing themselves.

By the 1800s such ideas were granted scientific respectability by the systematic measurement of physical features such as head shape, and this was used to justify the exploitation of colonial labor and the seizing of land by European farmers. The legacy of this era was a system of skin color hierarchies that stretched across the Caribbean and Central and South America. The subsequent discrediting of scientific racism by theories of genetics did not mean the end of racism. The immigration policies of such countries as Australia, Britain and the United States continued to be informed by racial thinking into the 20th century.

Racism disguised and challenged

In the period since World War II relations between nations have been regulated and improved by the setting up of the United Nations and the ending of colonial rule. Yet racism has come to occupy a more central place within individual countries. Immigrant communities may be racialized by the host nation – meaning that ideas of race seemingly serve to explain both their supposed distinct social identity and also their lack of economic success. In many liberal countries race is no longer a respectable term, but the term culture often replaces it and performs the same function: attempts to explain a person's behavior in terms of their culture may contain the germ of racism.

South Africa's apartheid policy has often been singled out for condemnation as abnormal and therefore immoral, but racism in various guises is equally apparent in many other parts of the world. Wherever immigration, education or other policies make assumptions of innate cultural difference, the possibility of racism exists. The experience of the United States, which has done more than most countries to combat racism through legislation but where prejudice and disadvantage nevertheless remain, suggests that legislative responses alone are inadequate to deal with it.

Racism has not gone unchallenged by racialized peoples. Slave revolts and the creation of the Brazilian *quilombos* – societies of runaway slaves – were early attempts to fight back against repression. In the 20th century black intellectual and artistic movements and the politics of Pan-Africanism have crossed national

THE CULTURAL POLITICS OF NAMING

Europeans devised new terms to describe the peoples they enslaved and conquered, such as "Negro" and "Indian". This suited their belief that humankind could be classified systematically, and established skin color as the criterion for categorization. In the United States, for example, the government at various times has used the terms "Colored", "Negro" and then "Non-White" to describe the descendants of enslaved Africans. Unlike Caribbean societies, no distinction was made between Africans and the children of Africans and Europeans. The right of a group to choose its own name has since become a central demand of many antiracist cultural movements.

During the 1960s civil rights struggle in the United States, Americans of African descent began to describe themselves as black, turning a pejorative term into a source of positive pride. Mexican-Americans did the same with the name Chicano, formerly a term of abuse. In 1980 the government replaced "Negro" with "Black" in its official census, but leading black politicians pressed for a new term, African-American. This would, it was argued, put blacks on a par with other ethnic groups and end the preoccupation with skin color as the means of defining a group.

The Ku Klux Klan parading in Washington DC in 1926. Originally founded in 1861 to oppose civil rights for freed slaves, the Klan gained strength after World War I and extended its activities to include Catholics, foreigners, Jews and organized laborers.

Germany, April 1991 *(above)* As the barriers between East and West officially fall, Polish citizens crossing the frontier into Germany are greeted by a group of protesters. Hurling abuse and taunts, they have borrowed the gestures and slogans of Germany's prewar antisemitic Nazi movement. The raised arm salute remains a powerful and chilling image of racial hatred.

Banners and swastikas *(left)* Nazi symbols are also used by these white demonstrators in South Africa. Extremists such as these are motivated by a belief that black South Africans are inferior to whites in ability and potential. The removal of the official apparatus of apartheid in the 1990s intensified such protests, especially among rural communities of Afrikaners.

boundaries by challenging the derogatory evaluation of blackness. Literature, arts and music are central to this positive self-evaluation. For example, in the 1920s the Harlem Renaissance in New York focused on the new writings of such influential thinkers and writers as the novelist Langston Hughes (1902-67) and the historian W.E.B. DuBois (1868-1963). They began to assert a positive pride in their culture in contrast to those black Americans who wanted a more complete assimilation with white-dominated society. Rastafarianism, a religion originating in Jamaica that links Africans throughout the world with their homeland in Africa, has been a recent expression of those attitudes of cultural pride.

Family and Kinship

Individuals derive a sense of personal identity and belonging through their place or niche in the social group. The most immediate and intimate sources of personal identity are generally those of family and kinship. These relations may be based initially upon biological ties, but their meanings are established socially and culturally. They take an enormous variety of forms; there is no such thing as a universally normal family or kin system.

Families are intimate kin-based groups usually, but not always, based on a parent-child pairing that share the activities of daily life. The nuclear family – two parents and their children – is the norm throughout the industrialized parts of the world, but elsewhere extended families of two or more parents and their children, and of two or more generations, or joint families based on brothers and sisters and their children, are common.

Kinship describes a set of rights and obligations that individuals have to others they identify as their relatives. It defines systems of inheritance and the duties of exchange or sharing. In much of the world kinship also determines the way farming or manufacturing industries are organized though in industrialized countries its influence is more usually restricted to the distribution of goods.

It is possible to classify peoples by their kinship system, distinguishing the ways in which lines of descent are defined. In some cases, an individual's obligations and rights are related to the mother's kin. Called matrilineal descent, this is common among small-scale agricultural societies in which there is a clear division of labor between men and women. Examples of matrilineal societies exist throughout Africa and were also found among Native-Americans before the European conquest. Identifying with the father's side, which is called patrilineal descent, is widespread among pastoral nomads and the majority of other traditional peoples. In Western societies, bilateral descent is more usual, in which individuals count as their kin persons on both their mother's and father's sides. In practice many peoples combine both patrilineal and matrilineal principles, or even widely ignore them.

These systems may also determine where individuals live. In matrilocal societies, a newly married couple is required to live with the bride's family, but in patrilocal societies they go to the groom's side. Neolocal residence is where a married couple set up their own household – a practice that is usual in industrialized societies and becoming more widespread elsewhere. The territorial organization of peoples may be founded on such kinship groups – this is common throughout Central Africa for example, where village hut clusters are traditionally organized around lineages (groups that share a common and known ancestor) or clans (groups that share an ancestor but are not fully aware of the exact links of descent).

Changing families

Although many societies have norms of family size and membership, in practice a range of family types is found. This is

especially so where there is rapid social and economic change. During the period of European industrialization, for example, the nuclear family became the norm among the urban middle class, who then encouraged the rest of society to emulate them. Yet, despite this group's influence over both government and the means of communicating ideas, the nuclear family never became universal. In fact, evidence from around the world suggests that urban living and industrialization are not necessarily associated with smaller, nuclear families, though they do seem to be becoming more widespread.

The nuclear family (*left*), thought to be the cornerstone of Western society, is fast becoming a minority institution. With more people choosing not to marry, and divorce on the increase, a growing number of households are headed by a single parent.

The extended family (*below*) includes grandparents, cousins, uncles, aunts, siblings and relatives by marriage as well as parents and their children. It may be organized in a number of ways. Young members of the nomadic Masai people of Kenya, for example, are cared for by several close female relatives living in the same hut group or kraal.

In Africa, the Caribbean and parts of South America – particularly where male labor migration is common – female-headed single-parent households are becoming more normal. Indeed, many women in such areas may be choosing not to marry. Households headed by females have become well-established among the urban poor.

In India, the desire of many women for greater control over their lives has led to a move from extended to nuclear families. By contrast, among many immigrant families in the industrialized world, the resources of an extended family – their labor, money and a place to live – may be a means of securing an economic foothold. In general, however, immigrant families after the second or third generation tend to resemble those of the host society. Income, the desire for personal autonomy and cultural background all influence the lines along which families are organized.

WHO IS KIN?

There are many ways of defining kin. Most cultures, however, distinguish between relatives of blood and relatives of marriage. In Hawaii all relatives of the same sex and generation are referred to by the same name. An individual's biological father as well as his brothers are all equally called "father". By contrast, the Sudanese system is far more complex. There, a range of names is used to distinguish between various categories of cousin.

Fathers do not always have to be male. In the ghost marriages of the African Nuers, for example, an older high-status woman adopts a younger woman and her child by an arranged lover. A more common form of kinship that is based neither on ties of blood nor marriage is the *compadrazgo* (godparent) relationship that is widespread in Christian, especially Roman Catholic, countries. For certain purposes the religious bond between the *compadre* and the child ritually adopted in the rite of baptism may be more important than biological parenthood.

Male and Female

GENDER REFERS TO THE CULTURAL MEANINGS assigned to the biological difference between the sexes. All peoples distinguish, to a greater or a lesser degree, between men and women in terms of the work they are supposed to do, their rights in the social system and the roles they are expected to perform. The terms masculine and feminine do not refer simply to people's roles. Through a range of symbolic associations they give meaning to personal characteristics, objects and places. The gender distinction is often central to a people's cultural or symbolic organization of the wider world. Although the form and meaning of the distinctions are diverse, in virtually every society women are accorded a secondary status in cultural values, though not necessarily always in practice. This may be one of the few universal cultural ideas.

Nature and culture

The reasons that women have acquired this secondary status are many and diverse. However, there is a common cultural association of women with nature and men with its opposite. Western cultures, for example, generally personify nature as female, as in such phrases as "Mother Nature" or "the virgin wilderness". Early European descriptions of the Americas used metaphors of the female body to describe the new lands.

Women and nature are viewed as sharing qualities of birth or fertility – whether of crops or children – and of being wild and uncontrollable, requiring male reason and technology to master them. Once dominated, women, like nature, may be romanticized in arts and poetry, and celebrated for their mysteries and universal characteristics as givers-of-life.

A Russian female worker (*above*) on a construction site in Moscow expects to work as hard and take the same risks as her male colleagues. Equality of opportunity in the workplace is one of the ideals of socialism. Yet women are still expected to take responsibility for most household tasks as well.

Unmarried girls (*below*) from a wealthy Pathan family in Pakistan do not venture into public without a chaperone, and keep their faces and bodies covered in front of strangers. In common with women in other traditional Muslim societies, they will have marriages arranged for them by their families.

Male preserve Farmers at a livestock sale in Britain exchange traditional wisdom and advice. In spite of extensive equal rights legislation in Europe and the United States, there are many areas of life where time-honored gender roles persist. Female engineers are still uncommon, as are male nurses. Cultural conditioning is hard to alter.

Other cultures share similar views. In Southeast Asia, for example, myths of the origin of rice – the staple food – commonly describe the miraculous plant springing from the ground where the body of a young woman, who has been ravaged by a man, lies buried.

The gender equation does, however, sometimes work the other way around. In certain islands of the South Pacific, the forest is commonly regarded as a masculine realm, while the cultivated clearing is feminine. Although the idea of the femininity of nature, and thus the more natural quality of women, is widespread, not all peoples devalue nature and seek to dominate it in the manner typical of Western societies. Nor are women necessarily identified principally by their potential fertility: among Australian Aborigines, the Ilongot of the Philippines and the !Kung people of the Kalahari, the ability to give birth to children is not regarded as an especially fundamental distinction between men and women.

Domestic and public

A second common distinction – closely related to that between nature and reason – is between the two realms of the domestic (or private) and the public. The idea that a woman's place is in the home, raising children, cooking and doing the housework, is widespread in all kinds of societies. Men, by contrast, are identified with the larger decision-making world of waged work and politics.

The justifications for such a division vary. In Islamic societies, the idea that family reputation and honor depends upon the chastity of women gives cultural sanction to female seclusion or "purdah". In Europe and North America, this idea was coupled with another, that women – by virtue of their supposedly greater proximity to nature, desire and passion – were incapable of reason and therefore could not be trusted to make decisions rationally. Western democracies gave women the right to vote and sit on juries long after men, and even today their representation in politics is low – in 1990 only 2 of the United States' 100 senators were women. Socialist countries were quicker to accord such rights, and the parliaments of the Nordic Countries are an exception to the rule of female exclusion. Women's rights to property are also frequently less than those of men, not only among those societies – in Africa for example – where only men can inherit. In many Western societies women still have to fight in courts for a share of property following divorce.

Patriarchy redefined

These cultural ideas together form the basis of patriarchy, the near universal domination of women by men. Its practices and ideas vary, and have long been challenged by women. The most significant world development affecting the relationship between men and women is the growing number of women in the waged workforce. Despite cultural norms, women have always been workers, though their labors very often went unremunerated. Women produce half or more of the world's food, and in African countries they still perform three-quarters of all agricultural work.

Women today represent about one-third of the world's workforce. The figure is as high as 50 percent in the Soviet Union and 40 percent in Western industrialized countries. Only in the Islamic countries of the Middle East and North Africa are the vast majority of waged workers male. In most societies men and women perform different jobs with unequal rates of pay. Women's jobs are frequently an extension of their traditional household roles of caring, cleaning and provisioning. This suggests that patriarchal relations are now maintained more at work than in the family.

Such developments have helped to break down the division between domestic and public roles. Yet even in societies such as the Soviet Union, where most women work, the evidence suggests that they also perform the bulk of household tasks as well. Perhaps only Sweden has gone any significant way toward reducing the dual burden by making provision for maternity leave without loss of job seniority, offering paternity leave, and making adequate childcare arrangements for working parents.

Stages of Life: Infancy to Adulthood

ALTHOUGH INDIVIDUAL DEVELOPMENT IS A gradual process, a person's life is punctuated by distinct moments: birth, the attainment of adulthood, marriage and death. These are collectively known as the rites of passage.

Many beliefs and taboos surround the actual moment of birth. In some societies, for example the traditional peoples of Melanesia, the mother is secluded away from the village since childbirth is seen as a potential source of pollution. In others, the father joins the mother in seclusion, and may even observe certain rituals of diet that serve to acknowledge his role in bringing about the birth. Among the agricultural societies of India and Africa the birth is usually attended by the mother's female kin, while in Western societies it has become the practice for the male-dominated medical profession to control the labor and delivery.

A newborn baby is given social status by a variety of practices such as the ritual naming of the child. Some cultures, for example Jews and Muslim Hausa, circumcise infant boys; sometimes girls may be circumcised. Such rituals distinguish between the biological process of being born and entry into the social group. Babies born to the !Kung of southern Africa may be disposed of without social sanction in the three days before naming since they have not yet become a person. Among traditional Hindus a number of rituals are carried out during a baby's first two years of life. These include naming, the first time the father views the child, and its first feed. They are among the first of 16 such rites of passage (*samskaras*) in the individual's life up to death.

Boys and girls

In agricultural societies children are a valuable source of labor, and the more children a family has the better its economic future. In cultures that place value on fertility, large families may also confer status on both father and mother. The average number of children per woman of child-bearing age is currently over six in Africa. As soon as children cease being a valuable source of income and start being an expense, as may happen when compulsory primary education is introduced, families tend to become smaller. In many industrialized societies children are not considered especially to be a sign of status and large families are exceptional. The average number of children per women of child-bearing age in Europe is less than two, which means that these populations are not replacing themselves.

Male and female children are often valued differently by their cultures. Male

Young Masai warriors (*below*) on a hunting expedition. Groups of boys of the same age are initiated into adult life together. They then move up within this group through the various stages of adult life until they reach the rank of senior elder.

THE PORO INITIATION RITE

The male initiation rite of the peoples of southern Liberia, Sierra Leone and the Ivory Coast is widely known as the Poro. It follows a threefold sequence of events that is echoed in the initiation rites of many other cultures. The individual is first removed from society, perhaps to a place used only for this purpose, put into an intermediate state in which the former identity is cast off, and then eventually readmitted to society as a new person.

At the prearranged time gangs of masked raiders seize all the boys of the village and tell their mothers that they have been killed, miming the act of slaughter. The boys are taken into the bush, circumcised and taught the various skills needed by men. During this process they are subjected to trials that occasionally result in death. After several weeks or months they are returned to the village, given new names and considered to be new people. The act of rebirth is signified by pulling the man from underneath a cover or blanket. The Poro acts thereafter as a secret society among males; there is also a secret society known as Sande among females.

At first communion (*above*) children are admitted to the communion, or mass, the church's central ritual when bread and wine are shared. It is one of the highlights of a Roman Catholic childhood, especially for girls who dress as miniature brides.

children may be preferred for their potential labor, the fact that when they marry they will gain property from their wife's kin (where dowry systems operate), and sometimes for religious and ritual purposes – in Hindu societies a son is required to carry out his father's cremation ceremony. Cases of valuing girls more are rare. Some peoples are known to have practiced female infanticide in the past, including the Inuit of Canada, Australian Aborigines and the Yanonami of the Amazon. Often this was the only way they had to control their numbers. Today, some societies, such as South Korea, have far more male than female children, possibly because medical technology can establish the sex of a fetus, enabling unwanted females to be aborted.

Entry to adulthood

The transition from childhood to adulthood is determined by cultural factors. It may coincide with the biological changes of puberty and menstruation or, as in many industrial societies, with certain customary events – the completion of education or the acquisition of the right to vote, for example. The average age at which females start to menstruate has been gradually falling in Western industrialized societies to 12 years old. Boys are sexually mature as young as 14 years old. Such biological signs of adulthood rarely coincide with the legal recognitions of adult status – voting usually begins at 18 while in the United States a person must be 21 to buy alcohol. This causes friction between children and their parents, and leads society to ignore or condemn sexual activity between young adults.

Shaving a baby's skull (*below*) is one of several Hindu rituals that are carried out to mark the beginning of the individual's life journey. This ceremony, in the Mumbaderi Temple in Bombay, takes place in the presence of the baby's parents.

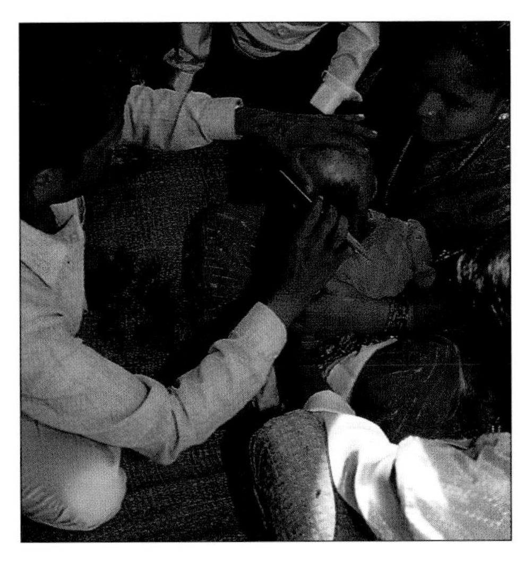

Among more traditional peoples there is usually a single undisputed moment at which the transition to adulthood takes place. In many societies a special initiation rite is accompanied by a marking of a part of the body, frequently the genitalia or the face. These are the public marks of the new status. The rites often take many weeks, and tend to be more elaborate for boys than for girls. Several world religions have an equivalent rite of passage. They include the Jewish "barmitzvah", which takes place for boys at 13 years old, first communion among Roman Catholic 6 to 8-year-olds and Protestant confirmation, generally at age 12.

Stages of Life: Marriage to Death

THE RITE OF MARRIAGE NOT ONLY INVOLVES the legal bonding of two people, but is also an exchange or contract between kin groups. At marriage, the rights over the sexuality, property and children of a man and a woman are established both between them and within society. The importance of this aspect of marriage is demonstrated by the various forms of property exchange, such as dowries and the giving of gifts, that accompany the rite. These tend to be more significant in societies where lineages are the main social, political and economic units. In such cases, marriages are usually between lineages, and if they are in a state of conflict with each other, they may be important acts of alliance.

In many such societies the choice of partner is so important for status that it is a matter for the kin group rather than the individual. Among a number of African peoples, the couple, after marriage, may not even be each other's principal personal companion. Arranged marriages are common in South Asia. They were banned in China in 1950, but the practice was so central to the structure of its peasant society that the law was widely broken. In Taiwan, by contrast, there has been an almost complete decline in arranged marriages in favor of the industrial world's practice of couples choosing each other.

In societies where lineage is less important, marriage may nonetheless involve a wider social grouping. In some European peasant societies this may include the village community, which expresses an opinion on the suitability of the match. This role is symbolically retained even in industrial societies by the Christian practice of announcing the proposed marriage at the local church before the event, and giving the opportunity for objections to be made to the marriage during the ceremony itself. In such societies, the role of kin is generally a reduced one, involving little more than separate seating areas during the ceremony, although it is still practice for a male member of the bride's family to "give away" the bride to the groom's family.

Marriages need not involve just two people. Polygamy, the practice of one individual (almost always the male) marrying more than one person at a time, is allowed by Islam and is common in the Middle East, Cental Africa and Tibet. It is becoming rarer, however, largely due to the expense involved in supporting several wives and their children. In truly matrilineal cultures, such as the Nayars of India, women may marry early but thereafter be free to enter into as many partnerships as they wish and end them with relative ease.

In industrial and secularized societies the practice of serial monogamy, having many partners one after the other, is increasing. This is accompanied by high rates of divorce, something which legislation has made much easier over the past 40 years in socialist countries such as the Soviet Union and Cuba as well as the predominantly Protestant countries of Western Europe. A third of all current marriages in Britain will end in divorce. However, religious injunctions against divorce still act as a restraining factor in Roman Catholic countries. Divorce is rarer among nonindustrial peoples, though high rates are found among some, such as the Muslim Hausa of Nigeria.

Attitudes to death

Almost all peoples have some conception of life after death, or the belief that death is not the end of being but a transition to another state. In the religions of the Judeo-Christian tradition, for example, death marks a one-way passage to a noncorporal or spiritual state; it is also the moment at which the individual's life is judged. In Hindu and Buddhist religions it is part of a cyclical process of reincarnation in which souls, including those of animals, are transferred to another stage in the hierarchy according to conduct in the former life. The ultimate aim is to be released from this cycle of rebirths.

In all cases death is a social event. The funeral or mortuary rites that accompany it serve to help society adjust to the loss of the individual. Hindus, Buddhists and followers of Shinto are among the people that cremate their dead, while Christians, Muslims and Jews bury them in special areas. The Parsis of India and Tibetan Buddhists leave the body in the open to be devoured by vultures, possibly because the ground is too hard to be dug up. In southern Europe, it is common to place bodies in walled graves above the ground

ZONPANTLI

Mexico's "Day of the Dead" is a fiesta rather than a time of mourning, celebrated on 1st November, All Soul's Day. It is an extraordinary mix of local animist belief and Christian ritual. Gifts of flowers and fruits are made to welcome the souls of the dead back to their living families.

BRIDEWEALTH AND DOWRY

In patriarchal, and therefore in most societies, marriage is essentially an exchange of women between male-dominated kin groups. Two forms of property exchange often accompany this transaction. In peasant societies, such as those of South Asia and the Mediterranean, where land is the basic unit of wealth and is passed on by inheritance, the "dowry" is the part of a woman's inheritance that she takes to her husband's kin, transmuted very often to a cash payment. In strongly patrilineal societies, including many, both pastoralists and cultivators, in Africa, women cannot inherit land from their families and "bridewealth" is a payment made to the wife's kin to compensate for the loss of her labor.

In practice, both systems usually result in the woman losing control of her property and wealth. With the integration of traditional societies into a cash economy such a transaction can become vital to a kin's well-being but detrimental to the woman's, as she becomes little more than a tradable commodity. Dowry was outlawed in India in 1961, though it is still widely observed, and opposition from the women's movement here now focuses on the disturbing practice of brides being burnt in cases where the payment is not fully met.

Spinning the corpse (*above*) is an essential part of burial rites in Bali. Male friends and relatives carry the body to be cremated in a litter decorated in white and yellow, the colors of mourning. On the way they whirl it around wildly to confuse the spirit of the deceased so it will not return to the body.

A white wedding (*left*) Many Western couples opt for an elaborate (and expensive) religious ceremony that involves the families of both partners, even though marriage is now a question of personal choice. Weddings have their own particular rituals. The bride wears white, and her father "gives" her to the groom.

so that the deceased can be visited by friends and relatives.

The mortuary rite may take a single day or many years, particularly if customary mourning or reburial is involved. In this period the living perform the rites that enable the deceased to complete the transition. Among Hindu communities, for example, the ashes of the cremated individual must be placed in a river, preferably the sacred Ganges, so that the soul can rejoin the cycle. The family, all of whom are required to attend if possible, must then observe certain purification rituals to rid themselves of pollution: the deceased's own sins are taken away by a Brahmin priest. By contrast, in industrial societies, in which the experience of death is rarer, professional funeral directors generally assume responsibility for the corpse.

The Role of Language

LANGUAGE IS THE PRINCIPAL EXPRESSION OF A culture – the beliefs and behavior that are shared by a particular ethnic, social or religious group. It defines a person's potential range of communication, and as the medium of ritual and literary tradition, it provides an intimate connection with the past. The central role played by language in establishing cultural identity means that struggles over the right to speak, to be taught and to use particular languages are often highly impassioned.

Language may also embody a view of the world itself, since its vocabulary contains the terms by which it is classified and the concepts by which it is given meaning. It has, for example, been suggested that languages such as Hopi, which is spoken by some indigenous American peoples and contains no terms for time and space, present insurmountable obstacles to translation. In such cases, language may provide an absolute barrier to the mutual understanding between different peoples.

It is estimated that there are between 3,000 and 5,000 actively spoken languages in the world, though problems of definition and classification make an accurate count impossible: Africa alone contains about 1,000 different tongues. The vast majority of languages, however, belong to a small number of language families that have common origins, and possess similarities of vocabulary and grammar. The largest, known as Indo-European, originated in central Eurasia about 3000 BC and includes all the main European languages, as well as those of the northern Indian subcontinent. At the other extreme, there are a few languages that appear to defy classification and have no close relatives. These include Euskera, the language spoken in the Basque region of northern Spain and southern France, the languages of the Andaman Islands in the Indian Ocean, and the Inuit and Aleut regions of the Arctic fringe.

Linguistic descent The Indo-European languages are all descended from a single unrecorded language, which is believed to have originated in the steppe regions north of the Black Sea. In time dialects were formed, and as these were spread by migration they developed into separate languages. Seven of the nine branches are found in Europe.

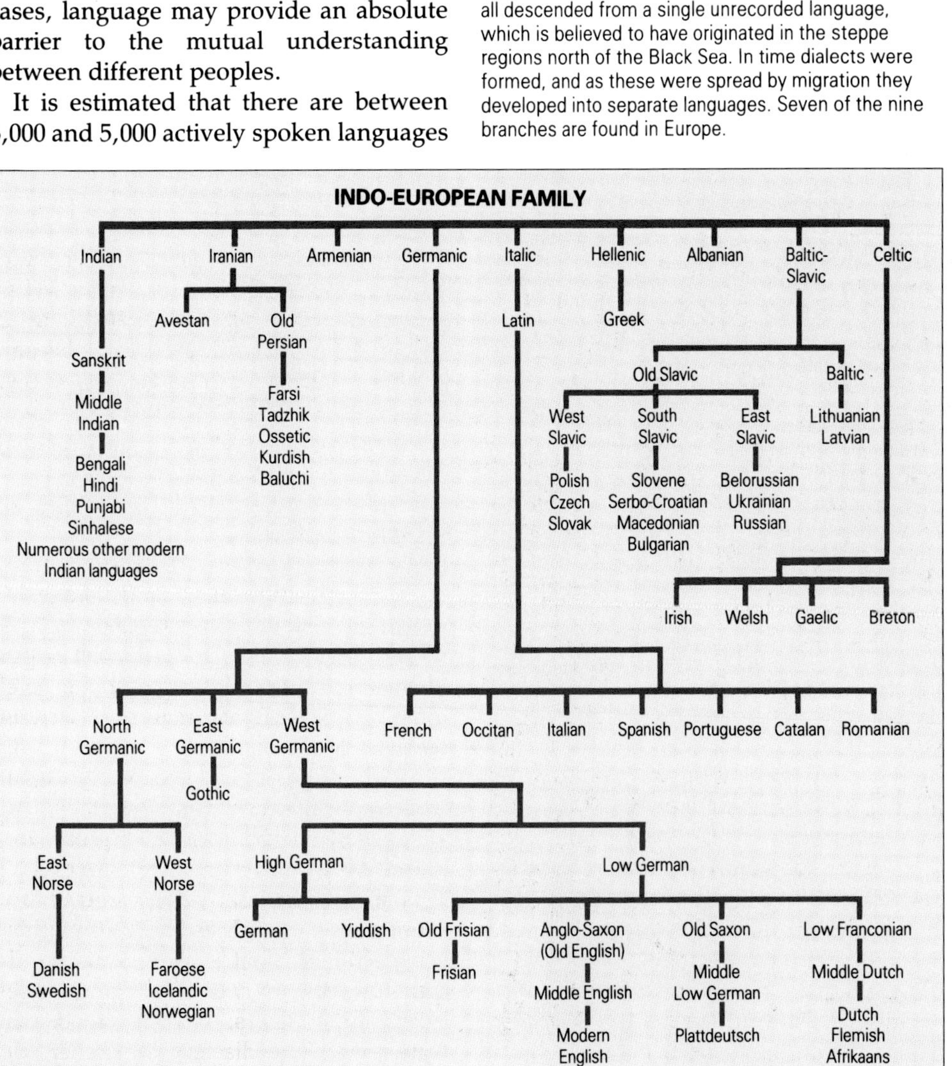

There are relatively few countries that have only one language. Japan, Uruguay and Iceland are among those that can claim to be virtually monolingual. In Papua New Guinea, by contrast, there are over 700 separate languages among a population of only 3.5 million. Most people in the world, therefore, need to possess at least one other language, in addition to their native tongue, in order to communicate with people within their own country, let alone outside it.

Official languages and pidgins

The domination of one language over one or several others is a major issue in many 20th-century nations, from India to Canada. The challenge of multilingualism has been met in a number of ways. In many countries, notably the former European colonies, national governments have adopted official languages for the conduct of administration, law and education. In post-independence India, English joined Hindi as the official language, though neither of these is the first language of a large number of the population. In Kenya and Tanzania, in eastern

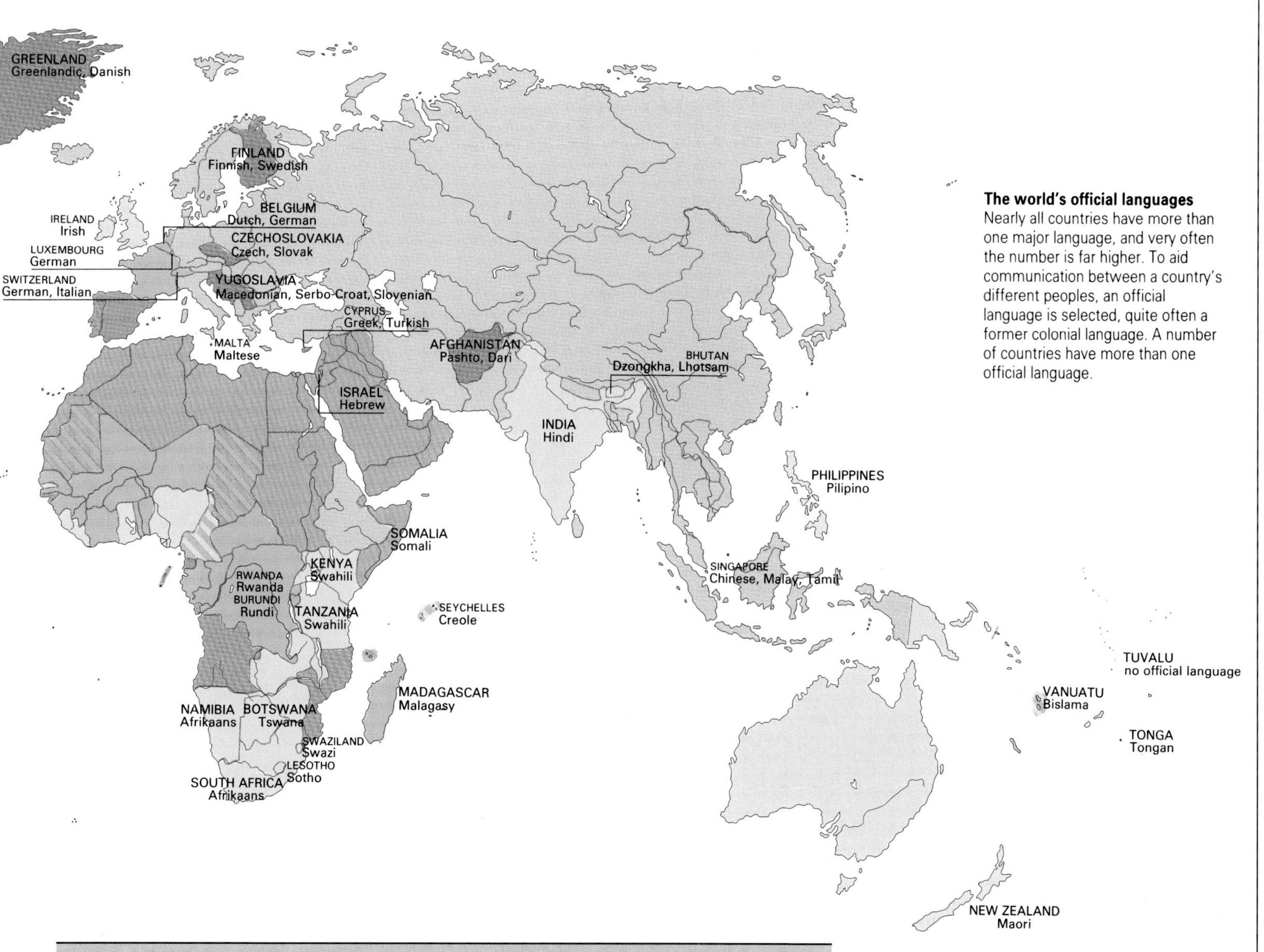

The world's official languages
Nearly all countries have more than one major language, and very often the number is far higher. To aid communication between a country's different peoples, an official language is selected, quite often a former colonial language. A number of countries have more than one official language.

POLYGLOT PHILIPPINES

With 56 million people spread over 7,000 islands, the Philippines represent an extreme case of linguistic diversity. Over 70 indigenous languages are spoken, in addition to the colonial languages of English and Spanish, which have both become modified as creoles. However, the islands are so scattered that not even these creoles have become generally established and there is no common tongue used by a majority of the population.

The national language is Pilipino, based on a local language, Tagalog. It is spoken as a mother tongue by less than a quarter of the population and was chosen as the official language because it is the dominant tongue of the capital, Manila. However, it is not very similar to many of the country's other indigenous languages, and government officials, doctors and teachers often cannot communicate with local people in some of the islands, thereby reducing the effectiveness with which they can work. Linguistic minorities, who may have difficulty anyway in finding a voice in national politics, have sometimes turned to armed resistance to make themselves heard.

Africa, Swahili is an official language alongside English. Swahili is a lingua franca – a language that was originally developed to ease trade between different peoples, and usually restricted to such contexts. Swahili, for example, is a Bantu language (the major language family of Central Africa) modified by Arabic, the language spoken by the coastal traders of eastern Africa. In some respects English, with 350 million speakers, has become a world lingua franca, being frequently used as a second tongue as well as for international transactions such as air-traffic control.

In other colonized societies languages known as pidgins developed as a way of enabling communication between peoples who had had little prior contact. These simplified tongues are widespread in the Caribbean, Africa and Oceania. Bislama pidgin is one of the official languages (with English and French) of Vanuatu, while Melanesian pidgin allows half the peoples of Papua New Guinea to speak to each other in a basic way. When pidgins become more elaborate, with an enlarged vocabulary, they may become mother-tongues. Such languages, known as creoles, are largely based on European languages and are common in the many islands of the Caribbean.

Dialects and slang

Regular variations of vocabulary, grammar or phrasing within a language are known as dialects. Such dialects may indicate the social or geographical origins of an individual and, despite the leveling or equalizing influences of writing, television and public education, still serve as badges of identity. Many subcultures or social groups develop their own specialized set of colloquialisms, or slang, one of the purposes of which is to exclude outsiders. From bankers to thieves and street gangs, speech as much as clothing is a flexible form of self-presentation. To speak is often a way of identifying oneself in relation to others, and individuals may move from one dialect to another, depending on the nature of the group they find themselves in.

The Written Word

The words of Allah *(above)* An ancient manuscript of the Qu'ran – the text from which all Islamic culture and belief springs – from Uzbekistan in Central Asia. Because Muslims believe that God dictated the Qu'ran directly to Muhammad, official translations of the sacred text were prohibited until quite recently.

Words and meanings *(right)* A Jewish scribe at his painstaking work in New York's East Side. Modern scholars keep up the centuries-long tradition of earlier scribes in reinterpreting Judaism's sacred texts and older commentaries in the light of new experiences and learning.

THE DEVELOPMENT OF WAYS OF COMMUNICATING, other than through face-to-face interaction, has facilitated the spread of cultural ideas through time and across distances. For thousands of years writing was almost the only non-oral method of communication, but as ever more sophisticated means of transmitting information electronically have been discovered and perfected, ideas are now able to reach a worldwide audience almost as fast as they are formulated.

The first forms of writing consisted of simple pictures (pictographs) of objects. Gradually, the direct visual link between a symbol and the object it represented disappeared, to be replaced by ideograms in which the symbol represents the idea of the object. One of the earliest such systems was used by the Sumerians of Mesopotamia (the area between the Tigris and Euphrates rivers in the Middle East) in the 4th millennium BC. The hieroglyphics used by the ancient Egyptians combined pictographs with ideograms. The Chinese script still used today, which can be traced back to the 2nd millennium BC, contains around 45,000 ideograms.

The alphabetic system of writing, in which a series of characters (letters) represent different sounds, and words are constructed from combinations of these characters, developed in western Asia about the same time. It subsequently evolved into a number of forms, including Arabic script, the Devangari script of the Hindus, and the Greek alphabet – the ancestor of both the Cyrillic (Russian) alphabet and the Roman script used throughout Western Europe.

Writing generally developed first among agricultural societies, as a means of recording crop surpluses. Although writing as such was not known in the Americas before the arrival of the Europeans, the Incas had developed a way of keeping tallies by a system of knotted cords (*quipus*). Writing in such early societies was a source of power, and was probably the preserve of a special priestly class. In about 1,000 AD the Chinese pioneered printing techniques using clay tablets, and were able to copy and disseminate the key texts of the philosopher Confucius. These writings formed the basis for the powerful bureaucracy that the Chinese empire developed. Writing was also crucial to the development of religious cultures. Although some Hindu sacred texts, the *Vedas* were transmitted orally with great accuracy for nearly 3,000 years, in general religious communities were held together by the written word. Holy books lie at the heart of the Jewish, Christian and Muslim faiths, and both devout Jews and devout Muslims carry miniature copies of their sacred texts on their bodies.

Writing and nations

Before movable-type printing presses were invented, written texts had to be copied by hand. This was time-consuming and allowed errors to be introduced. The first book to be mechanically printed in Europe, in the mid 15th century, was the Bible and from then on the text became mass produced and standardized. Printing also allowed translations of the Bible (until then known only in Latin) to be widely circulated and enabled the spread of ideas that challenged the authority of the established Roman

Catholic church. In early 16th-century Germany the works of Martin Luther, whose radical ideas led to the birth of Protestantism, accounted for a third of all books published. Between 1435 and 1500 more than 20 million printed volumes were produced in Europe, and in the 16th century perhaps as many as 200 million. This contrasts with China, where the imperial bureaucracy maintained control of books and writing, and so its monopoly of learning and literacy was perpetuated into the 20th century.

Writing brought about the standardization of languages. In late medieval Europe the range of dialects within a language meant that two people both speaking French, for example, could not necessarily understand one another's speech, but they could, if they were literate, read each other's letters. Certain forms of national languages became dominant and more or less stable over time. In Britain the dialect variant used in London and the southeast gradually gained acceptance over other forms to become the basis of standard English. The link between the written form of a language and national consciousness can be deliberately made. Koreans, Vietnamese and Armenians have all replaced an unwanted, imposed script with their own invented script as a symbol of their separate identity. In the 20th century, Joseph Stalin (1879–1953) standardized Russian on the Cyrillic alphabet and forced many nationalities to abandon their Roman script in order to create a distinct Soviet script.

So important is the written word that an inability to read or write in one's own language may bring considerable disadvantage, when dealing with the government or law courts for example. However, widespread literacy is a comparatively recent phenomenon. It was not until the 19th century, when primary education was made available to a broadening section of society, that it became common in Europe, the United States and eastern Asia. Today there are still large areas of the world with high rates of illiteracy. A quarter of the world's adults cannot read or write, a figure that rises to 54 percent in the case of Africa. In 1990 – declared International Literacy Year by the United Nations – over 100 million young children received no primary education.

The Idea of Religion

By addressing the nature of existence, religions describe the world and the position of the individual or a people within it, while also indicating how they may attain a place in some greater cosmological order. Although attempts to define religion by identifying its universal characteristics have mostly proved unsuccessful, most systems of belief generally recognize distinct realms of existence and some form of relationship between human and non-human beings, who are believed to intervene actively in human affairs. Humans can attempt to influence this intervention by the performance of particular rituals, prescribed personal conduct or through intercession by a priest or other mediatory agent, such as a shaman. Beyond this, religions have little else in common, though it is possible to distinguish broad types.

Universalizing religions are those that have historically sought new adherents by conversion, and therefore have worldwide significance. Christianity, with over 1 billion followers, has proved the most pervasive, though the number of active Islamic and Buddhist believers are growing worldwide. Cultural or ethnic religions are usually restricted to particular societies, such as Hinduism in southern Asia, Taoism in China and Shintoism in Japan. Finally, there are a myriad of local or traditional religions, often referred to as animist, in which the earth, sky, stars, fire and water are commonly worshipped, and the spirits of ancestors and animals are believed to intervene in daily affairs.

A further distinction is between monotheistic faiths that recognize one god, and polytheistic that worship many gods. The three monotheistic world religions – Judaism, Christianity, Islam – emerged one from the other in the Middle East between 4,000 and 1,400 years ago. They are "revealed religions", which means that they have sacred texts emanating from their respective God. Their adherents are divided into subgroups, or sects: notably Ashkenazy and Sephardic Jews; Roman Catholic, Orthodox and Protestant Christians; and Sunni and Shi'a Muslims.

The roots of Hinduism – the largest of the world's polytheistic religions – date back approximately 5,000 years to the early civilization of northern India. Other religions, including Buddhism, Jainism and Sikhism, also arose in the Indian subcontinent. Buddhism is unrevealed. This means its teachings stem from a human teacher and leader, Siddhartha Gautama, who lived in the 6th century BC. Confucianism and Taoism, followed in China, Korea and parts of Indochina, are also derived from the teachings of a particular individual; unlike Buddhism, they do not involve belief in an afterlife and are not, strictly speaking, religions but philosophies or ethical codes.

A mixing of religions

Wherever these religions spread, they introduced their own sense of sacred time and space. Religious calendars defined appropriate moments of ritual that often bore little relation to the natural rthythms of the seasons and the stars. Islam, in particular, involved a new sense of time free from the cycles of food provision and based on regular acts of observance. Christianity introduced an expectation of the imminent end of the world. This was spread by missionaries to the indigenous peoples of the Americas and Africa, who had not thought previously in such terms.

In very many cases, however, the imposed religion has not been entirely replaced, but has blended with local beliefs. The process of mixing one religion with another is known as syncretism. Local deities may be identified with outside ones and elements of local rituals transplanted into new ones. The mixing of African religion – a legacy of the slave trade – and Roman Catholicism in Brazil provides a good example of this process, but similar syncretic religions are found in Africa, Southeast Asia and Oceania. Buddhists in Japan will take part in local Shinto festivals; Sikhism contains elements of Islam and Hinduism.

The decline of belief?

Religions are sometimes mainly private beliefs, sometimes public ideologies – or a mixture of both – and require different levels and forms of observance from their adherents. Islam, for example, enjoins believers to pray together several times each day and at set hours. This emphasis on public worship means that the mosque is still the center of Islamic communities around the world.

With the separation of religious authority and government in much of the developed world and the waning of compulsory faith – a process known as secularization – some religions have

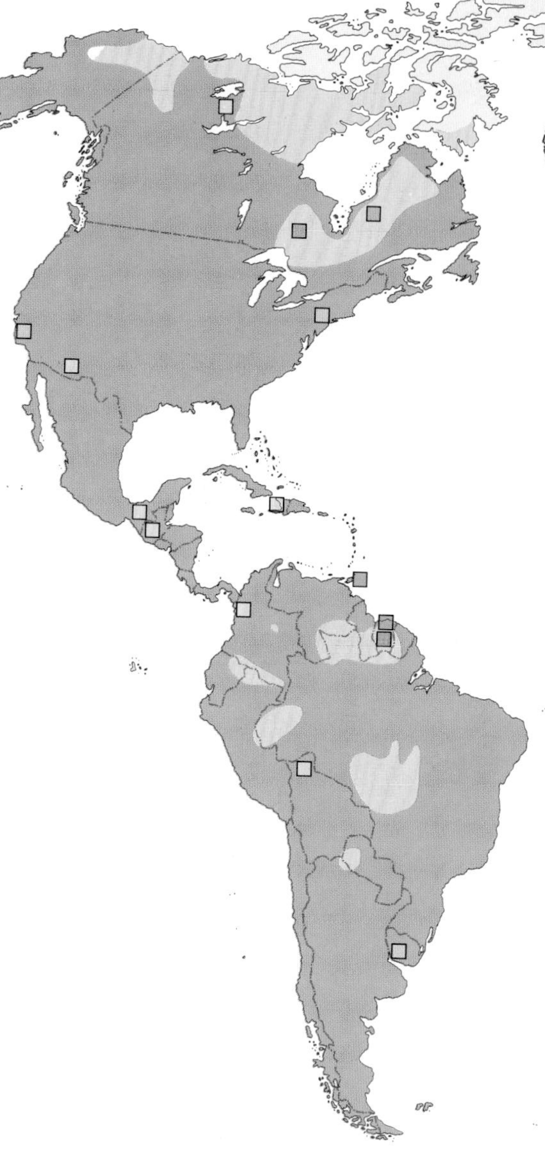

A world of many gods *(above)* Simple offerings adorn a shrine to a local deity in a rice field in Bali, an island where animist traditions and Hinduism blend. Local religions typically involve a belief in gods or spirits, often those of ancestors, that inhabit the natural world and have influence over human affairs, including crop success.

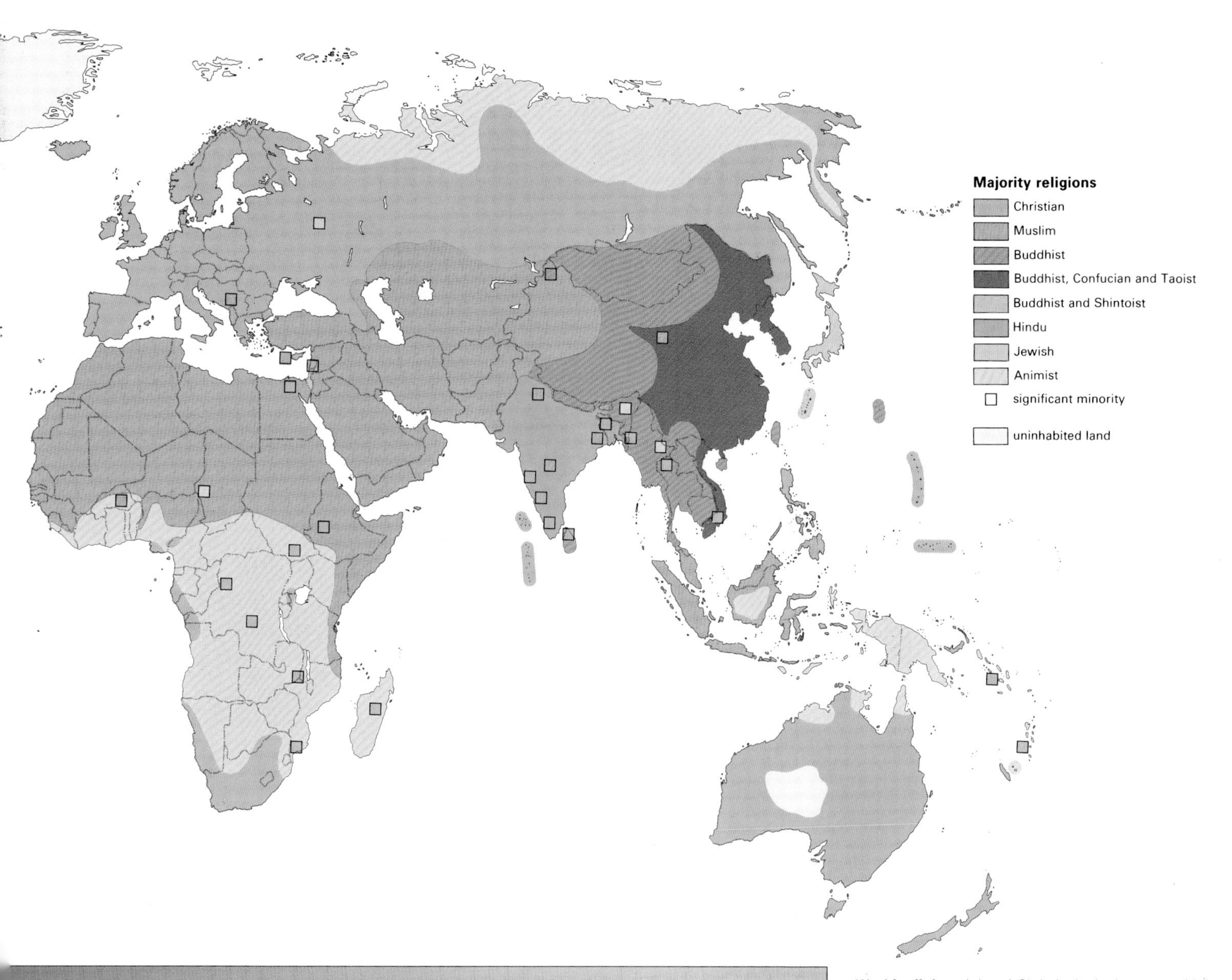

World religions *(above)* Christianity is the most widely distributed of the world's religions, reflecting the global impact of European colonialism. The picture is constantly shifting. With new patterns of migration, numbers of Muslims and Hindus are growing in Europe and the USA, though they are not yet significant enough to show up here.

Desert prayer *(left)* Obeying the Qu'ran's injunction to pray at five set times daily, a Kuwaiti driver has descended from his truck in the midst of the desert to perform his sacred duty as a Muslim. Islam in the 20th century has been the fastest growing of the world's great religions.

shifted more into the private sphere. This has particularly affected Christianity, especially in liberal Protestant countries such as the Netherlands, where the process of secularization is most advanced. However, the number of regular churchgoers is not falling in the United States, suggesting that a decline in observance is not inevitable in modern industrial societies. Even in countries where it has fallen, most people still claim to believe in God. In the Soviet Union and Eastern Europe, where secularization was imposed by the state, the collapse of communism has seen a revival of religious belief.

Ethnicity and Tribalism

POSSESSION BY A GROUP OF PEOPLE OF A common language, religious tradition, particular customs and a common place of residence may all be signs of an enduring collective identity, generally termed ethnic or tribal, that distinguishes it from other social groups. Ethnicity and tribalism place special emphasis on sharing time (rather than space) through the idea of common descent – a shared past as a distinct people. They therefore draw upon an expanded notion of kinship, and distinct ethnic groups are typified by a high degree of intermarriage (endogamy).

Two broad processes are involved in establishing ethnic and tribal identity. One is communal or inclusive, with the emphasis on what is shared and who is alike. This defines ethnicity at its core. The other is categorical and exclusive, indicating who is different and establishing ethnicity at the boundaries of a social group. Many small groupings within industrialized societies – from youth subcultures to the fervent followers of sports teams – behave in ways that correspond to those that define an ethnic group. Indeed, the term "tribe" is often used popularly and pejoratively to describe both urban gangs and sports fans.

Tribes of the soccer stadium *(above)* Hundreds of soccer supporters surge forward as one, raising their fists in celebration of a goal by their chosen side. The group behavior of urban sports fans – often reinforced by the wearing of team colors – establishes their differences from other groups, or tribes.

Ethnicity in cities

When immigrants arrive in the cities of their own country, or in a new land, they will usually become more conscious of what makes them different from the host society and from other immigrant groups. Local identities based on residence in a particular village or membership of a clan may take second place to a wider ethnic identity, particularly when many groups of immigrants are present. Chicago in the 1920s, for example, had areas known as Little Italy, Little Poland, Pilsen (a Czech enclave), as well as districts for Irish, Jewish and Greek ethnic groups. Common residence and common workplaces can reinforce such identities, especially as people are likely to find marriage partners among those they live and work with.

Chain migration – the process by which migrants from particular areas tend to follow their friends and relatives to the same cities or even neighborhoods – replenishes the group. Neighborhood institutions such as schools, temples, mosques and churches are all places where people can meet, socialize and practice their religion together to reinforce their separate cultural identity. Stores selling special foods and events such as saint's day festivals or parades may serve to mark out the boundaries of an ethnic area

Ethnic pride (*above*) In the heart of Chicago, an enormous Polish flag flies proudly during an annual Polish Day parade. Chicago's many immigrant communities have preserved their ethnic identity by settling in distinct areas and maintaining national customs and traditions.

Urban multiculturalism (*left*) A new mosque in London's East End attests to the increasingly varied composition of modern urban societies. Local places of worship are an important focus for immigrant communities, bringing people together and reinforcing their sense of cultural identity.

– New York has several such occasions throughout the year. Neighborhoods may retain these customs as a symbolic ethnicity, long after most of the original group has moved out. This has happened in several North American Chinatowns.

Geographical concentration of groups within particular areas can be the basis for ethnic political representation if elections are based on city districts. Many governments, from the United States to the Netherlands, thus face the dilemma of whether to encourage the dispersal and integration of ethnic minorities into the wider society, or to facilitate their continued separation.

The colonial context

Colonialism produced a "top-down" version of this ethnic or tribal process. The historical record is uncertain, yet it seems probable that among precolonial and smallscale societies group identity was frequently flexible, with individuals having only a loose sense of wider collectivities outside their clans or kin groups. Often a people's name for themselves simply means "human" or "people" in their own language – the Ainu of Japan and the Inuit of the Canadian Arctic are both examples of this. The European colonialists, however, brought with them their own conceptions of ethnicity, which assumed that all peoples of a region belonged to a tribe, and to one tribe only, and invented separate names for them: Eskimo, the name commonly given to the Inuit, for example, is French in origin.

It suited the colonial administrators of Africa, as well as the settlers of North America, to classify people in tribes and rule them indirectly through tribal leaders. In missionary schools Africans were taught that they belonged to discrete tribes. This lesson, well learned, benefited the postcolonial elites struggling for power, enabling them to legitimize their leadership. Current divisions in South Africa and Zimbabwe partly stem from these practices. The United States government used tribal identities to establish reservations for the indigenous peoples, though such categorizations may have meant little to the individuals concerned.

In many formerly colonial countries, such as the United States, Australia and Canada, as well as some older imperial powers such as Britain, the concept and practice of multiculturalism is often a means of perpetuating earlier ethnic divisions. Such policies, which are aimed at providing equal treatment for all groups in society, assume that everyone belongs to a distinct group, and that each individual is simply a reflection of that group's wider identity. But in reality intermarriage blurs group boundaries, while individuals of mixed ethnic origin may choose to associate themselves with one or the other group, depending on their circumstances.

Nation and Nationalism

THE TERM "NATIONS" IS DIFFICULT TO DEFINE. However, the political culture of nationalism runs through the modern relations of peoples and societies. Nationalism is the creation of a sense of common political and cultural identity among a people. It is wider than kinship or ethnicity, though it shares some of their aspects.

Nationalism combines two broad types of identity. One is a formal sense of belonging to a political body, revealed in rights of citizenship, adherence to a common law and often involving a common education. The other is a less defined sentiment, expressed in patriotism (love of one's country), the belief in common origins and shared destiny, and the reverence for national symbols.

One of the key characteristics that distinguishes nation from ethnicity is attachment to a particular territory or land. This may be the land actually occupied by a people, exemplified by the Promised Land of the Jewish nation of Israel, granted by God in a covenant with the Hebrews. Or the land may be anticipated, something that is hoped for, a sentiment felt by stateless peoples such as Kurds, Armenians and Palestinians. In the nationalist imagination there can only be one people in any given territory – the people who truly belong there. There is consequently a close association between nationalism and racism. The demand for national purity can excite tremendous passions, often leading to violence as in Sri Lanka and the Transcaucasian republics of the Soviet Union. But in a world of immigration and migrant labor the expectation of national cultural homogeneity is increasingly unrealistic.

The mismatch between the formal and emotional character of nationhood is highlighted by attitudes to immigration. Some countries, notably the United States, whose nationhood is founded on the principle of political freedom and equality, and forged by immigration, are relatively open and generous in their award of citizenship. Others such as Japan and Britain are more exclusive, restricting citizenship but also acknowledging a difference between it and emotional nationalism. Although formally recognized as citizens, immigrants may not be popularly regarded as real members of the national community.

Nationalism may be based on the struggle for independence against wider empires. This characterized the 19th-century movements in Latin America and Eastern Europe and the 20th-century demands for autonomy from colonies in Africa and Asia. In such cases attempts to find and create a common culture based upon popular traditions of art, music and literature were often significant. By contrast, the nationalism of Russia, Britain and Germany was forged through the creation of empires. Such nations often experience difficult adjustments following the loss of their empires.

National myths

Nearly all nations tell stories about themselves. They dramatize their past, either selecting a coherent sequence of key events or untangling certain events from others that happened in the same

MARIANNE

For new nations or those with a past of deep divisions, the selection of an appropriate national and traditional symbol can be important. In 19th-century France, for example, there had been several revolutions and changes of government that divided the country along both political and religious lines. The Third Republic, founded after defeat by Germany in 1871, tried to establish new national traditions that were associated with neither the church nor the monarchy, including a new flag, a new anthem and the adoption of Bastille Day as a national festival. The figure of a woman, Marianne, was chosen as an icon of French republicanism. This followed the common European practice of using women as general symbols not associated with particular events.

Busts and statues of Marianne appeared on town halls throughout the country, though there was little official compulsion to do so. Over the years a number of French women have served as models for the figure of Marianne, including in recent times the actresses Brigitte Bardot and Catherine Deneuve. When the model Ines de la Fressange was chosen in 1989 she was immediately sacked by the French parfumiers Chanel. The head of the company regarded Marianne as a petty provincial symbol that could have no association with international fashion.

The massive stone walls *(above)* of the Zimbabwe Ruins assumed great cultural significance in the era of nationalism as a monument to the country's precolonial achievements, and on independence gave their name to the new state.

People without a land *(right)* Kurdish refugees in flight from Iraq after the 1991 Gulf War. Aspirations for a homeland produce a strong sense of nationalism among peoples such as the Kurds, often settled for centuries in the same place, who are yet without one.

The great tartan myth *(below)* The connection between certain tartans, or plaids, and particular "clans" – now an indelible symbol of Scottish nationhood – dates back only to the 19th century, and has a spurious history at best.

place. Some national myths concern origins, the moment of a nation's birth, such as the American Declaration of Independence, Poland's conversion to Christianity in 966, or the courage of Australia's troops at Gallipoli in 1915. Others speak of spatial origins, such as the widespread myths of migration among African peoples or the founding myths of the Sinhalese concerning the kings Vijaya and Dutugemenu. Many contain ideas of golden ages, or moments of national rebirth or decline, that do not necessarily distinguish between legend and historical fact. National myths are animated by the heroes who gave life to the nation and supposedly embody its natural and particular qualities. Some are historical figures, such as Simon Bolivar, the liberator of several South American states, after whom Bolivia is named. Others are more legendary, such as Ireland's King Cuchulain.

The main ways that continuity with the national past is maintained are through tradition, repeated actions or ceremonies, or through reverence for objects invested with a similar aura of permanence. The 19th century was a period of widespread invention of such traditions, including flags, national anthems, national symbols and military uniforms.

Landscapes and Monuments

LANDSCAPES – THAT IS, THE NATURAL ENvironment as transformed by the working of the human imagination and technology – are historical records, expressions and often icons of peoples and cultures. Building styles, the layout of settlements and fields, and the prominence given to certain features reveal cultural values, systems of property ownership and the significance of particular social activities.

Although most landscapes emerge from the unplanned activities of people over time, many are the direct expressions of a people's concept of their own origins. Ancient Chinese cities were laid out as models of the Earth: square, oriented along cardinal directions and located by reference to the principles of *fengshui* – the view that the landscape contains poerful spirits of ancestors, animals and mythic beings. Aboriginal Australian landscapes are organized along songlines, the paths taken by clan ancestors as they came out of the Earth and called the features of the landscape into being. Conflict occurs when landscape ideas such as these come up against secularized, rational views that regard the land as inert. In Australia, European settlement paid no regard to Aboriginal landscapes, and even today mining companies disrupt songlines and destroy sacred sites.

Europeans also have landscape principles. In Italy and France in the 16th century a combination of revived Greco-Roman ideas of harmony and geometry, together with the Renaissance view of a universe following mathematical laws, created a view of the ordered landscape that came to dominate European town planning and is found in city layouts from the Vatican City to Leningrad and the Midwestern United States. In the 20th century modern architects adapted these principles to create highly-planned cities such as Chandigarh in India and Canberra in Australia.

Landscape may also assume deeper significance as a national symbol or icon. The American west, the Russian steppes and the Japanese rice fields are examples of widely celebrated and romanticized landscapes that are seen as somehow intrinsic to a people. It is ironic that among the most urbanized societies the rural landscape is often viewed as the most authentic reflection of national culture. Yet this piety may be expressed for a landscape long since despoiled by the modernization of agriculture.

The heart of the nation

Within landscapes there may be sites of special significance – monuments and public spaces – that locate a people both within a spatial area and within a past. They give to that people not only a sense of spatial continuity, but also confirm their right to be there. Such sites may be natural features, such as Mount Fuji in Japan and Ayers Rock in Australia, or created by humans, such as the ruins of Great Zimbabwe.

Monuments often bear witness to the heroes or heroic deeds of a people's past. War is the most universal theme. From Moscow to Canberra the tomb of an unknown soldier stands as a monument to all those who perished in the nation's wars. Sometimes an empty tomb, or cenotaph, symbolizes those who have died. The state and people meet from time to time in ceremonies at these memorials that confer a sacred status on the act of war, and reemphasize the theme of national dedication.

Often there are public spaces in front of these monuments where crowds of loyal spectators can be assembled to witness the rituals that affirm a nation's identity. In Moscow's Red Square the ruling figures of the day at the annual Soviet parades stood on the tomb housing the body of Lenin, the founding figure of Soviet communism, that symbolized the power of the state. Many public spaces began as the visible expression of autocratic power. But they may also hold crowds of disaffected citizens. In Beijing's Tiananmen Square in 1989 democracy demonstrations were ruthlessly suppressed by the army, and in Bucharest President Ceausescu was refused a hearing by the Romanian people gathered in the square he had himself created in his monumental rebuilding of the city.

The sacred mount (*right*) Two Buddhist monks pray at a shrine overlooking Mount Fuji, Japan's national symbol. This natural feature is considered sacred and holds a powerful place in Japanese culture. Every summer thousands of people undertake the long climb to visit the shrine on its peak.

SADDAM'S MONUMENTS

Many men have rebuilt cities as their own memorial. The latest in the line that includes Peter the Great of Russia, who created St Petersburg (Leningrad), and the Emperor Napoleon III who rebuilt Paris, is President Saddam Hussein of Iraq. He has sought to unite a divided country and impose upon it his Baathist party's own version of history by claiming Iraq's ancient heritage and the ideas of Islamic revolution in a series of monuments. These include a reconstruction of King Nebuchadrezzar's legendary city of Babylon, as well as the Shaheed (Martyrs') Monument and the imaginative Unknown Soldier's Monument.

The most remarkable monument is the Victory Arch that commemorates the Iraq–Iran war of 1980–88. This consists of two pairs of arms rising from the ground and holding swords that meet to form two archways at either end of a huge square. The arms have been modeled directly from Saddam's own, complete in every detail, while the swords are replicas of those of Sa'ad ibn-abi-Waqas, the commander of the Muslim army that defeated the Persians in 637. They have been cast from the weapons of Iraqi soldiers. At the point where the arms spring from the ground there are metal nets holding the helmets of Iranian soldiers. Thus Saddam Hussein has identified himself with a past hero in the war for Islam against Persia (now Iran). The irony of the monument is that it was conceived as a victory celebration before the war – which did not end in victory but in stalemate – had finished.

Reflected glory *(above)* The Martyrs' Monument in the center of Baghdad is one of the impressive monuments built by Saddam Hussein as a means of identifying his political authority with Iraq's historical heritage as a champion of Islam.

Revival and Survival

THE 20TH CENTURY HAS WITNESSED AN unparalleled acceleration in the diffusion of ideas, technologies and commodities across the world. Some observers feared that the dominance of Western economies would lead to a leveling of cultures – a homogenizing process sometimes termed "westoxication" or "cocacolonization" – though not everyone viewed this as necessarily a harmful development. Some Western economists argued that the modernization of traditional cultures was necessary to increase standards of living. The aim of international socialism was to sink national and ethnic identities in a larger socialist identity. In the event, homogenization did not occur, in many ways global economic changes established the conditions that allowed cultural diversity to flourish.

The Western countries failed to produce uniform national cultures. Their cities are the sites of a variety of new lifestyles, from San Francisco's gays to the alternative communes of Europe. Ecology and feminism are examples of cultural movements that, though not community-based, question deep-seated assumptions about humanity, nature and personal life. In a climate of increasing religious and cultural tolerance, today's immigrants are often able to create vibrant and distinct communities.

The revival of European regional cultures after World War II, which challenged the integrity of the older powers, often drew on colonial struggles across the developing world for their inspiration. Basques, Corsicans, Quebecois and others have combined demands for regional autonomy with a renewal of their distinct languages, arts and ceremonies. Many use increasing access to communications media – once thought by authoritarian states to be a force for control – to express their identity. These movements indicate a plurality of cultures that questions the desirability of national cultural homogeneity. Spain, Belgium and Canada are among those countries whose constitution has been amended to facilitate diversity.

Outside the West, capitalism did not totally destroy local cultures, though the majority have undoubtedly been transformed. Regional peasant societies may

Peoples under threat Indigenous (so-called tribal) peoples have escaped cultural assimilation and retained their centuries-old ways of life only in remote, inaccessible parts of the world. Today even these survivors are threatened with extinction, particularly those inhabiting the tropical rainforests.

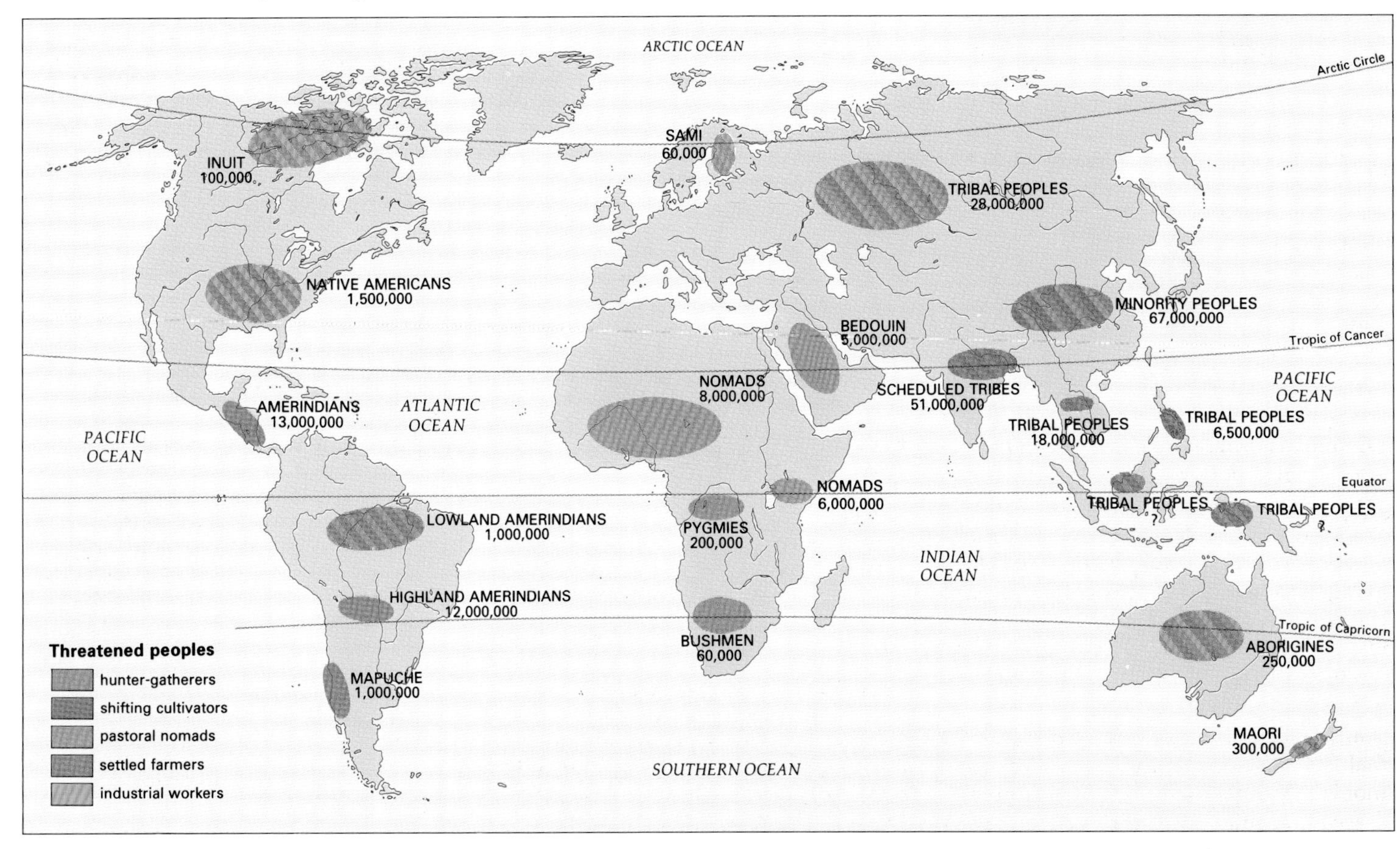

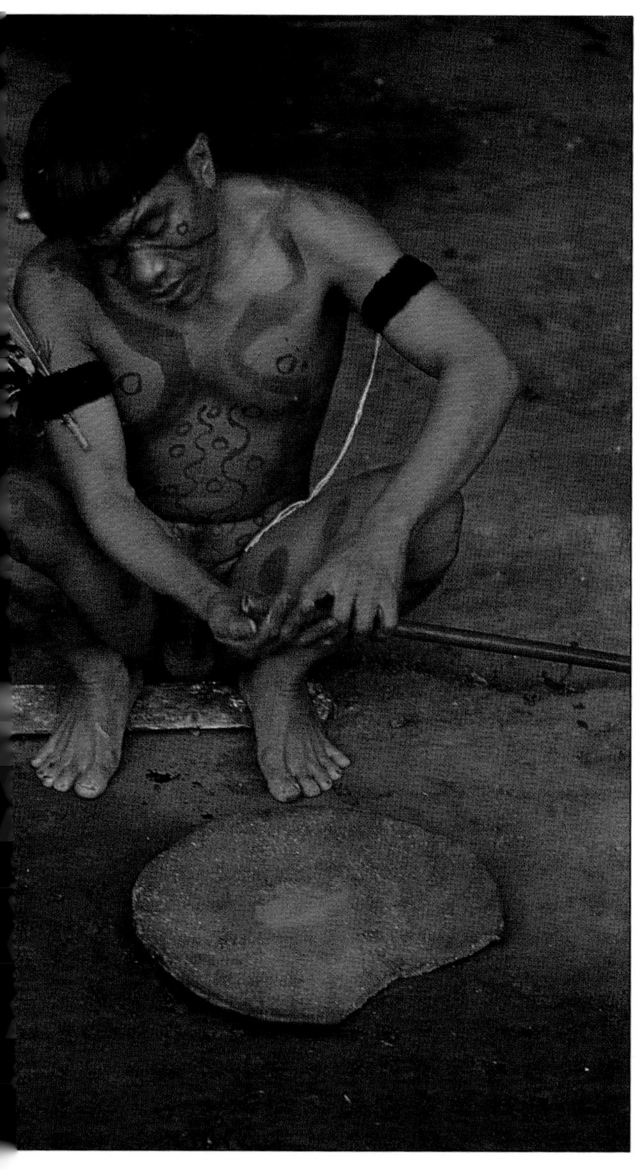

Glad to be gay *(above)* Members of San Francisco's large gay community make their own particular statement at the annual Lesbian and Gay Freedom Parade. Alternative lifestyles have found room to flourish in the cosmopolitan and culturally diverse atmosphere of many larger towns and cities in the West.

Fighting for survival *(left)* The warlike Yanomami people of Venezuela and Brazil live in scattered villages in the forests of the Orinoco River basin. For all their ferocity, they have been powerless to resist the diseases and forest clearances that threaten their way of life and very existence.

buy Western products but they also maintain local traditions, even when their members migrate to expanding urban centers. Bonds of kinship and exchange networks may in fact be useful resources of economic survival.

Indigenous peoples in danger

Peasant societies have historically proved both resilient and adaptable, able to weather revolutions, land reform and other dramatic changes. The same cannot be said for many small-scale hunter–gatherers, shifting cultivators and pastoral nomads. There are some 200 million so-called tribal peoples, three-quarters of them in Asia, whose lifestyles are under threat as the very ecological bases of their cultures are being destroyed. Mining, logging and the flooding of valleys to form reservoirs displace them; smallpox and venereal diseases, introduced by migrant laborers, reduce their numbers; while in some places evangelical missionaries seek to overturn their traditional beliefs. A third of Brazil's Yanomami people died in the 1970s from disease and battles with gold prospectors. The San (bushmen) of Botswana are being displaced to create nature reserves. The numbers of the Onge of the Andaman Islands, the Junira of Brazil's Mato Grosso and the Waorani of Ecuador have fallen so much that these peoples face complete extinction. For others, such as Mexico's Lacandan peoples, or the Papuans of Irian Jaya, government hostility comes very close to ethnocide, or the complete eradication of their culture.

For larger indigenous groups in the liberal democracies, the 1960s brought promising signs in the form of land rights and limited political autonomy. In 1971 the United States government created a number of tribal-run land corporations in Alaska, granting them mineral rights, a move that Canada later adopted. South of the Canadian border, Native-American reservation lands are also held by trusts that retain the land in communal rather than individual ownership. The Australian Aboriginal land rights battle began in the 1960s as an attempt to recover land from both the state and a variety of trustee organizations. Since then, many lands have been returned, including in 1988 the sacred site of Ayer's Rock to the Pitjantjatjara. Some Australian states, however, have so far granted only 99-year leases.

THREATENED PEOPLES OF IRIAN JAYA

The 1 million or so indigenous inhabitants of Irian Jaya, part of the island of New Guinea in Southeast Asia, are divided into 250 or more distinct linguistic groups. Many are isolated by the island's mountainous terrain. Some groups in the interior valleys were not visited by outsiders until the 20th century. Most maintain a traditional lifestyle, based on animistic beliefs, They practice hunting–gathering or shifting cultivation.

Annexed by Indonesia in 1969, the sparsely populated lands of Irian Jaya are being used to relieve the overcrowding of the island of Java through a government resettlement program. Out of a planned total of 4 million, so far 300,000 settlers have arrived. Land has been seized by the government, often without compensation, and cleared, adding deforestation to displacement.

A resistance movement claims that the resettlement program amounts to the total destruction of their culture, since it severs people from their land and lifestyles. Christian missionaries have added to this cultural extinction by discouraging nakedness, polygamy, stone fetishes and other elements of traditional culture. Once contact with the outside world was made, however, it is difficult to see how change could have been avoided.

World Culture, Global Spectacle

Where in the world? *(above)* This traffic policeman in Bangkok would not look out of place on the streets of Miami. The age of mass communications and air travel has brought peoples from all around the globe into contact with one another, exposing them to new cultural influences and reducing the differences between them.

THE FIRST PICTURES RETURNED TO EARTH BY the Apollo and Soyuz spacecraft in the 1960s made real what until then could only be imagined and projected mathematically. The dramatic image of the planet isolated in black space gave a global dimension to the relationship between humanity and nature. A new set of questions were raised. Concerns with overpopulation, food supply and environmental pollution also focused on the ethical issue of human and planetary survival. From space, the idea of the world as a mosaic of discrete human cultures recedes into the reality of One Earth. The rapidity with which the 1987 stockmarket collapse sped around the world demonstrated how the communications revolution has shrunk the globe.

The world is growing smaller in other ways, too. By 1990, 400 million tourists were traveling to other countries, while another 200 million or more journeys abroad on business. By the year 2000 tourism will be the world's largest industry. Tourists are principally spectators, consumers of other people's cultures and landscapes, capturing them in photographs and souvenirs. But in trying to encounter others, their concentrated numbers inevitably change them.

Cultures on view and for sale

Tourism places national monuments and icons in a new global context, partially dispossessing local peoples of their history and heritage. Traditional peoples who have been drawn into modern economies have realized that they can stage their ceremonies for tourists. Japan's Ainu, for example, a despised minority, have turned their difference to an advantage by recreating traditional villages and inviting outsiders to watch them follow their crafts. In Borneo, Papua New Guinea and North America, other formerly tribal peoples have turned their tradition into a marketable commodity.

In recent decades traditions and the past have also become more commercialized in developed countries. The heritage industry – the packaging and selling of the past for touristic consumption – suggests an ambivalence toward culture, as being both essential to a people's identity and a sellable product, as well as a nostalgia for the past that can be taken to indicate uncertainty about present identity.

A counterflow of foreign foods, artifacts, music and literature enables some citizens of the wealthy world to consume elements of other cultures, but only at the

WORLD RITUALS

When the astronaut Neil Armstrong walked on the Moon in 1969 he was watched on television by 723 million people in 47 countries. Television enables the world to create its own rituals and cultural events. By far the most elaborate of these are the Olympic Games: the Los Angeles games in 1984 attracted an audience of 2,500 million. However, since the 1980 Moscow games they have been split by nationalistic rivalries based on the Cold War and the politics of apartheid. The 1990 soccer World Cup in Italy was watched by 1 billion people across the world. It, too, is a spectacle infused the nationalism. The defeat of national teams prompted suicides in Brazil and riots in some small English towns.

Perhaps only charity events, such as the Live Aid concert in 1985 – seen by 1.6 billion people live in 100 countries – come close to achieving a really distinct world cultural event. Even then, it took the suffering of one country, Ethiopia, to establish briefly a worldwide community of strangers.

cost of uprooting them from their proper contexts. Archaeologists and explorers in the 19th century removed cultural objects from many places in the world to fill the great museums of Europe and the United States. Anything from the smallest artifact to entire buildings, such as the Egyptian temples reconstructed in New York and Madrid, can enter the global circulation of culture. Many peoples are now demanding the return of such treasures: Australia's Aborigines, for example, have regained some of the remains of their ancestors from British museums. Even developed countries fear the loss of their cultural objects, particularly paintings and sculptures that have entered the world art market.

With an audience of millions (*left*) The spectacular opening ceremonies of the 1990 World Cup in Italy were staged less for the spectators in the stands than for the worldwide television audience. The ubiquity of the television set has created a new cultural phenomenon; that of a world event that briefly links its many different peoples.

"Cocacolonization" (*below*) One indication of how far the world has shrunk in cultural terms is afforded by the sales performance of certain Western consumer goods. Where once the British empire painted the world map red, today it is the international corporations. In 1990 Coca Cola found markets for its products in countries in every continent, including the Soviet Union and China.

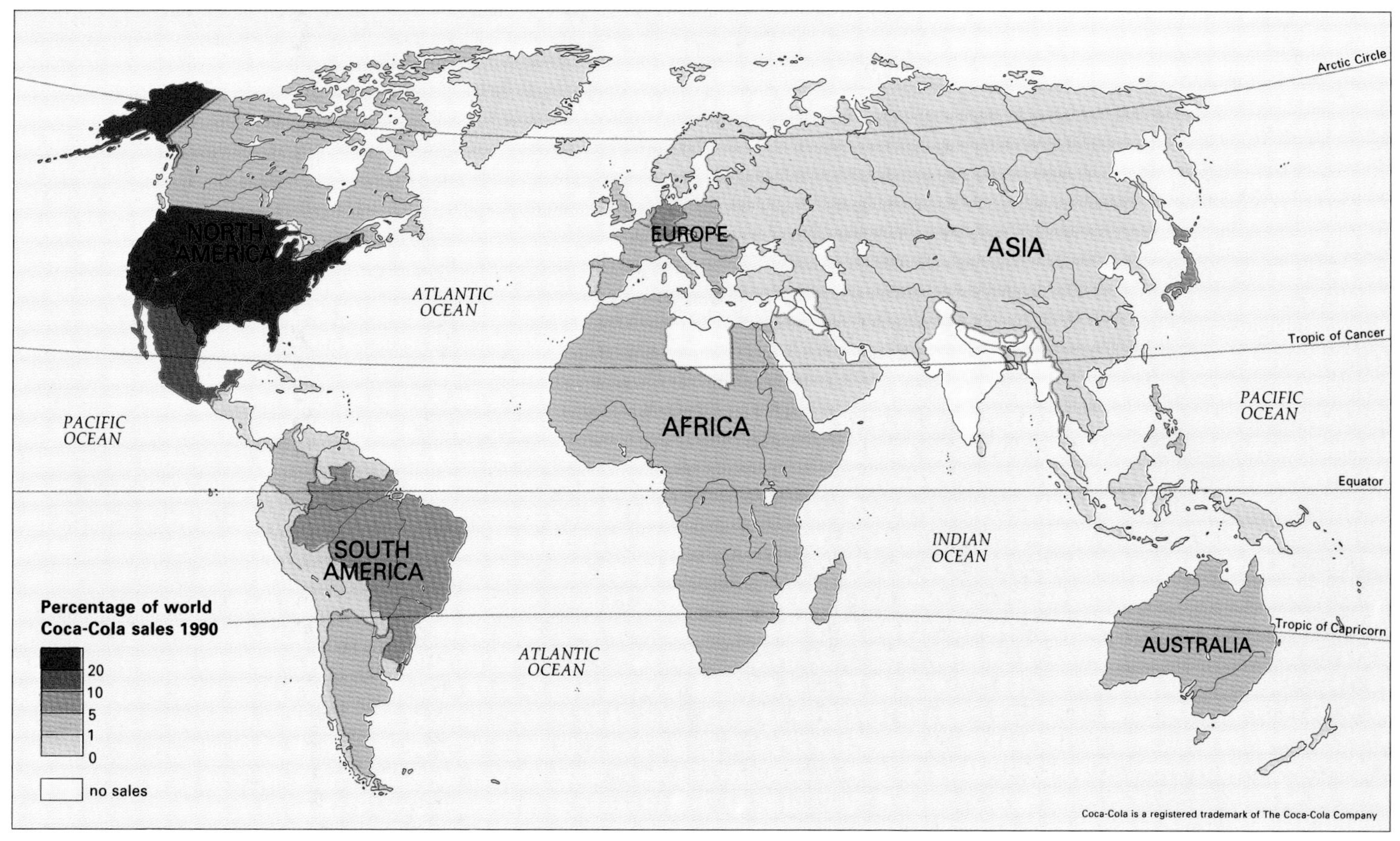

REGIONS OF THE WORLD

CANADA AND THE ARCTIC
Canada, Greenland

THE UNITED STATES
United States of America

CENTRAL AMERICA AND THE CARIBBEAN
Antigua and Barbuda, Bahamas, Barbados, Belize, Cosat Rica, Cuba, Dominica, Dominican Republic, El Salvador, Grenada, Guatemala, Haiti, Honduras, Jamaica, Mexico, Nicaragua, Panama, St Kitts-Nevis, St Lucia, St Vincent and the Grenadines, Trinidad and Tobago

SOUTH AMERICA
Argentina, Bolivia, Brazil, Chile, Colombia, Ecuador, Guyana, Paraguay, Peru, Uruguay, Surinam, Venezuela

THE NORDIC COUNTRIES
Denmark, Finland, Iceland, Norway, Sweden

THE BRITISH ISLES
Ireland, United Kingdom

FRANCE AND ITS NEIGHBORS
Andorra, France, Monaco

THE LOW COUNTRIES
Belgium, Luxembourg, Netherlands

SPAIN AND PORTUGAL
Portugal, Spain

ITALY AND GREECE
Cyprus, Greece, Italy, Malta, San Marino, Vatican City

CENTRAL EUROPE
Austria, Germany, Lichtenstein, Switzerland

EASTERN EUROPE
Albania, Bulgaria, Czechoslovakia, Hungary, Poland, Romania, Yugoslavia

THE SOVIET REPUBLICS AND THEIR NEIGHBORS
Estonia, Latvia, Lithuania, Mongolia, Union of Soviet Socialist Republics

THE MIDDLE EAST
Afghanistan, Bahrain, Iran, Iraq, Israel, Jordan, Kuwait, Lebanon, Oman, Qatar, Saudi arabia, Syria, Turkey, United Arab Emirates, Yemen

NORTHERN AFRICA
Algeria, Chad, Djibouti, Egypt, Ethiopia, Libya, Mali, Mauritania, Morocco, Niger, Somalia, Sudan, Tunisia

CENTRAL AFRICA
Benin, Burkina, Burundi, Cameroon, Cape Verde, Central African Republic, Congo, Equatorial Guinea, Gabon, Gambia, Ghana, Guinea, Guinea-Bissau, Ivory Coast, Kenya, Liberia, Nigeria, Rwanda, São Tomé and Príncipe, Senegal, Seychelles, Sierra Leone, Tanzania, Togo, Uganda, Zaire

SOUTHERN AFRICA
Angola, Botswana, Comoros, Lesotho, Madagascar, Malawi, Mauritius, Mozambique, Namibia, South Africa, Swaziland, Zambia, Zimbabwe

THE INDIAN SUBCONTINENT
Bangladesh, Bhutan, India, Maldives, Nepal, Pakistan, Sri Lanka

CHINA AND ITS NEIGHBORS
China, Taiwan

SOUTHEAST ASIA
Brunei, Burma, Cambodia, Indonesia, Laos, Malaysia, Philippines, Singapore, Thailand, Vietnam

JAPAN AND KOREA
Japan, North Korea, South Korea

AUSTRALASIA, OCEANIA AND ANTARCTICA
Antarctica, Australia, Fiji, Kiribati, Nauru, New Zealand, Papua New Guinea, Solomon Islands, Tonga, Tuvalu, Vanuatu, Western Samoa

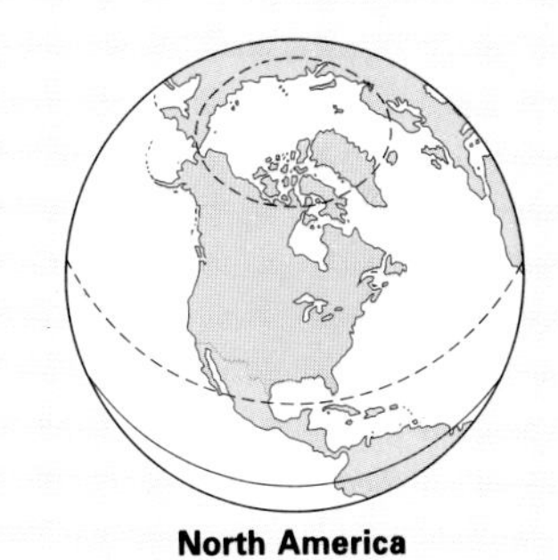

North America

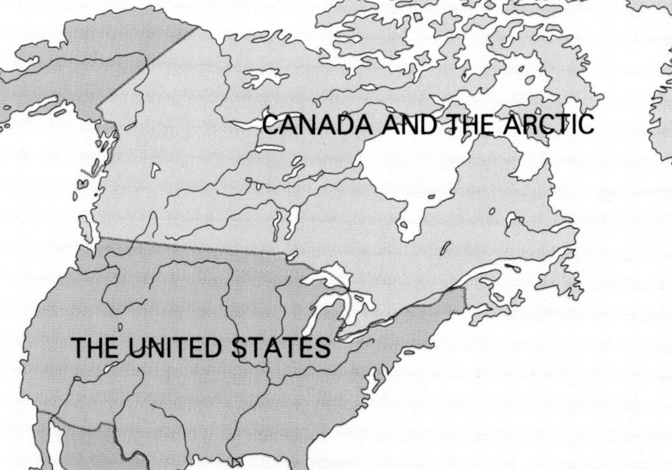

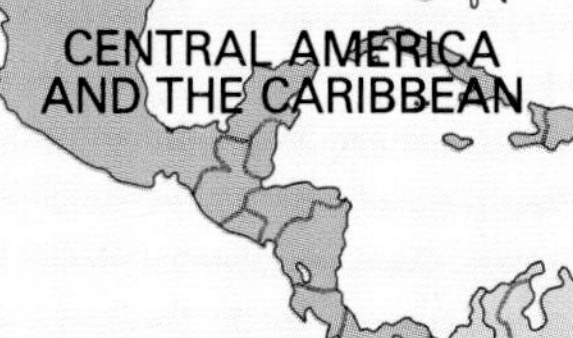

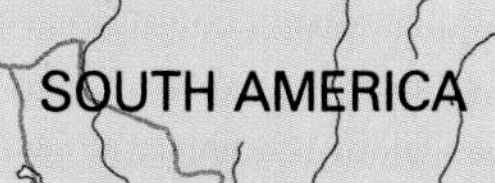

Central and South America

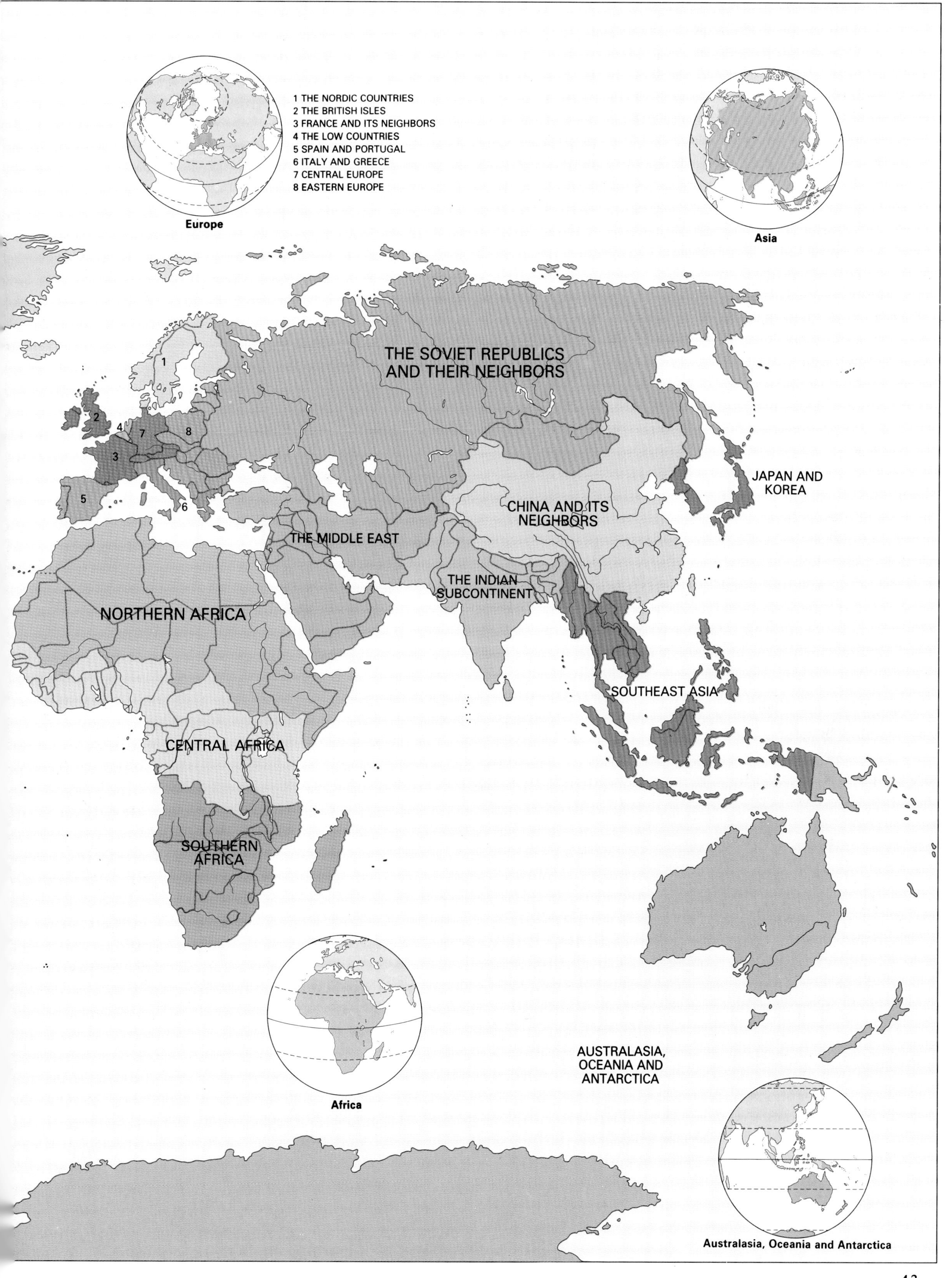
1 THE NORDIC COUNTRIES
2 THE BRITISH ISLES
3 FRANCE AND ITS NEIGHBORS
4 THE LOW COUNTRIES
5 SPAIN AND PORTUGAL
6 ITALY AND GREECE
7 CENTRAL EUROPE
8 EASTERN EUROPE
Europe
Asia
THE SOVIET REPUBLICS AND THEIR NEIGHBORS
JAPAN AND KOREA
CHINA AND ITS NEIGHBORS
THE MIDDLE EAST
THE INDIAN SUBCONTINENT
NORTHERN AFRICA
SOUTHEAST ASIA
CENTRAL AFRICA
SOUTHERN AFRICA
AUSTRALASIA, OCEANIA AND ANTARCTICA
Africa
Australasia, Oceania and Antarctica

THE ENIGMA OF CANADIAN IDENTITY

TWO CULTURAL TRADITIONS · A DISTINCT NATION · TOWARD A MULTICULTURAL SOCIETY

Canadians are not easily defined. The population of 26 million, living in scattered pockets stretched out in a narrow belt along a vast east–west territory that covers almost 90 degrees of latitude, is made up of a collection of diverse ethnocultural groups, the result of Canada's history of welcoming immigrants from Europe, and more recently from Asia and Latin America. Yet these groups are accommodated within a fundamental cultural dualism that divides the country into an English-speaking and a French-speaking community. The question of national identity has inevitably been an abiding concern: are national icons English, French, neither or both? Are dominant loyalties regional or national? How are the indigenous populations of Native Indians and Inuit (Eskimos) to be integrated into a modern state?

COUNTRIES IN THE REGION

Canada, Greenland (part of the Danish realm)

POPULATION

Canada	26.2 million
Greenland	52,940

LANGUAGE

Countries with two official languages (English, French) Canada; (Danish, Greenlandic) Greenland

Percentage of population by first language (Canada) English (61%), French (24%), other (11%), bi- or multilingual (4%), indigenous languages (0.3%)

RELIGION

Canada Roman Catholic (46.5%), Protestant (41%), nonreligious (7.4%), Eastern Orthodox (1.5%), Jewish (1.2%), Muslim (0.4%), Hindu (0.3%), Sikh (0.3%)

Greenland Protestant (97.8%); other (2.2%)

TWO CULTURAL TRADITIONS

The original inhabitants of Canada made their way over a land bridge across the Bering Strait from northern Asia about 30,000 years ago; to the far north were Inuits belonging to the group of hunter–gatherers that lived along the edge of the Arctic ice from Siberia to Alaska. The first contact with Europe came about 1000 AD when Norse explorers established a brief settlement at L'Anse aux Meadows in Newfoundland – an event that is recorded in the Icelandic sagas. Nothing more occurred for more than 500 years until, following the discovery by English and French sailors of the Newfoundland coast and the St Lawrence river, fishing fleets from European ports started to exploit the rich North Atlantic fisheries and established seasonal settlements along the coast. Gradually, as some people decided to spend the winter there, these became permanent settlements.

The making of French Canada

At this time, the region was inhabited by between 220,000 and 250,000 indigenous peoples belonging to one of 12 major language groups. Apart from the Iroquois, who were settled as farmers in the area around Lake Ontario, most survived by hunting and fishing, moving around with the seasons through much of the continental interior. Trading contacts were established, first by the French and then by the English, with these migratory groups, who exchanged furs, highly valued in Europe, for a variety of goods. Competition for trade between European rivals was fierce. The introduction of firearms brought substantial changes to the traditional Native Indian patterns of hunting and warfare. Gradually they were displaced by European settlement, and were compelled to abandon their seasonal migratory way of life.

The French were the first Europeans to open up the interior to settlement. Progress, following the course of the St Lawrence river, was slow, limited by the climate and the volatility of relations with indigenous groups. Quebec, founded in 1608, was by 1800 the largest settlement away from the coast, and the center of French-speaking Canada, yet its population was only about 8,000 – roughly the same size as Halifax, the capital of Nova Scotia, which was the largest coastal town. After 300 years of European contact, the total European population of the seven colonies of British North America, which would come together in the course of the 19th century to form the Canadian confederation, was no more than 340,000 – and 60 percent of these inhabitants were French-speaking.

The British influence

The French-speaking people, however, were living within a territory that since 1763 had been British-owned. The British element in Canada (at this time predominantly Scots and Irish) was strengthened by the addition of thousands of loyalist emigrants from the newly independent United States of America to the south, and the division of the colony into English-speaking Upper Canada and French-speaking Lower Canada (corresponding to Ontario and Quebec today) confirmed the existence of two parallel cultures. The English-speaking community looked to Britain for its laws, its models of government and its Protestant Christian culture; French-Canadians retained the use of their own language, their distinctive system of land-holding, and their Roman Catholic religion.

Expansion westward across the continent did not fundamentally alter this cultural division. In 1901 only 10 percent of the population fell outside the Anglo-Canadian or the French-Canadian communities, or were not members of the indigenous peoples. These included the *Métis*, people of mixed Indian and French or British ancestry, who had developed a distinctive way of life and thought of themselves as a nation with rights in the area that was to become Manitoba, west of Hudson Bay.

Control of political and economic resources made the British influence culturally dominant in most of Canada. Although there were some local concentrations of minorities – for example, Asian settlers, predominantely Chinese and Japanese, accounted for 11 percent of the population of British Columbia in 1901, and there were communities of European immigrants (particularly Scandinavian, Dutch, Italian, German, Polish and Ukrainian) across the continent – they were islands in a much broader ocean of Anglo-Canadian culture.

Even in Montreal, the commercial center of the province of Quebec, the English-speaking minority successfully

dominated local business and politics – a change that had started to come about when the French fur trade was taken over by British interests. They came to form a ruling elite that was segregated by language, religion, class and residence from a largely rural, Roman Catholic French-Canadian majority.

British influence in Canada attained its highwater mark between 1914 and 1918. Many Canadians fought on the British side in World War I – the Molsons, one of Montreal's leading industrialist families, sent 30 members to the war in Europe, more than half of whom were killed or wounded in action. However, by the

1920s United States' investment in Canada had outstripped British investment and in the period since then its popular culture – spread through newspapers and journals, television, movies, and the increased movement of population – has become a pervasive element in Canadian culture. Even to the people of Quebec the United States' border has become increasingly permeable. Over a quarter of a million older Quebecois are today permanently resident in Florida.

A Quebec street scene (*above right*) highlights some of the influences at work in Canadian society. French shops and chateau-style architecture reveal European links; the rise in popularity of American sports shows the impact of the United States.

Urban convenience (*above*) Huge American-style shopping malls, where shoppers can park and purchase all their needs under one roof, are well suited to Canada's often rigorous climate.

A sign of belonging (*left*) Canada's national symbol of a red maple leaf helps to unite its diverse population. It is, however, relatively recent: the Canadian flag bearing the maple leaf only became official in 1965.

A DISTINCT NATION

The establishment of a coherent cultural identity has been an elusive goal in Canada. This is common to most former colonial nations, but in Canada the problem is compounded by the distances that divide its centers of population, the bicultural nature of its society, and the fact of its living next to an economically more powerful neighbor. Canada has, in many respects, substituted one dominant presence for another. Sharing an immense land boundary and with an economy and population some ten times smaller than the United States, it has a major task in defining its distinctiveness.

One example of how this distinctiveness expresses itself at the popular level is found in the area of sport. Hockey (which in Canada means ice hockey) is a major national interest, and a powerful symbol of national unity. For decades one of the most popular television programs has been the Canadian Broadcasting Corporation's *Hockey night in Canada*. Yet even in the context of sport, competition from the United States is evident. Baseball is growing in popularity, and Canadian baseball teams compete in the United States' leagues.

The threat from the south

Canada's cultural tug between its own identity and that of the United States makes itself felt in a number of ways. For example, there is confusion over the style of spellings used in written English – although the government and large business corporations (including United States-owned multinationals) will usually follow the British spellings of words ("neighbour" rather than "neighbor", for example). At the popular level, including newspapers, an inconsistent mixture of British and American spellings is used.

During the 1970s and early 1980s, concerns over Canada's cultural and economic independence led to the then Liberal administration's program of "Canadianization": attempts were made to nationalize the economy, and the constitution was freed from British parliamentary control by the passing of the Canada Act. These moves stimulated a wide-ranging debate on national identity, which came to a head with the vehement discussion over the signing of the Free Trade Agreement (FTA) with the

National passion (*above*) Ice hockey, developed in its present form in Canada during the 19th century, is the country's most popular sport. Children are encouraged to compete for the big teams from an early age. Thousands of amateurs play in leagues across the country, and top professional players have the status of national heroes. Team loyalties are affected by ethno-religious boundaries.

Canada's changing face (*left*) The image that many Canadians have of themselves is colored by the country's recent pioneer past, but belies the fact that theirs is now a highly urbanized society. Highrise commercial buildings dominate the center of Edmonton, capital of the western prairies, dwarfing its earlier, more modest architecture.

United States in 1988, opponents (who failed in their political attempt to prevent the agreement) arguing that it would erode Canadian identity by threatening its cultural and social traditions.

Among the people who spoke most vociferously against the FTA were artists and other members of the cultural media. The arts, broadcasting, publishing and film industries have always enjoyed a privileged position in Canada, provided with subsidies and tax incentives to advance the cause of Canadian distinctiveness. There are, for example, specific guidelines to regulate the degree of Canadian content in broadcasting: the 1968 Broadcasting Act identifies the need "to safeguard, enrich and strengthen the cultural, political, social and economic fabric of Canada" through the Canadian ownership of broadcasting.

Canadian artists and intellectuals consequently possess significant material grounds for resisting Americanization, in addition to their own individual emotional response to the issue. They have played an important role in contributing to Canada's images of nationhood. In the 1920s and 1930s, the Toronto-based Group of Seven established the first major school of Canadian art. Their paintings, which represented "the true north, strong and free", portrayed natural landscapes; typically, though not always, they showed primitive yet colorful wilderness scenes devoid of settlement.

Rural pioneers to urban dwellers

These images of northern austerity and purity were offered as icons of Canadian

THE HUTTERITES – A RURAL COMMUNITY

While most of the European immigrant groups to Canada have sought to preserve some elements of their old-world culture, they have been assimilated to a greater or lesser degree into Canadian society. A few, however, have sought complete seclusion in rural settlements to maintain their distinct ways of life free from contamination from mainstream cultures.

One of the most interesting of these groups is the Hutterites. Originally from Moravia (today part of Czechoslovakia), they take their name from Jacob Hutter (d. 1536), a religious leader who was burned as a heretic. Taking their authority from a strict interpretation of the Bible, they hold all goods in common, and are the oldest and the largest communal group in the Western world. On emigrating from Europe in the 19th century, they first settled in the United States and only moved into the prairie provinces of Canada in 1940, during World War II, to avoid persecution for their pacifism.

Living on collective farms, Hutterites pursue a modern commercial agriculture, and have a reputation as innovative and successful farmers. Their farming colonies increased rapidly from 52 communities in 1940 to nearly 247 a little more than 25 years later, reflecting their high birthrate: colonies divide once the population reaches about 150. This is done to safeguard the financial wellbeing of the parent colony, and it is the Hutterites' successful pursuit of a modern commercial economy that enables them to preserve their culture.

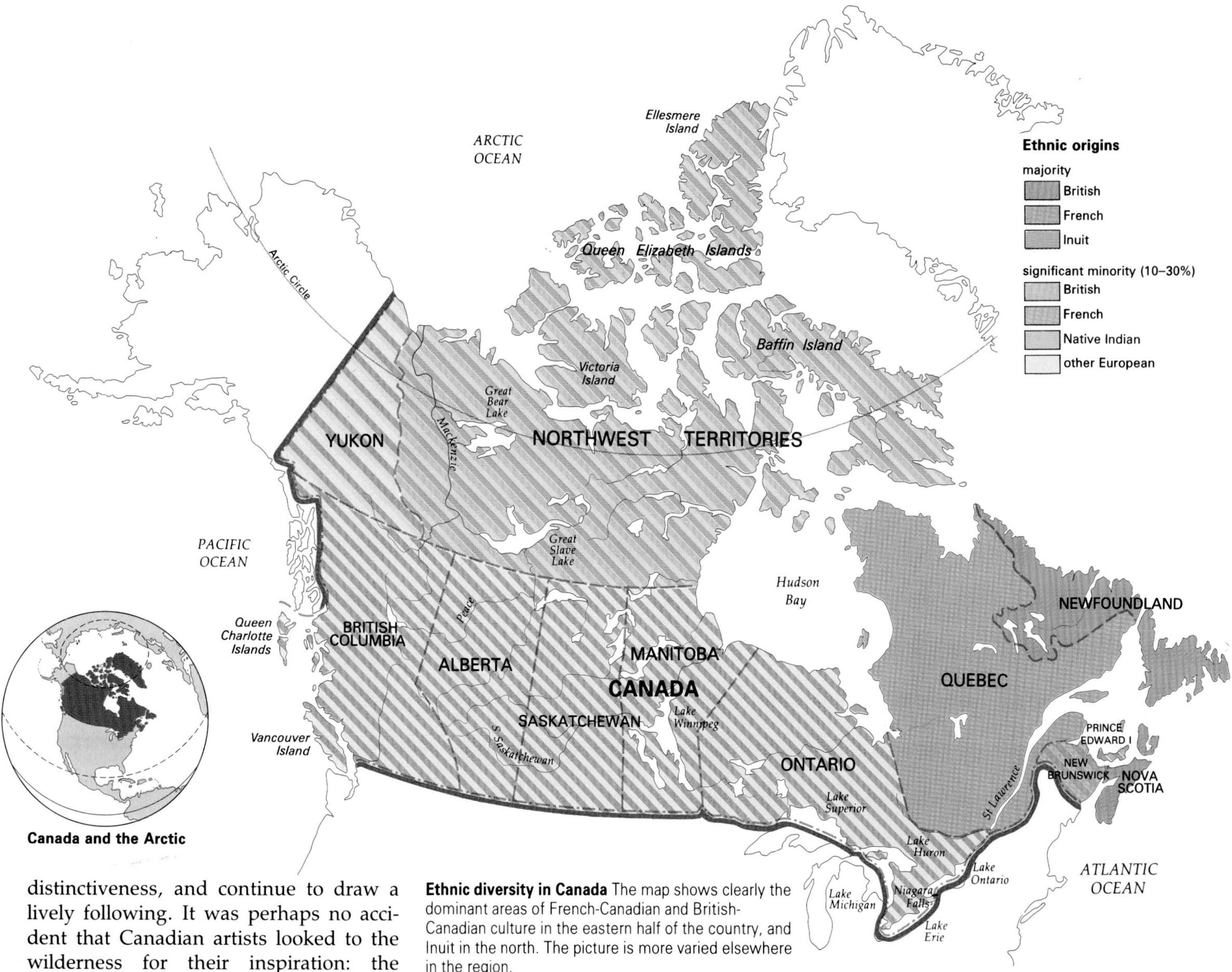

Ethnic diversity in Canada The map shows clearly the dominant areas of French-Canadian and British-Canadian culture in the eastern half of the country, and Inuit in the north. The picture is more varied elsewhere in the region.

distinctiveness, and continue to draw a lively following. It was perhaps no accident that Canadian artists looked to the wilderness for their inspiration: the choice of a maple leaf to symbolize Canada's nationhood on the country's flag points to a deep-rooted identification with the natural environment. But such an image today overlooks the fact that Canada is now one of the most urbanized of nations. Canadians may seek to recreate their pioneer past by weekending at lakeside cabins and taking part in log-rolling competitions, and the northern wilderness may offer a useful symbol of national cohesiveness, but it is far removed from the everyday life of most Canadians working in office blocks in large urban centers.

For these people, the rural conservatism and traditional religious values of the past (whether the strict Protestant Christianity of those of Scottish, German and Scandinavian descent, or the devotional piety of French Roman Catholics) have mostly been replaced by urban, secular values. This transformation has been most marked in Quebec, where the role of the church as the dominant institution in the province was successfully challenged and displaced in the 1960s by the "Quiet Revolution" of the Liberal administration, led by Jean Lesage. In this period of rapid secularization and modernization, Quebec's birthrate plummeted from being one of the highest to one of the very lowest to be found among Western societies.

The division between urban and rural values has been at the root of two bitter ideological struggles that have divided Canadian society in recent years. The base of support for the "pro-choice" lobby in the debate over abortion, enunciating the right of women to control their own fertility, has been in the major urban centers, while "pro-life" advocates have drawn their strength from the suburbs and small towns. Similarly urban opinion has been unequivocally expressed on a range of environmental questions, in particular the culling of seals on the Atlantic coast, while small town residents balance these arguments against their own economic livelihood.

TOWARD A MULTICULTURAL SOCIETY

From the earliest days of British power in Canada, the interests of the distinct French cultural community in Quebec were legally safeguarded within the constitution. This has not been the case with Canada's other cultural groups. For example, would-be Chinese immigrants to British Columbia at the turn of the century had to pay a head tax. In 1895 this was rated at $50 per immigrant, but it increased dramatically to $500 in 1905 – a figure that reflected contemporary anti-Oriental sentiment.

British Canada's prevailing rhetoric of race and Empire established a clear ethnic pecking order among immigrants, and many groups faced a marginal existence, whether in Irish Catholic slums in Montreal, deprived Eastern European districts in Winnipeg's North End, or in the barely tolerated ghetto of Vancouver's China-

Kensington market, Toronto, is a lively meeting place for people of many different cultural traditions and from many parts of the world. One-third of recent immigrants to Canada, particularly from Hong Kong, Southeast Asia and India, have settled in the city.

town. However, as the numbers of various groups increased, their social status slowly improved. Canada's Jewish population – principally made up of immigrants from Eastern Europe – had grown from 16,000 in 1901 to 126,000 in 1921. Most were initially employed within the garment industries of Toronto and Montreal, and as they grew they prospered to form a significant merchant class. Progress among the Asian communities, however, was delayed by a lingering anti-Orientalism. During World War II this found political expression in the decision to intern Japanese-Canadians and confiscate their property.

More liberal policies in the postwar years have seen the flow of immigrants increase in national diversity. In 1957, 95 percent of immigrants were from Europe or the United States, by far the greater number of them from Britain. Ten years later this figure had fallen to 80 percent, declining still farther to 47 percent in 1977 and 24 percent in 1987. The leading countries of origin in 1989 were Hong Kong, Poland, the Philippines, India and Vietnam. Before the war most immigrants were unskilled laborers; but since then greater numbers of middle-class immigrants have been admitted, a trend that the Canadian government's policy of welcoming political refugees encouraged.

Pressure from ethnic groups

By far the greater number of new arrivals seek a life in the cities. In 1988, more than a third of all immigrants settled in the Toronto metropolitan area. This has had a significant influence on the nature of small businesses setting up in commercial streets, and has led to changes in domestic styles of architecture.

There has been among these groups a growing sense of exclusion from Canada's bicultural society, particularly in the light of the debate about national identity that the Quebec separatist movement has stimulated. Under pressure from ethnic groups, positive attitudes to multiculturalism have been encouraged by providing funds for the promotion of ethnic festivals and community centers. A growing consciousness of ethnic rights is reshaping public policy in other ways, too. Japanese-Canadians have launched a campaign to gain redress for wartime losses, and the issue of multiculturalism looms large in debating current immigration policy.

French voices

In Quebec, the ideals and policies of multiculturalism collide uncomfortably with regional goals. The province's falling birthrate means that immigrants are needed to sustain its economy and labor force. But few are native French-speakers, and since 1977 French has been the language in which all work and education are legally conducted: immigrants are required to send their children to French-language schools.

The increasingly protective nature of legislation to preserve the French language in Quebec has led to some unusual situations. In 1983 the Language Commission had to pass a ruling on a case that involved the daughter of a woman who had died in hospital. She complained that her mother "didn't die in French", and the Commission – which heard evidence from a doctor and three nurses – concluded that the woman had died only 66 percent in French, since she had been treated by English speakers for 34 percent of her last months in hospital.

As a result of the campaign for French language rights, there is today a fuller and more confident consensus in Quebec society than ever before. That Quebec separatism continues to pose a most serious challenge to national unity was made clear in the constitutional crisis that erupted in 1990 following the emergence of opposition to the Meech Lake Accord and the subsequent failure to secure Quebec's standing as a "distinct society" within Canada. Acceptance of the accord might have been forthcoming if some provincial governments had not pressed for an equal recognition of distictiveness, in effect suppressing Quebec's claim of a unique cultural status.

MONSTER HOUSES

The ever-growing multiculturalism of Canada has led to some difficult adjustments on all sides. In some major cities, particularly Toronto and Vancouver, which both contain sizable Chinese communities, controversy has raged since the mid 1980s over the building of "Monster Houses" – large, sometimes ostentatious new houses on cleared lots. Special incentives for business immigrants have encouraged wealthy Asian households – prominent among them families from Hong Kong who have moved to Canada to avoid the feared consequences of the colony's reunion with China in 1997 – to take up residence in upper middle-class neighborhoods. Such districts usually have a well-established Anglo-Canadian appearance, with houses built in Tudor or other European architectural styles discreetly hidden behind mature trees and shrub fences.

Chinese residents have very different values, based on the principles of *feng shui*, which take into consideration how propitious the site of a building is and direct buyers to particular favored locations and street numbers. New structures are preferred to old, and sun-blocking trees are removed to achieve the right balance between light and shadow – consequently large and often ostentatious houses on cleared lots are greatly favored.

This clash between the traditional Canadian landscape values and a no less venerable Asian style has led to several confrontations and poses a dilemma for planning departments, which are thrust into the role of cultural adjudicators. In one celebrated conflict, the destruction of two mature sequoia trees on a street in Vancouver by a Chinese resident generated a public row that was reported on the front page of newspapers in Hong Kong.

Northern frontier, northern homeland

When the French established their trading post at Quebec in 1608, the surrounding area was home to the Huron-Petun group of the Iroquois family of Native Indians, settled agriculturalists who then numbered between 20,000 and 30,000. Outbreaks of measles, influenza and smallpox, all introduced by the Europeans, had immediate and devastating effect upon them: by 1639 only some 12,000 survived. This pattern of events was repeated many times across the continent in the next 300 years.

The combined effects of disease, enforced relocation to accommodate European settlement, and wars aided by the use of European arms brought widespread destruction to Canada's indigenous peoples. Their numbers reached their lowest point in about 1920. Since then greater immunity from disease and the provision of improved health care have contributed to a rapid growth in numbers, and by 1981 the populations of Native Indians, *Métis* and Inuit had climbed to threequarters of a million. Most of these groups today are found in the Yukon and Northwest Territories.

Even in Canada's most remote areas, their way of life is under threat. The Inuit people of the Arctic inhabit one of the most inhospitable environments on Earth. Survival in these severe conditions was dependent on a seminomadic hunting–gathering culture – winter settlements of between 500 and 1,000 people

Campaigning for lost rights (*above*) Mohawks and other Indian nations have become increasingly politicized in their pursuit of cultural independence and restoration of their territory.

Under pressure (*left*) Inuit communities have been severely disrupted by the commercial exploitation of their land. The breakdown of their traditional way of life has led to serious social problems and alcohol abuse.

Artistic heritage (*below*) Inuit craftsmen continue the skill of their ancestors in soapstone and bone carving. Craftwork is now an important source of income.

broke up into small hunting bands in summer, and social life was communal, with strong family bonding.

Today, the far north is Canada's last frontier. Exploitation of its mineral resources has exposed the Inuit to modern pressures, and their traditional way of life has been almost entirely abandoned. Survival, materially and culturally, is pursued in other ways. Inuit art, particularly carving and printmaking, has become a significant economic activity, and cultural integrity is protected by the Inuit Tapirisat, an umbrella organization that safeguards Inuit interests, including the Inuktitut language. But changing conditions bring new threats and pressures from outside. The campaigns of animal rights' activists around the world have limited demand for furs and skins, the trapping of which provided a means of support for many Inuit communities. Rapid cultural change leads to demoralization and displacement – Inuit suicides are four times the national average, and violent crimes, almost invariably associated with dependency on alcohol, are also high.

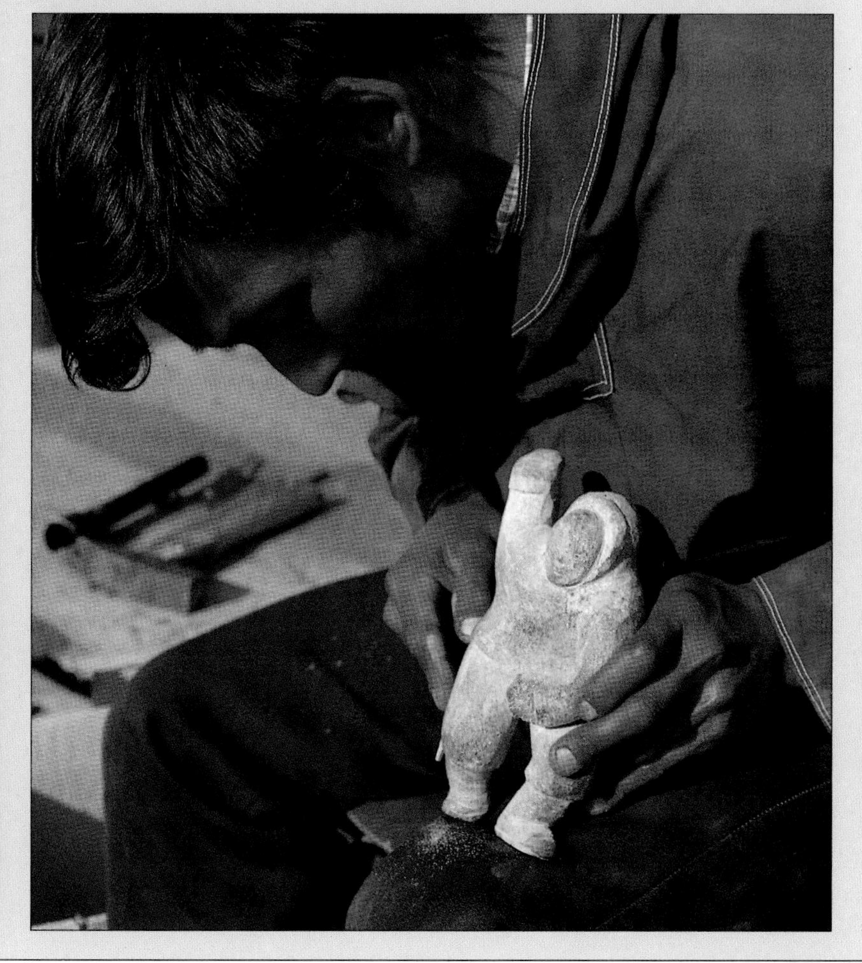

A culture in transition

A study of a remote village in the Yukon, carried out by Robert McSkimming over a long period of time, shows how a traditional indigenous economy and culture has been affected by European contact. The Kutchin people who live today in Old Crow were originally fishermen and caribou hunters – the village became a permanent settlement only when a trading post was established there in 1912. The villagers quickly turned to trapping for their main means of support. The acquisition of firearms meant that hunting could be carried out quickly, leaving more time for trapping. Their communal hunts were abandoned, and by the 1940s perhaps three-quarters of the adult population were trappers; their trap lines reached 240 km (150 mi) south of Old Crow.

By 1973 all this had changed. With a declining market for furs, only 10 percent of Old Crow adults remained trappers, and few young people knew how to set trap lines. There was widespread unemployment. Nevertheless, a quarter of the village's income, and over half of its food needs, were still derived from the land.

"Without land, Indian people have no soul, no life, no identity, no purpose. Control of our own land is necessary for our cultural and economic survival." These words of the Yukon Native Brotherhood lie at the heart of the land claims movement. For Canada's indigenous peoples, possession of their land not only has symbolic meaning, but fundamental material importance. Substantial land claims offer an opportunity for economic and cultural survival. The first agreement, made with the Council for Yukon Indians in 1990, included a cash settlement of $232 million and ownership of an area of land covering 41,000 sq km (15,830 sq mi). Such land claims are likely to increase in the future.

The Mounties

The Royal Canadian Mounted Police – commonly called the Mounties – are an important symbol of Canadian nationhood. Founded in 1873, they began as a small mounted force charged with keeping the peace between the indigenous peoples and European settlers in the vast territories acquired from the Hudson Bay Company. From a series of posts and forts they maintained order first in western Canada and then later in the northwestern gold fields and inside the Arctic Circle. Canadian folklore contrasts the peace and stability brought by the Mounties to these frontier lands with the violence and disorder that ruled south of the border.

In their scarlet tunics and blue pants – the uniform they still wear – Mounties were used on immigration pamphlets and tourist advertisements as early as the 1880s. Through novels, paintings and movies, they acquired an almost mythical status that was, in some ways, reminiscent of the European tradition of the romantic knight. The image of the calm, dutiful, honest and courageous individual provided an archetypal hero for a society undergoing rapid change. The heroes were, however, strictly English, men from yeoman and aristocratic backgrounds bound together by ties of mutual loyalty and respect, engaged in tackling the threat from outsiders, usually American or Chinese. The Mounties were never so popular among either French–Canadians or Native Indians.

The legendary character of the Mounties was such that they were able to withstand a series of scandals in the 1970s over their involvement in illegal surveillance and bomb-planting activities, much of which was directed against Quebec separatists. They are now a force of 20,000 men, and training on horseback is no longer essential, although their Musical Ride remains a popular show and an important source of publicity.

Mounted cavalcade Mounties are no longer required to be skilled horseback riders – except those that take part in the famous Musical Ride.

THE PEOPLING OF A CONTINENT

THE EARLIER AMERICANS · UNITY IN DIVERSITY · THE REGIONAL EQUATION · THE FOURTH WAVE · FROM UTOPIA, TO SUBURBIA · LIVING BEHIND BARRIERS

The United States, the world's fourth largest country, is culturally one of the most varied. Within a span of barely 500 years, a land of great natural diversity has been occupied and transformed by 60 million immigrants and their descendants from virtually all continents. In a grand political and cultural enterprise European ideals of individual freedom were transplanted to a new land. Their influence can be seen at work in the earliest colonies, through the westward advance across the continent, to the global projection of American power in the 20th century – a process that involved the conquest and dispossession of indigenous peoples and the enslavement of Africans. Maintaining such a diverse society has encouraged among its citizens a particularly self-conscious and critical analysis of society.

COUNTRIES IN THE REGION

United States of America

POPULATION

246 million

LANGUAGE

Official language English

Percentage of population by first language English (79%), Spanish (4%), German (3%), Italian (2%), French (1.3%), Polish (1.2%)

Over 90% of the population can speak English

IMMIGRATION

Percentage of foreign born	6.2
Total immigrants 1987	601,516

Countries sending most immigrants (1987) Mexico (72,351), Philippines (50,060), China and Taiwan (37,772), Korea (50,060), Cuba (28,916), India (27,803); refugee arrivals 70,000

RELIGION

Protestant (55%), Roman Catholic (29%), nonreligious and atheist (6.8%), Jewish (3.2%), Eastern Orthodox (2.3%), Muslim (1.9%), Hindu (0.2%)

THE EARLIER AMERICANS

The first inhabitants of North America are believed to have crossed the Bering Strait from Eurasia some 10–30,000 years ago, if not earlier. Their ways of life developed according to the range of natural environments. When the first Europeans arrived in the 16th century there were at least three distinct centers of civilization: town-dwelling Pueblos practicing irrigated farming in the arid southwest, the village communities of shifting cultivators in the southeast, and the Iroquois confederacy, a powerful military presence in the northeast. In the Great Plains and Pacific coast areas, other peoples lived traditionally by hunting and fishing.

The arrival of European fishermen, fur traders and farmers from 1500 onward gradually changed the lifestyles on both sides. The Europeans adopted the cultivation of indigenous crops such as maize (corn), beans and squash, while the Native-Americans acquired manufactured goods. The exchange of guns and metal knives for furs and meat created a continent-wide trading system that altered the patterns of indigenous lives well in advance of European settlement itself, bringing with it disease and intergroup warfare over hunting grounds. As horses were obtained from the Spanish in the south, the Native-Americans began to hunt with guns from horseback across the Great Plains in the center of the region. This entirely new way of life, pioneered by the Dakota, was eventually to become the European stereotype of all so-called Indian behavior.

A people conquered, a land changed

At first Europeans treated the indigenous peoples as if they belonged to separate nations, and entered into agreements and alliances with them during their wars to win possession of the continent. But as the British established control they began to change the country. Hungry for land, they came to regard the Native-Americans as an uncivilized people, first defeating them in war, then seizing their land and finally deporting them westward, leaving the land of the eastern seaboard free for settlement.

In the lands they took over, nature was transformed in line with European ideals of improvement: forests were cleared, swamps drained and grasslands plowed up. Small farms dominated in the north of the settled area, while in the south large numbers of slaves were imported from Africa to work on the large cotton and tobacco plantations belonging to an aristocratic, landowning elite.

The frontier of settlement marked a division between garden and wilderness, with indigenous peoples banished to the latter. The final act of displacement was the imposition by the new American federal government of a regular geometric system of land survey and individual property ownership, which was completely alien to indigenous conceptions of communal land use. Culminating in the Homestead Act of 1862, this grid-based system ignored topographical features such as rivers and uplands and left a legacy of linear state boundaries, straight roads and regular settlements across the heart of the continent.

"Go west young man"

The westward movement of the frontier was encouraged by the over-farming and consequent exhaustion of land and by rising rates of immigration. Although begun by American-born settlers, the invasion and settlement process of new lands farther west was continued by these newcomers. They included small farmers and their families from northern and western Europe who had been displaced by famine, land clearances or agricultural mechanization. In general, particular nationalities tended to settle particular areas. They maintained a gradual movement west across the vast expanses of the interior by following variations in the terrain and natural environment.

After the 1860s the national economy became self-sustaining. Immigrants – now mainly from the impoverished parts of southern and eastern Europe – were more likely to be single and consequently found work as urban industrial workers. A major cultural and political divide gradually emerged between the Old Europeans, who, generally, tended to be Protestant, rural-based, skilled workers or business owners, and the New Europeans, who were predominantly Roman Catholic, urban and less skilled. In the second half of the 20th century the "fourth wave" of immigrants, mostly from Latin America and eastern Asia, together with large numbers of refugees, have redrawn these lines of cultural difference, particularly in urban areas.

A cultural melting pot (*left*) Players in an open-air chess game in New York's Washington Square reflect some of the city's ethnic mix. American society is one of the world's richest in terms of ethnic diversity; colonization and an open immigration policy have brought new citizens from virtually every part of the globe.

Spirit of self-sufficiency (*below*) The picturesque villages of New England were founded during the 1600s by European settlers – many of them English Puritans – who came to create independent communities free from religious persecution. Each small village had its own church, public buildings and elected officials.

UNITY IN DIVERSITY

Despite its great cultural diversity, the United States has, for various reasons, never experienced the levels of sectarian conflict found in other parts of the world. After independence in 1776 the new nation's identity was forged around political ideals rather than cultural similarities, and citizenship was open to anyone prepared to abide by them. This has remained the case. Some proficiency in English and a knowledge of the constitution are all that is required after five years' residency to become a United States' citizen. Government has played an active role in managing cultural diversity, restricting immigration when it was unpopular, assisting oppressed minorities to participate in community and national affairs and generally avoiding excessive attempts to create a culture of uniformity. The absence of an official state religion and the fact that cultural groups do not dominate particular regions have also lessened the basis for conflict.

Patriotism has been encouraged by stressing the ideals of liberty, democracy and self-government, by maintaining that anyone can succeed in the United States, and in the decades since World War II, by emphasizing the United States' world role in defeating communism. In few nations is the flag and the national symbol (the Bald eagle) so prominently displayed in private homes and businesses.

Ethnic distinctions

The proliferation of pizza restaurants, delicatessens, and beers with German or Czech names points to the fact that the United States has absorbed many things from immigrants and transformed them into a national culture. Yet ethnic distinctions are made both in everyday life and in public institutions such as schools and workplaces. People still tend to marry others from a similar ethnic background (although this tendency is progressively weakening), while colleges and employers may take ethnic identity into account when deciding on university places or jobs. Eight out of every nine white Americans (frequently termed Anglo-Americans) identified themselves as having one or more European ancestries in the 1980 census. English, German and Irish are the most common self-descriptions, although it is thought that many people now simplify their ancestry by choosing one from among a number of national identities in their families' past.

Sometimes this means little more than a faint pride in the homeland, but among Italian-Americans, for example, ethnic rituals retain symbolic importance. In New York's Little Italy, where even the fire hydrants are painted in the Italian national colours of green, white and red, the festival parades of San Antonio and San Gerrano are still observed in the area, even though the majority of residents are no longer of Italian origin. Nevertheless, traditional Italian attitudes to family or the role of women appear to have converged with the rest of society. Similarly, among Jewish-American families, rites of passage such as *bar mitzvah* (for boys) and *bas mitzvah* (for girls, a ritual that was invented in the United States), are popular but have lost much of their religious significance; today they are primarily observed as an excuse for a family and social occasion.

A greater distinction exists between Anglo-Americans and the fifth of the population that is usually classified as Latino (or Hispanic), African-, Asian- and Native-American. These terms are not necessarily used by the people so described and conceal both linguistic diversity and a variety of national origins. Latinos may be said to include the descendants of Spanish colonialists, Mexicans conquered by the United States and several generations of subsequent immigrants and refugees from Cuba and Central America, as well as Puerto Ricans, who are United States' citizens. Asian-Americans include the descendants of 19th-century Chinese, Japanese and Filipino immigrants, Indochinese refugees and the more recent immigrants from countries such as Korea, Taiwan, and India who are also widely differentiated by language, religion and custom.

Still a religious society

The constitutional separation of church and state in the United States has allowed a great number of religions and churches to flourish. Contrary to the trend of secularization in other Western countries, religious observance is not declining and is strong across all regions and sections of society. Some 40 percent of Americans attend church regularly, while smaller denominations have recently been growing at the expense of larger ones.

Home from home (*above*) Immigrant communities have recreated the atmosphere of their home countries in many city districts. These women are serving pasta and sauce at a festival in an Italian neighborhood where many of the store signs are in Italian.

Praising the Lord (*below*) A preacher delivers a rousing sermon at a tent revival meeting in Atlanta, Georgia. Unlike many Western countries, religious belief is not declining; evangelical organizations raise millions of dollars from their faithful.

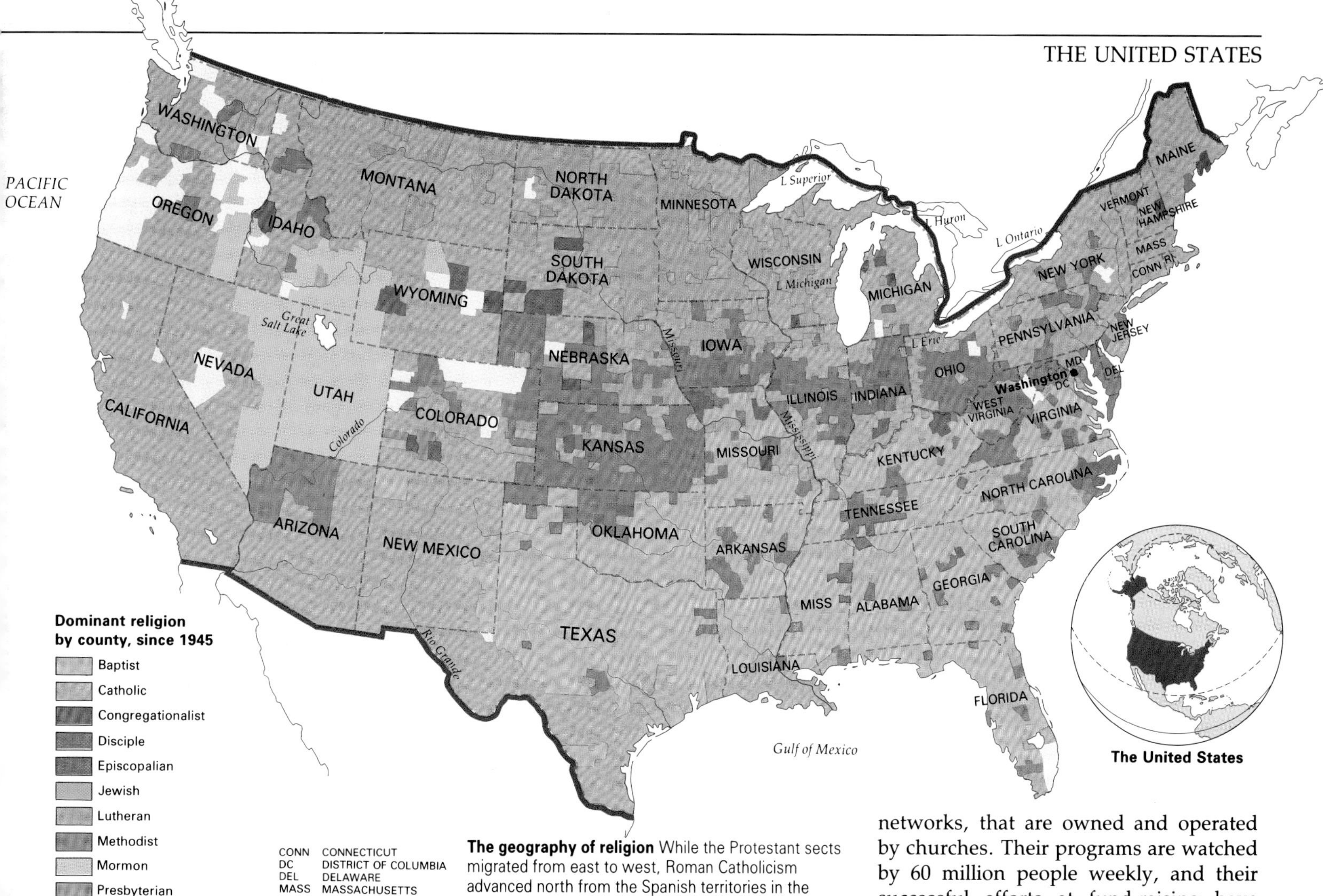

The geography of religion While the Protestant sects migrated from east to west, Roman Catholicism advanced north from the Spanish territories in the southwest. Immigrants from eastern and southern Europe swelled its numbers in the northern cities.

About 50 million adults claim to have been "born-again", meaning that they have had some direct spiritual experience that has renewed their Christian faith. One factor that contributes to the continued vitality of religion is the existence of more than 200 television and 1,300 radio stations, as well as three television networks, that are owned and operated by churches. Their programs are watched by 60 million people weekly, and their successful efforts at fund-raising have enabled some evangelical churches to build hospitals, universities (such as Liberty College in Virginia) and even theme parks such as the Praise The Lord ministry's Heritage USA. The Protestant churches of the United States were among the first to ordain women priests, while priests and nuns of the Roman Catholic church still play a role in mediating between new immigrants and society.

FLYING THE FLAG

"I pledge allegiance to the flag of the United States of America and to the Republic for which it stands, one nation under God, indivisible, with liberty and justice for all." Each morning at school American children repeat these words. With no monarchy or state religion, the United States' flag (known familiarly as "Old Glory" and "the Stars and Stripes") is a powerful secular symbol of nationalism.

The national anthem, "The Star-Spangled Banner" commemorates the flag. It is displayed in every courthouse across the region, while politicians habitually pose for photographs with the flag in the background. It embellishes many family homes, and elaborate regulations surround when and how it must be flown.

The burning or desecration of the flag, common during anti-Vietnam war protests in the 1960s and 1970s, is seen as deeply unpatriotic or subversive. Laws forbidding it have been passed in 48 states. When, in 1989, the Supreme Court ruled that banning flag-burning violated another American ideal, the right of free speech, President George Bush campaigned to protect the flag permanently by amending the constitution itself. This would have been a far-reaching step – only 26 amendments have been added to the constitution since 1787 – and Congress persuaded him to try to pass a federal law instead, though the Supreme Court rejected even that in 1990. In few other countries is the national flag held in such high esteem or accorded such protection – only Germany approaches this degree of reverence.

THE REGIONAL EQUATION

The history of the United States, with its distinct regional economies, political traditions, and patterns of migration and settlement, has led to the creation of a number of different regions, with which people identify strongly. It is an indication of how far regional differences still matter that presidential candidates generally choose a running-mate from another part of the country.

Regional variations in attitudes are reflected in the range of states' laws, covering such things as alcohol use, gambling, women's rights, unions and the environment. For example, northeastern states have the highest church attendance, midwesterners vote the most, while southerners drink the least (many southern states ban alcohol sales altogether). Local identity appears in the lack of any widely read national newspapers – the United States has over 1,700 dailies – and the local coverage of the 9,000 radio and 1,000 commercial television stations. Each state has its own flag, song, animal and flower, and their nicknames (which often refer to events in the early history of the state) appear on car license plates: for example, Tennessee is the "Volunteer State" and Oklahoma the "Sooner State".

North versus South

The American Civil War of 1861–65 was more than a conflict between economic systems. It was also a clash between a northern, individualistic, industrializing and urbanizing culture with a history of radicalism, and a southern land-owning culture based on hierarchy, paternalism and racial order as established in the slave plantations. There are still traces of these differences and of the rivalry associated with them.

Northern states are known for their liberal attitudes on social and political issues, while the south has tougher laws against trade unions, and is traditionally opposed to such things as abortion, women's and gay rights. The symbol of southern resistance, the Confederate battle flag, still adorns homes and car bumpers, and is even flown above the Union flag on the statehouses of Georgia and South Carolina. Since for African-Americans it has a more racist meaning, controversies over its display in public indicate that old racial tensions remain unsettled. Yet the mixing of African and European cultures made the south one of the world's great musical heartlands, being the origin of, for example, blues, jazz, rock-and-roll, bluegrass, country, zydeco and other styles of music.

Middle America

Middle America, a region that has come to typify perceived American characteristics, is the site of many sociological studies: its large cities are often used by political pundits to gauge national opinion. Peoria, Illinois, is where firms often test their products, while the state of Iowa holds the first run-offs in the presidential elections.

The Midwest was the first truly American landscape to be cleared and stamped with the geometry of the federal land surveys. Within this framework a series of distinct ethnic islands arose. The cultural landscape still bears the influence of Germans, Scandinavians, Dutch and others in the form of settlement patterns, placenames and vernacular architecture. Many small towns employ their ethnic heritage as a selling point for tourists by holding annual fairs; for example Wilber, Nebraska, has a Czech fair and New

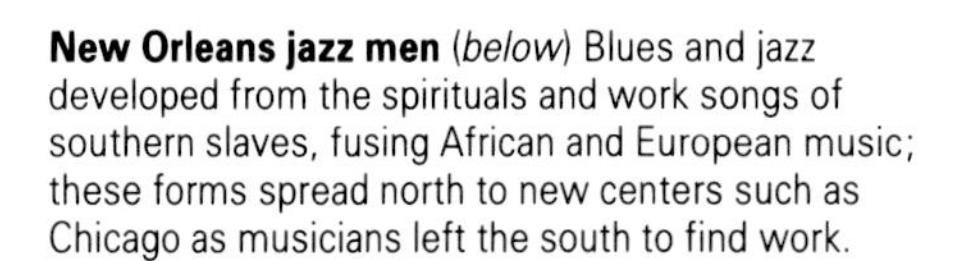

New Orleans jazz men (*below*) Blues and jazz developed from the spirituals and work songs of southern slaves, fusing African and European music; these forms spread north to new centers such as Chicago as musicians left the south to find work.

The Lone Star State *(left)* Frontier Day in Eastgate, Texas. The sense of the past, and of regional difference, is strong among Texans, who inhabit an area larger than most European nations. They are intensely proud of their state's independent history, and of their heroes such as Davy Crockett (1786–1836) and Jim Bowie (1796–1836) who fought in the war against Mexico.

Sliding into base *(below)* Regional rivalries are seen in the battle of the big cities to own a major baseball or football team. There are too few teams to go round, and so they are poached with the lure of new stadiums and other benefits.

Glarus, Wisconsin is a Swiss theme town with an annual William Tell pageant.

The region is also marked by a northern European political tradition based on community ideals and compassionate, progressive government. Wisconsin for example, whose population is mainly of Scandinavian, Polish and German descent, was the center of the Progressive Party in the 1920s – the last serious challenge to the two-party system. It was also the first state to ban child labor and pass laws safeguarding gay rights.

Way out West

As the cultural and political traditions forged in the early centers of settlement were taken westward, eventually crossing the Rockies during the 19th and early 20th centuries, they combined to produce a region of unstable politics, new ideas and frequent dissent. Some traditional practices did not endure in the arid, sparsely populated west. Water laws, for example, which had previously respected the rights of those owning the land through which it flowed, switched to give exploitation rights to the first users. Water scarcity was central to the struggles between homesteaders and ranchers on the Great Plains and between farmers and cities, later giving impetus to the environmentalist movements that are such a feature of life along the western seaboard. The states of the northwest have led the nation in ecological law-making. Oregon, for example, was the first state to ban drinks cans with detachable ring-pulls and throwaway bottles, and gives tax credits for alternative energy schemes.

The Sierra Club, the country's largest and earliest conservation lobby, was founded in 1892 in San Francisco, California, over the issue of creating the first National Park in Yosemite. More recently, that city's restless population has provided the seedbed for various "countercultures", such as the Beat Generation of the 1950s, hippies in the 1960s and the New Age movement, with its eclectic mix of mysticism and holistic mind and body healing techniques, in the 1980s. The west's demographic momentum is maintained by high levels of immigration, particularly to California. In addition, migration of young people to the Rocky Mountain states from other parts of the country, attracted by its clean air and attractive outdoor lifestyle, make the west the youngest and fastest growing region.

BALL GAMES AMERICAN-STYLE

The United States is credited with the invention of three major sports – basketball, baseball and American football. Over half the population regularly follows such sports. In 1987 they drew 49 million, 53 million and 16 million spectators respectively, while millions more watch on television; entire channels are devoted to sport. Americans support their local teams with extreme fervor. Every big city wants to house a major team for reasons of civic pride and for the boost that they are assumed to give to local economies.

With the exception of basketball, however, these sports have not traveled well. Baseball has caught on in Japan (where it was introduced by American servicemen during the postwar occupation), Cuba, Mexico and parts of Central America, but has not spread farther. American football, which began to diverge from its antecedents in British rugby football in the sporting encounters between the Universities of Yale (Connecticut) and Princeton (New Jersey) in the 1880s, has made the leap to certain European cities, such as Barcelona and London, but remains definitively American.

Foreigners find the slow pace of both games, the labyrinthine rules and the obsession with statistics a deterrent to their enjoyment. Yet both have acted as integrating forces within United States' society. A large number of baseball stars are Latinos, while sports scolarships at the major colleges provide potentially lucrative opportunities for African-American football players. This unique link between higher education and professional football may be the reason that this most American of pastimes has proved so hard to export.

The frontier myth

All nations and peoples create their own myths. By selecting and dramatizing historical events they try to say something about the forging of group identity and its perpetuation through time. The frontier is one of the more enduring American national myths. In 1893 the American historian F.J. Turner (1861–1932) wrote that the frontier was "the meeting place of savagery and civilization" and that "in the crucible of the frontier the immigrants were Americanized, liberated and fused into a mixed race". In other words, man and nature wrought changes in each other and in doing so created a new form of person and society, that of the independent individual and of the post-European democracy.

Even by this time the West and the cowboy-hero had been romanticized in songs and dime novel characters such as Deadwood Dick. In his traveling shows William F. Cody (1846–1917) featured himself as Buffalo Bill, blurring the line between fact and fiction in the dramatic reconstruction of his exploits. It was however, the new cinema industry, which began to produce hundreds of short action-features in southern California in the 1900s, that gave substance and body to the myth. Many of the stars and stunt riders had been cowboys in real life, driven to the movies by the collapse of the ranching industry in the 1880s.

These films provided the staple diet of American history for the thousands of newly arrived immigrants. Throughout its history, Hollywood has turned to the "Wild West" far more than to any other American theme. The fullest elaboration of the movie myth was worked out between 1939 and the early 1960s, a time in which war and postwar uncertainties created a need for nostalgia and stories of national identity.

The classic western movie revolved around a number of common themes already established in the works of 19th-century authors such as James Fennimore Cooper (1789–1851). They included the conquest of nature, the creation of a stable moral community and the heroic role of the individual, almost invariably male. Nature and land were represented as entities to be possessed and altered by males. Both films and histories of the frontier tended to ignore women and their views of nature as a garden to be nurtured. Often a woman symbolized the new land itself, fought over by suitors

Marketing the American way (*above*) Advertisers have long realized the commercial benefits of associating their products with the powerful myths and values that inform American society. Campaigns such as this, linking a brand of cigarette with the heroic appeal of the cowboy, have helped to spread the frontier myth around the world.

Celluloid hero (*left*) The development of the movie industry gave new force to the frontier myth in such films as this 1925 western *No Man's Gold*, starring Tom Mix. Contemporary westerns generally try to paint a more realistic picture of what life must have been like for the men and women who migrated west to a harsh and unfamiliar land, and are more sympathetic to the role of Native-Americans.

An enduring fantasy (*right*) A couple at a livestock show in Kansas City wearing the modern-day, luxury version of cowboy gear, complete with stetson and hand-tooled boots.

representing opposing forces. Civilization, the East, community and order battled against savagery, the West, individuality and violence.

The problem of how to create order through acts of violence (against outlaws and "Indians") often required the death of the heroic or dangerous individual, leaving the community to celebrate his acts but also to go beyond them. In the myth, Native-Americans generally had a minor role, Mexicans went unacknowledged, and the ethnic diversity of the settlers was often forgotten. Also missing was the role of government, railroads, banks, corporations and land speculators. By neglecting the larger society, the impression that the frontier was opened up by pioneering individuals, who then made themselves into new Americans, was falsely perpetuated. At the same time, the polluting and destructive aspects of settlement, including the extermination of the buffalo and the overexploitation of marginal farmlands, was also conveniently forgotten.

The new frontier

The frontier myth proved adaptable. In 1960, future President John F. Kennedy (1917–63) took as his campaign theme the New Frontier to persuade American taxpayers that an ambitious, risky and costly space program was a continuation of their national destiny. Aspects of Kennedy's liberal vision, which included the United States' world role of peace and understanding, found their fictional expression in the 1960s cult television series, "Star Trek". Some 20 years later President Ronald Reagan spoke of a High Frontier, a network of space-based weapons' systems that would finally enclose the world in peace and civilization. Science and fantasy blended in his new vision of "Star Wars".

The myth moves on. Always at the disposal of advertising agencies to sell a variety of goods from cigarettes to fashion denim clothing, it has most recently entered the jargon of the real estate and urban planning world. Renewal of old housing for middle-class residents is often described as the pioneering and homesteading of an urban frontier. The myth retains its ability to dramatize and remodel both history and the present day, to elevate the heroic individual and to draw boundaries between savagery and civilization.

THE FOURTH WAVE

The waning of the Cold War climate and the growing liberalization of society under the Democrat administrations of Presidents John F. Kennedy (1961–63) and Lyndon B. Johnson (1963–69) encouraged the revision of immigration laws in the 1960s. There was a renewed flow of people to the United States from around the globe: two-thirds of the world's emigrants now settle in the United States. Sometimes termed the "fourth wave", these new arrivals come mostly from Asia and Central and South America, rather than Europe. Mexico, the Philippines and Korea alone account for a quarter of their numbers, and thousands also arrive from small Caribbean countries such as the Dominican Republic and Jamaica.

Among their numbers are a growing proportion of the skilled and professional workers that the new immigration laws were designed to attract. In addition, its tradition of welcoming refugees has made the United States an obvious destination for people escaping political repression. American involvement in wars and civil unrest in Indochina and Central America in recent decades has caused large numbers from those regions (particularly Vietnamese and Cubans) to seek refuge in the United States. No other country accepts so many refugees. In 1989 more than 110,000 entered the country, including over 40,000 Jews who were finally able to leave the Soviet Union.

Unlike 19th-century immigrants, the fourth wave immigrants tend to concentrate in the western and southern states, and in urban areas. Over 40 percent go to just three states, California, Texas and Florida, while a further 20 percent head for New York. The cities of New York, Los Angeles (California), Miami (Florida) and Chicago (Illinois) are the chosen destinations of over a third. They encounter a cultural climate that is more tolerant of ethnic and religious diversity, and the conditions for maintaining the traditional cultures of their homelands seem more favorable than in the past. For example, public schools now provide bilingual education in more than 80 languages, while many larger cities have Spanish- and Asian-language television and radio stations.

Niches and neighborhoods

The character of many of the larger cities has been transformed by the fourth wave. Over a fifth of the people living in Miami, Los Angeles and New York were born abroad, and the newcomers will often come to dominate one particular niche within a city's economic structure, either taking it over from a previous group or reviving a declining industry. The availability of cheap, often female, labor facilitated the recovery of Los Angeles' manufacturing economy. Its downtown and surrounding neighborhoods are today full of small garment workshops staffed by women from Mexico, El Salvador and Vietnam, who often work in conditions and for wages that are below those acceptable to longer established groups. Similarly, in New York the Chinese community has restored the garment trade, Koreans have taken over the independent grocery stores while Asian Indians and Pakistanis run the majority of the newsstands. Asian Indians have moved into California's motel business and Cubans into Miami's construction and banking sectors. Often working long hours and investing their savings in their businesses, immigrants are held up as exemplary workers.

Specific neighborhoods are also emerging. Asian immigrants, many from urban middle-class backgrounds, often have large sums of capital to invest in property, and this creates new commercial and residential districts such as New York's Chinatown or Los Angeles' Little Tokyo. Displaced refugees will reside close to one another in an effort to recreate their lost communities. For example, in California, Little Saigon has been created among the Vietnamese of Orange County and Little Phnom Penh among Long Beach's Cambodian population.

Perhaps the most unusual refugees are

BROOKLYN'S MELTING POT

Of Brooklyn's 2.3 million inhabitants, one in seven was born abroad. This borough of New York has long been a port-of-entry for immigrants, and newcomers to the area today occupy neighborhoods that have enjoyed long histories of cultural diversity. The Crown Heights and East Flatbush communities, for example, were once home to communities of Jewish, Italian and Irish people. Now there are more West Indians in Brooklyn than in any other city in the world, and Jamaican and Haitian Creoles mix on the streets, while migrants from St Vincent and the Grenadines add calypso, steel drum orchestras and cricket to the sounds of Jamaican reggae. On Labor Day the Eastern Parkway hosts crowds of 800,000 at the West Indian carnival.

In Sunset Park, once home to Finns, Jews and Puerto Ricans, Chinese from Hong Kong are creating a rival to Manhattan's Chinatown with money withdrawn from the British colony before its handover to China in 1997 and in Brighton Beach, an aging Jewish community has been revived by the arrival of 20,000 Soviet Jews from the Ukraine and Black Sea area. As a result, the neighborhood has come to be known as "Little Odessa", and Yiddish has been replaced by the sound of the Russian language in the streets and by notices written in Cyrillic script. Free to practice their religion openly, many Soviet Jews are now beginning to attend neighborhood synagogues and to observe previously unfamiliar Jewish holidays and religious rites.

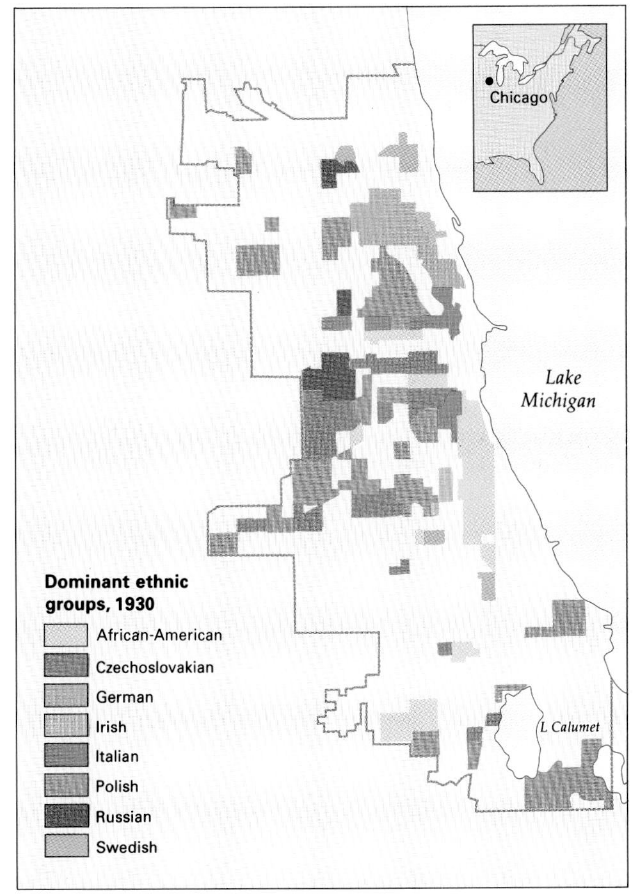

the Hmong. These hunting–gathering people from the hills of Laos were dragged into the Indochinese conflict and forced to flee from genocide to safety in the United States. Numbering some 97,000, most now live in just two places, in California's Central Valley and in Minneapolis-St Paul in Minnesota. Many refugees have difficulty in adjusting to the cultural ethos of the United States, but for the Hmong the transition from a premodern to a modern society has been especially rapid; they suffer from high rates of depression and alienation.

Closing the Golden Door?

As immigration reached its highest postwar levels in the 1980s, the majority of American citizens expressed their opposition to unrestricted entry. Popular misconceptions that immigrants live off welfare benefits, take jobs from citizens and increase crime rates fueled some people's resentment. In particular, public anxiety was focused on the 2–3 million so-called "illegal aliens" resident in the country, half of them believed to come from Mexico. Such is the degree of poverty experienced by many Mexicans

An ethnic neighborhood *(above)* Several of America's larger cities have a Chinatown such as this. Although new immigrants to the United States derive a sense of identity from living in closeknit groups, they share a pride in American citizenship with the rest of society. The Stars and Stripes flies alongside signs in Chinese in this busy street.

The changing inner city *(above right and right)* Chicago has one of the most ethnically diverse populations in the United States. The maps of ethnic neighborhoods in 1930 and 1980 show considerable change over 50 years. Large numbers of Hispanics have moved into the center areas. The African-American community, restricted to two small neighborhoods in 1930, have moved eastward and southward to dominate most of the downtown area of Chicago's Southside where urban dereliction is greatest. Here a distinctive African-American culture has developed, with its own style of "House" music. The Polish and Italians have relocated on the city fringes. Some of the small communities in the north, such as the Swedish, had all but vanished by 1980 and been replaced by Asians.

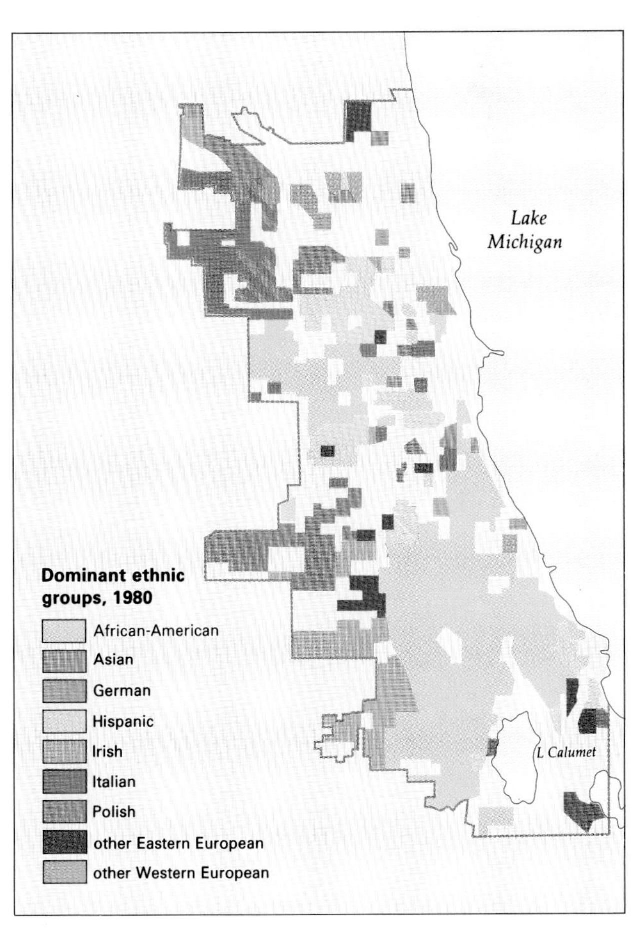

Suburb in the desert The American penchant for suburban living has created some strange juxtapositions of landscape and development. This suburb of Las Vegas is little more than a single strip of land bisecting the empty desert.

that even the lowest paid jobs north of the border can provide a better living. The existence of family and friends in the United States makes it easier to find work and worth running the risk of crossing over. In 1986 the Border Patrol made 1.6 million arrests on the United States–Mexico border. New laws passed in that year, however, provided an amnesty for 1.5 million undocumented workers who were already living in the United States.

There were also signs that the continued granting of entry to refugees was raising concern. Each person costs $7,000 to resettle, and consequently there were pressures to cut numbers. Yet even apparently successful immigrant groups, such as Asian-Americans, face hostility. An Asian-American is twice as likely to go to college as the rest of the population – they make up over a quarter of the students at the University of California at Berkeley for example – and this remarkable progress and adaptability causes jealousy. Perhaps those who feel the rub of their success most are the African-Americans, who have for years struggled against discrimination only to see others gain the benefits. Miami was the scene of riots throughout the 1980s, fueled by frustration and resentment against the successful newcomers.

FROM UTOPIA TO SUBURBIA

Individualism, the idea that individuals should work principally for themselves and lead private lives, is a hallmark of American society. Yet throughout its history there have always been currents of communalism, very often based on religious idealism, that stress the importance of a community living and working together for the good of all. From the 17th to the early 20th century, the availability of land for settlement, and the prevailing climate of political and religious toleration, encouraged numerous groups of religious dissenters to escape from persecution in Europe in order to establish utopian communities where they could practice self-sufficiency and preserve their distinct cultural identity.

Hundreds of place names, from Harmony, Maine to Philadelphia, Pennsylvania ("the city of brotherly love"), are relics of past utopias. Although most of them eventually collapsed under the pressures of individualism, the tradition continued in the 1960s' hippy communes, and in urban squatters' movements and New Age communes such as Rajneeshpuram in Oregon. Founded by an Indian spiritual leader in the early 1980s, this community had become a small and economically dynamic town before collapsing under the weight of internal feuding and local hostility in 1985. Biosphere 2, a self-contained ecosystem in a dome in the Arizona desert was home to eight scientists from 1990 to 1992, in an ecological utopian experiment.

Only two historic utopias still thrive: the 80,000 strong Old Order Amish, who speak German, avoid all modern technology and run successful farming communities in Pennsylvania, Indiana and Ohio; and the Mormons. This religious sect, which originated in the United States in the 19th century, now has 5 million members worldwide, a third of whom live in Utah, the state they first settled in 1847 and still govern.

Californian affluence The irrigated yards, swimming pools and individually designed houses in a wealthy district of Los Angeles are status symbols in a society that places high value on material success.

COMING OUT IN SAN FRANCISCO

Homophobia, or the hatred of homosexuality, is as pervasive in the United States as in other western countries. Gay identity as the basis of an explicit lifestyle is a recent phenomenon. Homosexual relations between adult males are still illegal in several states, particularly in the south, Midwest and Rocky Mountains; gay lifestyles consequently tend to be most prominent in the more tolerant ambience of the cities. In San Francisco, where gays and lesbians number over 100,000, a public gay identity has emerged alongside the development of a distinct territory. The Castro district, an area of gentrified older houses, has become a place to "come out". Beginning as a network of bars and meeting places, it has grown as gay-run businesses were created to meet the needs of a gay market, and is today the focus for gay culture, arts, literature, an annual parade and political power within the city. In 1977 Castro elected the country's first gay politician, Harvey Milk, who was assassinated by an embittered fellow councillor in 1978.

In the 1980s the spread of AIDS led to a process of evaluation and choice-making among gays, dividing the community between those who endorsed a promiscuous lifestyle, and those who believed the disease could only be held in check by restricting sexual activity to one partner. The AIDS crisis created anguish for bereaved partners: gays have had to fight for the rights normally extended to spouses such as making funeral arrangements and the right to stay on in rent-controlled apartments.

Class in the classless society?

For most Americans, utopia extends no farther than the boundaries of home, and homes tend to be grouped in neighborhoods distinguished by class. No Western nation has so persistently denied the significance of class in shaping society, and the development of separate class cultures has perhaps been complicated by the prominence given to ethnic or racial identity and a strong ideal of class mobility. Yet, although there is no aristocracy, a distinct upper class perpetuates itself through restricted marriage, the choice of select schools for its children, and the continuation of exclusive occasions such as debutante balls. Elite institutions such as country clubs may still exclude women, African-Americans and Jews if they wish.

This elite culture borrows heavily from the varied landscape designs and codes of behavior of the English land-owning gentry. Tudor, Georgian and Victorian Gothic architecture all typify country houses, parklike suburbs and the "ivy league" universities such as Harvard, Massachusetts. Fox-hunting is practiced in Virginia and Maryland, while polo is played in the rural suburbs of New York, Chicago and Los Angeles.

Most Americans would identify themselves as middle class, and the suburb became the epitome of the middle-class community. It combined urban decentralization with the rapid diffusion of automobiles, refrigerators, telephones and televisions: by 1956 the population was spending more time watching television than engaged in waged work. In 1960 the average American watched five hours a day, while now it is over seven hours; over half of all homes have video cassette recorders and cable channels, giving possibly the greatest choice of viewing in the world.

While dress and accent are rarely reliable guides to a person's status, suburban landscapes are typified by the care given to the external and internal appearance of dwellings. Neat lawns and tidy rooms are marks of respectability. In this individualized landscape of small, nucleated families, a sense of community is provided by the shopping mall, which combines every element of a village except a church. Since the first shopping mall was built in Kansas City, Missouri, in 1920, they have become widespread, growing ever larger in design. Today they are significant social centers. In particular, their cinemas, restaurants, ice rinks and shops provide the focus of suburban teenager life.

Beyond the suburbs

After nearly half a century in which suburban life was equated with the American dream, there are signs of a reaction against it. Recent television programs and movies have stressed the banality and uniformity of the suburbs, and suburbanites are increasingly aware of the spread of crime, gangs and child abuse, things they had assumed were confined to the cities. For some middle-class white parents the suburbs had offered the promise of safe surroundings for the education of their children, enabling them to avoid schools with high minority concentrations. Throughout the country, many such communities have resisted school desegregation programs, which would produce mixed race schools, while others have fought the construction of low-income or rental dwellings.

For younger, wealthier and childless Americans an alternative to suburban life can be found in renewed and gentrified city housing. Others have left the big cities altogether to settle in the small cities and towns of the more remote areas such as the Rocky Mountain states and the Pacific North West. For the older and wealthier, specialized retirement communities have sprung up, notably in the warmer states such as Florida.

LIVING BEHIND BARRIERS

The United States has a violent society, if crime statistics are taken as a reliable guide. Murder rates are 4 to 5 times higher than those of Western European countries, rape 7 times higher and robbery between 4 and 10 times greater. Various explanations have been given, including the country's relatively recent past of violence when the land itself was seized and taken, and the celebration of violence in cinema and on television. The outlaw and the gangster are well-established cultural heroes.

Contrary to popular opinion, however, television violence has not increased much since the 1960s: there are still 5 to 6 acts of violence an hour and the average viewer can expect to see 15 murders a week. Perhaps the most obvious reason is the ready availability of guns. There are thought to be 60 million handguns and 120 million long-guns in the country, or two guns for every three people. Handgun murders exceed 10,000 a year. Although the importation of some assault weapons has been banned, efforts to curb gun ownership run up against the powerful National Rifle Association lobby, whose 3 million members uphold the right of an armed citizenry as a democratic counter to the power of the state.

With some 500,000 prisoners, America's penal confinement rate is more than double that of Britain or Japan. Alone among the Western nations, the United States executes criminals, and support of the death penalty can be an effective way of winning votes in elections. In 1987 25 people were put to death and another 90 awaited execution. Only nine states do not have the death penalty, while Texas, Florida and Louisiana have carried out a third of recent executions.

In the cities, where violent crime more than doubled in the 20 years between 1970 and 1990, the fear of crime underpins segregation. Those who can afford it protect themselves by defending their buildings with barricades and alarm systems, creating neighborhood watch groups and employing private security patrols. Forty percent of prisoners are African-American, and among white Americans the fear of crime is inextricably linked with race and racism. Murders are committed overwhelmingly within ethnic groups, however. Confined to ghettoes, African-Americans bear the heavy burden of crime itself.

The first and last Americans

Native-Americans were the first, but in many ways are also the last, Americans. Their status in society has yet to be fully settled. Should they be treated as separate sovereign nations governing their own land, or simply as another ethnic group? Their numbers have now recovered to 1.5 million, the level they were before coming into contact with Europeans. There are 150 recognized "tribes" – as the government still calls them – a third of which live on the 270 or so Native-American reservations. The majority belong to 10 large tribes, which are led by the Cherokee and Navajo. Tribal self-government allows traditional languages to be taught in schools (there are 14 language groups, many now with written forms) and religious rituals to be continued, but years of hardship and poverty have produced high levels of alcoholism and low life expectancy.

The ghetto and the barrio

Among the African-Americans, and to a lesser extent the Latinos, the residential concentrations typical of all immigrant groups upon arrival have persisted. Although all forms of racial zoning of neighborhoods were illegal by the 1940s, levels of African-American segregation are comparable to those of South Africa. The Southside of Chicago, Harlem in New York and Watts in Los Angeles are all ghettoes with hundreds of thousands of inhabitants. Throughout the southwest, Mexican-American barrios like East Los Angeles are the centers of equally solid communities. Outsiders may stigmatize them as places of poverty, crime, gangs, drugs, graffiti and AIDS, but from the inside, they are the site of political power and of community pride.

For many African- and Mexican-Americans the choice to reside in such areas is a positive one, to remain near family, friends and churches. East Los Angeles' annual Cinco de Mayo parade and its world-renowned murals, its lowriders (classic old cars) and *mariachi* bands attest to the vibrant culture of a city within a city. The ghettoes have produced much of the world's most popular contemporary music, from Detroit's Motown to Chicago's House style. African-American styles of dress, slang and even

Women against crime (*above*) The police force is one of the areas where women have made significant progress in a previously male-dominated profession. In a country where guns are widely owned – and used – both female and male officers are armed.

Planning the campaign *(above)* Native-Americans intently discuss strategy before a political meeting held at the Cheyenne River Reservation. A process of politicization and cultural renewal has brought Native-American demands for full self-government into the arena of public debate.

In the ghetto *(left)* A crumbling inner city district of Philadelphia populated almost exclusively by African-Americans. The handsome architecture makes areas such as this ripe for middle-class gentrification, pushing the existing population into new ghettoes on the city outskirts.

the distinctive hand-greeting are mimicked across the globe.

Working mothers

Many women are also constrained, not in neighborhoods, but in homes and workplaces. As in many industrial countries, women are expected to be both wage-earners and child-carers. Women form over 40 percent of the workforce and 60 percent of new workers. Over half of married mothers with young children do waged work.

Some states have encouraged women to work by requiring large employers to provide childcare, but restricted access to cars, discrimination that holds back career advancement, and the demands of combining work and childraising frequently confine women to lower-paid work close to home. Divorce rates rose 150 percent between 1960 and 1980 and pregnancy among unmarried women is increasing rapidly, so female-headed households are among the newest poor.

Increasing levels of divorce may reflect the growing unwillingness of women to suffer the kinds of abuse and neglect from their husbands that previous generations endured in relative silence. But society still makes this decision difficult. Only 7 percent of divorced women receive regular alimony payments and in many states they are not legally protected in housing transactions.

Despite a strong and vigorous feminist movement, women in the United States still come up against the "glass ceiling", a hidden barrier of discrimination. Only 3 percent of top management in 1989 were women, showing little gain over 10 years.

IROQUOIS CULTURAL RENEWAL

Although the so-called "Indian Wars" belong to the last century, many of the sources of conflict remain unresolved. Clashes in 1990 between Canadian and United States' authorities on the one hand and the residents of a Mohawk reservation on the other provided a case in point.

Mohawks, along with the Oneida, Onondaga, Cayuga, Seneca and Tuscarora, are one of the six nations of the Iroquois. In total they number almost 40,000 people and are scattered throughout 17 communities living in reservations mostly in New York State and Ontario (Canada). These are their original homelands, where they lived as hunters, farmers and traders. Almost all adults speak Iroquoian languages as well as English. A matrilineal people, roughly half practice the Longhouse religion founded in 1800 in reaction to domination by the whites. This Native-American version of Christianity not only contains a moral code, but also envisages a heaven that only Native-Americans can enter.

Depopulation and cultural attrition followed conquest, but the threat to their land from hydroelectric and flood-control schemes in the 1950s led to a reassertion of cultural identity and renewed interest in Iroquoian history. Attempts were made to prevent further robbing of ancestral graves, and to recover the ceremonial *wampum* belts now distributed among museums.

Today as part of their campaign to obtain full recognition as a self-governing people the Iroquois issue their own passports. The Mohawks' Akwesane reservation straddles the United States–Canada border and a treaty made in 1794 gives them rights of movement across it. However, clashes between factions involved in casino gambling on the reservation resulted in several deaths in 1990, with subsequent pressure on both governments to intervene and close the border.

The graying of America

Rising levels of affluence and advances in health care have combined to produce a new generation of the elderly, and possibly a new cultural lifestyle, in the United States as in many other Western countries. In 1900 only 4 percent of Americans were aged over 65; by 1990 the figure had risen to 12 percent and by 2030 will be 21 percent. These 30 million individuals are charting new territory, as the former definitions of old age are being refashioned. The cycle that previously linked the termination of paid work to poverty and ill health is being broken in an ever-increasing number of cases. For more and more Americans, retirement at 65, far from marking the end of an active life, now heralds a new beginning.

Today's senior citizens differ in one very significant way from those of a generation ago. Medical advances have reduced the incidence of debilitating illnesses and diseases, so that after 65 they may have a longer life-expectancy. This is particularly true of women. By 2010 it is thought that three out of every four people over 85 will be female.

If the elderly worked and were well paid earlier in life, then they are likely to possess not only income from private pensions and savings, but also property. Over 75 percent own their homes. Even those on state pensions are better off than their parents were, as social security benefits have gradually shifted from helping the young to assisting the old. And men, though not women, are leaving work earlier; two-thirds of males with private retirement benefits stop working before their 65th birthday, and the trend is likely to grow.

The new old no longer live with their children. In 1900 two-thirds of them did so, but today the number has fallen to less than one-fifth. Many stay in the cities and neighborhoods they lived in during their middle years, some in specially designed retirement homes and sheltered accommodation. These are being built at an ever-increasing rate.

Cheering for old age (*above*) Seeing retirement as a fresh start, many senior citizens take up new leisure activities, with more time on their hands to enjoy themselves than ever before.

Migrating south (*right*) Leaving their former life in a northern city far behind them, a Jewish couple relax – suitably protected against the sun – outside their retirement home in Miami.

A place in the sun

A significant minority, mainly the younger, healthier and richer retirees, have moved away from their families to start new communities. Florida (where 17 percent of the population in 1990 was aged 65), Arizona, New Mexico and California are major destinations for retiree migrants seeking a leisured life in warmer surroundings and cheaper housing. Most of these people, often known as "snowbirds", come from New York and Illinois; a significant number of widows return to their places of origin after the death of their husbands.

Some of these move into mobile-home parks, of which there are now over 700 in Florida alone. But others have founded new settlements – retirement communities that have been developed since the 1950s for the more affluent retired. The largest of these, Sun City, Arizona, has over 60,000 inhabitants, but there are many other smaller communities dotted around Phoenix and St Petersburg in Florida. These places, with their distinctive bungalows, attractive but easily maintained gardens, golf courses and leisure centers, are unique developments of Western lifestyles. Few such communities have cemeteries, as a preference for cremation rather than burial seems to be on the increase among the present generation of the elderly.

Despite the forecasts of many professional social commentators, however, a distinct retirement culture has not fully emerged. Differences in class, voting and age itself cut across the communities so that the anticipated "gray vote", which was expected to exert a considerable influence over local political issues, has not developed as a significant force.

לידער פון היינט און פון נעכטן

A MEETING PLACE IN THE TROPICS

CONQUESTS AND COLONIZATION · THE CULTURAL MOSAIC · SOCIAL STRATIFICATIONS

Central America and the Caribbean islands originally supported a culturally diverse set of peoples who had lived there for at least 12,000 years before the European discovery of the "New World" in 1492. This event sparked off a sustained invasion of the region by Europeans, with devastating effects on the indigenous peoples. The Spanish quickly seized control of Mexico and the Central American isthmus, and their language, religion and customs continue to prevail there today. The scattered islands of the Caribbean were colonized by a number of European nations, but the predominant cultural influence is African, derived from the millions of slaves shipped there to work the sugar plantations. Asians and impoverished Europeans, both brought as laborers, added to the cultural amalgam.

COUNTRIES IN THE REGION

Antigua and Barbuda, Bahamas, Barbados, Belize, Costa Rica, Cuba, Dominica, Dominican Republic, El Salvador, Grenada, Guatemala, Haiti, Honduras, Jamaica, Mexico, Nicaragua, Panama, St Kitts-Nevis, St Lucia, St Vincent and the Grenadines, Trinidad and Tobago

POPULATION

Over 83 million Mexico

5 million–15 million Cuba, Dominican Republic, El Salvador, Guatemala, Haiti

1 million–5 million Costa Rica, Honduras, Jamaica, Nicaragua, Panama

250,000–1 million Barbados, Trinidad and Tobago

Under 250,000 Antigua and Barbuda, Bahamas, Belize, Dominica, Grenada, St Kitts-Nevis, St Lucia, St Vincent and the Grenadines

LANGUAGE

Countries with one official language (English) Antigua and Barbuda, Bahamas, Barbados, Belize, Dominica, Grenada, Jamaica, St Kitts-Nevis, St Vincent and the Grenadines, Trinidad and Tobago; (Spanish) Costa Rica, Cuba, Dominican Republic, El Salvador, Guatemala, Honduras, Mexico, Nicaragua, Panama

Country with two official languages (Creole, French) Haiti

Other languages spoken in the region include Carib, Nahua and other indigenous languages; creoles and French patois; Hindi (Trinidad and Tobago)

RELIGION

Countries with one major religion (P) Antigua and Barbuda; (RC) Costa Rica, Cuba, Dominica, Dominican Republic, El Salvador, Honduras, Mexico, Nicaragua;

Countries with more than one major religion (P,RC) Bahamas, Barbados, Belize, Grenada, Jamaica, St Kitts-Nevis, St Lucia, St Vincent and the Grenadines; (P,RC,V) Haiti; (H,M,P, RC) Trinidad and Tobago

Country in which religion is officially proscribed Cuba

Key: H–Hindu, M–Muslim, P–Protestant, RC–Roman Catholic, V–Voodoo

Colorful culture *(above)* Maya women from Guatemala bring textiles and other handicrafts to sell in the markets. Virtually all Amerindian communities have their own style of colorful clothing that distinguishes one from another.

Pride in the past *(right)* The great murals painted by the Mexican artist Diego Rivera (1886–1957) after the 1910 revolution signalled a rebirth of nationalism by celebrating the country's native civilizations and rejecting European ideals.

CONQUESTS AND COLONIZATION

Christopher Columbus (1461–1506) and the Spanish explorers and conquerors of the New World encountered a variety of indigenous peoples (whom, without distinction, they termed "Indians") practicing many differing ways of life. Among these peoples, chiefdoms existed in what are now the countries of Panama and Costa Rica, seminomadic societies inhabited eastern Nicaragua and Honduras, and a succession of complex empires (notably the Olmecs, Toltecs, Mayans and Aztecs) – each renowned for features such as hieroglyphic writing, advanced calendars, mathematics, monumental buildings and marvellous sculptures – left a cultural legacy in Mexico and Guatemala. On the many islands, settled agricultural villages existed in the Greater Antilles: Cuba, Hispaniola (Haiti and the Dominican Republic), Jamaica and Puerto Rico.

The conquering Spanish immediately and ruthlessly imposed their language, religion, bureaucratic forms of government, architecture, town planning and economic system upon all of these indigenous Amerindian peoples. The Spanish regarded the New World as a wild and hostile environment needing to be tamed, the fruits of which were there for the picking. They extracted all the wealth they could from the land and its people, shipping it back to Spain in fleets of treasure-bearing ships. From the Old World of Europe, new animals, new crops and new techniques for building, mining and agriculture were introduced to Central America and the Caribbean. All these radically altered the traditional ways of life of the original inhabitants.

In the course of being colonized, the Amerindians were treated so harshly that their numbers rapidly plummeted. From a total of somewhere between 50 and 100 million before Columbus's arrival, the indigenous population fell to only some 3 million within the first 100 years of Spanish colonization. Many devastating diseases (smallpox, measles, typhus and

yellow fever) were inadvertently introduced, to which the Amerindians had no immunity. Forced labor under harsh conditions and systematic slaughter also took large death tolls. The Amerindians who survived mainly did so because they lived in isolated pockets in remote areas.

By 1600, some 300,000 colonists had migrated to the region, and many mixed with the Amerindian population. In rural areas of Central America a rapidly growing population of racially mixed people (*mestizos*) took the place of the declining Amerindians to supply labor on the vast agrarian estates (*haciendas*), owned by a small elite of Spanish descent.

The Caribbean experience

The Spanish could not prevent other European nations from laying claim to the numerous islands of the Caribbean; many changed hands several times in the two centuries after Colombus. Although Spain continued to control the mainland, England, France and the Netherlands were the nations that were principally involved, with some incursions from Denmark and Sweden, and later from the United States. Languages, buildings, public works, laws, elements of popular culture and other features imported by these nations are still found throughout the Caribbean.

The indigenous peoples of the Caribbean islands – Caribs and Arawaks – were practically wiped out by disease, slavery and warfare within a few years of colonization. During the following two centuries, therefore, more than 5 million slaves from Africa were transported to meet the labor needs of the emerging, European-imposed plantation economy; here, many were kept in degrading conditions. After the abolition of slavery in the 19th century, laborers from the Middle East, Europe, China and India came to bolster the labor force. Many, particularly the Asian Indians, arrived as indentured workers who were bound by contracts to work for as much as 5 to 10 years, virtually as slaves themselves.

Following the Spanish–American War at the end of the 19th century, the United States seized Cuba and Puerto Rico, and later became heavily involved in Haiti and the Dominican Republic. This powerful neighbor has continued to have a significant influence – culturally and economically, as well as politically – over most other islands in the region.

THE CULTURAL MOSAIC

The mark of the Spanish empire permeates virtually all of Central America and the Spanish-speaking islands of the Caribbean (Cuba, Puerto Rico and the Dominican Republic). This is especially noticeable in the layout of towns and cities, the elaborate architecture of colonial buildings and churches, the heritage of certain forms of music, and even, in places, the popularity of bullfighting. Yet the greatest remaining influence of Spanish rule is found in the continuing role of the church in these countries.

In the early years of the conquest, the mission to convert the Amerindians to Roman Catholicism motivated the Spanish perhaps as strongly as their lust for gold. In addition to the exercise of political control and enforced labor, the religious conversion of the Amerindians was another way of dominating them. However, over many years, the blending of many pre-Colombian beliefs, practices and symbols with those of the Spanish has given the Roman Catholicism of Central America a character and flavor of its own. Among the Otomi of Mexico, for example, ancient religious shrines are still honored, and shamans intercede with spirits on behalf of villagers who are, nonetheless, devout Roman Catholics. In some places today, Roman Catholic influence is being challenged by North American-backed Protestant evangelical missions. This is especially evident in Guatemala, where already up to half the population has converted to such fundamentalist forms of Christianity in relatively recent times.

Amerindian survivals

Although in most of Central America *mestizos* are in the majority, pockets of Amerindian predominance are also to be found, and Amerindian languages, lifestyles and customs persist in one form or another throughout the region. There is great variety in racial composition from country to country: from Honduras, where *mestizos* account for 92 percent of the population, to Costa Rica, which is 92 percent European, to Guatemala, which has 62 percent Amerindians, the highest proportion in Central America. In Mexico, only 10 percent of the population is recorded as being of pure Amerindian descent, yet there are few Mexicans without some Amerindian blood, and Nahuatl, Mayan, Zapoteco and Mixteco languages are widespread, especially in southern parts of the country.

While European aspects of society and culture are generally widespread across the region, some facets of Amerindian life nonetheless survive from place to place, especially in the more remote rural districts. Such traits include (in addition to language) types of family and social relationships, patterns of collective land use, beliefs and ritual practices devoted to a series of supernatural beings, the honoring of ancestors, cultivation habits and food preferences, and styles of dress and textile production.

In Belize and along the Caribbean coasts of Nicaragua and Costa Rica, another unique cultural tradition is to be found. Here, the descendants of escaped African slaves form the majority of inhabitants. They mainly speak a Caribbean dialect of English, and many features of their society and culture are also closer to those of the Caribbean than to their Hispanic and Amerindian neighbors. This kind of heterogeneity found along the Caribbean coast is a mirror of the complex makeup of societies found on islands in the Caribbean Sea itself.

A multicultural region

More than 50 nationalities or groups, many with their own art forms, religions, languages, and social institutions, comprise the multicultural societies of the Caribbean islands. Language diversity is the most immediate sign of this – throughout the islands English, French,

A Voodoo priest (*above*) Haiti's folk religious cult is a mixture of Roman Catholicism and African animism and magic. A priest or priestess leads devotees in a ritual involving song, drumming, dance, prayer and the preparation of food.

Day of the Dead (*left*) Mexico's version of All Souls Day is widely observed. Offerings to ancestors' souls are made at a candlelit vigil, and picnics are held at their graves. Shops sell chocolate skulls and other souvenirs of the day.

Dutch, Hindi and Chinese, as well as a wide variety of *creole* or *patois* variants that have evolved through the blending of these with different African languages, can all be heard. The creation of these Caribbean languages coincided with the emergence of a colorful tradition of oral literature, based upon folkloric stories told from generation to generation. Out of this tradition has come, in turn, a number of important contemporary writers, in both English and French, who have explored the heritage of Caribbean peoples in evocative and entertaining ways.

The Caribbean is also renowned for its unique syncretic religions, such as voodoo in Haiti, pocomania in Jamaica and shango in Trinidad. These bring together the beliefs and practices of Roman Catholicism with facets of traditional West African religions, often involving the inducement of ecstatic states of consciousness and trance, rituals of healing, and the identification of Christian saints with West African (especially Yoruba) gods or supernatural powers.

Because societies in the Caribbean were created by colonial powers wholly for the production of export crops, their populations grew well beyond the capacity of each island to support them. After the ending of slavery a tradition of migration between the islands developed, widening to countries farther afield as regional economic problems intensified. Following World War II, Britain was the principal destination for immigrants from Jamaica, Barbados, Trinidad and other islands then belonging to Britain. Today, the goal for most is the United States.

THE VIRGIN OF GUADALUPE

Perhaps no other symbol encapsulates the identity of a nation so well as Mexico's patron saint, the Virgin of Guadalupe. Her image preceded armies of insurgents in the War of Independence from Spain in 1821, and today it is found everywhere: in homes and churches, taxis and buses, restaurants, bullrings and brothels.

According to legend, at Tepeyac, to the north of Mexico City, in 1531, the Virgin appeared to an Amerindian convert named Juan Diego, asking that a shrine be built in her honor. Diego petitioned the local archbishop unsuccessfully until the Virgin allegedly worked a miracle: producing roses where they could not have grown and marking her image on Juan Diego's cloak. Today, the famous cloak hangs above the altar in the church where his vision was believed to have taken place, and it now attracts hundreds of thousands of pilgrims each year.

In Tepeyac, in pre-Columbian times, there had stood a temple in honor of Tonantzin, the Aztec mother goddess. The pagan Tonantzin was replaced by the Christian Virgin of Guadalupe, thus linking Amerindian beliefs with those of their Spanish conquerors. The Virgin has thus now become the symbol of a single Mexican people.

SOCIAL STRATIFICATIONS

The longest-lasting legacy that European colonialism has given to Central America is a rigid, pyramidal social structure, with a powerful local aristocracy at the top, *mestizos* filling the middle classes, and Amerindians at the bottom. Attempts to wrest social and economic equality from the predominantly white elite have contributed to the endemic political instability of the region, resulting in a history of left-wing revolutions and right-wing military takeovers.

The struggle to survive

Regardless of their ethnic background, impoverished peasants make up the bulk of Central American society. For many, the struggle for survival is hard, whether as subsistence farmers on land where the soil has been eroded and degraded by deforestation or living in the crowded slums of Mexico City. It is among such people that the ideas summed up in the "liberation theology" movement, which calls for direct action to free the poor from dehumanizing social structures, have taken strong hold – as evidenced by the presence of Jesuit Roman Catholic priests in the left-wing Sandinista government of Nicaragua. Others seek to escape from poverty by migrating. Since the 1970s there has been a massive, sometimes illegal, exodus of peasants from all over Central America to the United States.

Social causes and movements will often harness local indigenous symbols to rally popular support. But indigenous culture itself is vulnerable to all kinds of pressure from above. For example, the military-controlled government of Guatemala has pursued an overt and hostile program of "cultural genocide", seeking to drive into extinction all traces of indigenous culture – and often its practitioners as well. Similar policies have already led to the virtual extermination of traditional indigenous cultures in Costa Rica, El Salvador and Honduras.

Shades of color

In the Caribbean, the plantation took the place of the *hacienda* in determining the social structure. The local plantation represented a complete social and economic unit that controlled practically every aspect of the resident workers' lives, from housing and work routines to marriage patterns and leisure activities. French, British and Dutch colonial rule created a strict hierarchy. In most places, whites formed the ruling and wealth-owning class, the numerically dominant blacks the laboring class, and those of mixed race (variously called mulattos, quadroons or mustees) were sandwiched in the middle – a pyramidal structure not dissimilar to that found on the Central American mainland.

However, in ways unique to the Caribbean, a separate "brown" racial category came to have great importance in determining social status and acceptability. Depending on the society and its par-

More than a game *(above)* The West Indies cricket team unites the islands of the Caribbean. Cricketing prowess represents black pride and independence from former British colonial masters.

A new identity *(below)* A Rastafarian family in Jamaica. Rastafarianism replaced Western religion and culture with a specifically black creed that turned to Africa for its inspiration.

RASTAFARIANISM

During the 1930s a religious movement was born in Jamaica that offered black people a cultural identity and called for an end to the inequalities that determined their place in society. Rastafarianism preached that the only way for black people to escape their poverty was to return to Africa. Adopting the symbolism of Judaism (and holding that the Bible has been deliberately mistranslated by whites), Rastafarians believe that the Africans of the Caribbean are in Babylon, the place of exile, from where a Messiah, or leader, identified as the late emperor of Ethiopia, Haile Selassie (1892–1975) – known as Ras Tafia, the Lion of Judah – will lead them to a promised land in Africa.

Rastafarians use a distinct dialect and obey Biblical injunctions not to cut their hair, wearing it long in dreadlocks; they adopt the distinctive colors of Ethiopia: red, green and gold. The movement's promotion of black pride and black power, and criticism of imperialism and social inequality, mean that its symbols and rhetoric have been borrowed by political parties at home and farther afield.

Rastafarianism has had widespread cultural influence – adherents are found around the world, particularly in the United States and Britain, and many of its usages have been absorbed by urban youth subcultures. This is largely due to the worldwide audience for reggae, which has become the Rastafarians' most successful method of communicating ideas, especially through singers such as Bob Marley (1945–80).

icular ethnic makeup, Chinese, Arabs (invariably called Syrians regardless of national origin) and Jews spread through the middle classes. Sometimes Portuguese and other groups of poor whites maintained a vague status apart from the browns and blacks, but were nevertheless placed in society's lowest ranks. Asians, too, were usually relegated to the bottom rank, largely because of their position in the labor market and their alien language, religion and culture. Only in Dominica and the Netherland Antilles do pockets of full-blooded Caribs and Arawaks survive; inevitably, these also occupy a place at the bottom of society. Since independence, most countries have been struggling to overcome this legacy of colonial rule. However, nearly everywhere, the white rulers of the past have been replaced by a brown elite, while whites often still control some of the most important economic assets.

Trinidad – colonized in turn by the Spanish, French and English – has perhaps the most diverse and complex of all Caribbean societies. Most of the ethnic groups of the Caribbean are represented here. In addition to English, three creole and several Asian languages are spoken, and both the mixed race and African-origin groups, including those that have migrated in recent times from nearby islands such as Barbados and St Vincent, are further subdivided on the basis of whether they are Roman Catholic or Protestant. Those of Asian descent (approximately half the population) are similarly differentiated by religion into Hindus, Muslims and Christians.

These deep-rooted ethnic and cultural differences within the population means there are few common symbols to unite the country. Social and cultural fragmentation is a problem that affects all Caribbean societies. Even though Caribbean peoples share a similar heritage and general social and cultural patterns, most attempts to arrive at some sort of common regional identity have been unsuccessful, partly due to size; the peoples of the small islands insist on maintaining their social and cultural independence.

Yet, especially in the fields of music (particularly reggae, soca and calypso), sport (cricket and soccer), theater and literature, Caribbean peoples are increasingly finding mutual expressions of their common historical roots. The existence of Caribbean communities in other parts of the world, such as New York, Canada and Britain, means that aspects of this distinctive culture are also found worldwide.

Hindus in the Caribbean

Hindu and Tamil languages, spiced Asian food, Indian music and decorated Hindu temples may seem out of place in the accepted picture of Caribbean life, with its dominant creole culture based on the merging of Roman Catholic and African traditions, but they are all to be observed within a variety of thriving Indian communities throughout the islands.

Between 1838 and 1917, more than half a million Indians were brought to the Caribbean. Most came from northern India and settled in the British colonies of Trinidad, Jamaica and some of the smaller islands. Others, from southern India, were transported to the French colonies of Guadeloupe and Martinique. Most of them were Hindus, adherents of the dominant religion of the Indian subcontinent, which consists of a complex system of rites and ceremonies.

Within the first years of their coming to the Caribbean, Indian Hindus built temples and organized the celebration of major festivals. However, in the new environment in which the immigrants found themselves planted, many aspects of Indian culture that had distinguished groups and communities at home, such as languages and their dialects, customs, religious practices and art forms, merged with one another. The result was a diminution in cultural vigor. The most significant feature of Indian culture to become weakened was the caste system. Inextricably linked with Hinduism, it was the traditional method of maintaining social, economic and political distinctions between groups within society – but, transferred to the plantation economy of the Caribbean, it ceased to operate in an effective way.

There were regional variations in the vigor with which Indian cultural identity was preserved. In the smaller islands, where Indian immigrants were few in number, their social and cultural habits were more easily superseded by the local creole ones. Where Indians were more numerous and densely settled, however, many characteristically Indian ways of life were strongly upheld.

Cultural reinvigoration

In recent years, many religious and cultural practices have been revitalized and embellished, largely through the growth of local and national Hindu organizations. These include *yagnas* (week-long sets of ceremonies surrounding readings of the Hindu sacred books), and the annual Ram Lila, a religious play concerning the life of Lord Rama, one of the main gods of Hinduism. Performances may last for several days, and sometimes involve more than 100 participants. Hindus have also reestablished the performance of important rituals, such as *pujas* (ritual offerings to deities) and the rites-of-passage, which are celebrated to mark the stages of an individual's life.

Strong as their sentiments and activities are, many Caribbean Indians feel alienated and discriminated against. In Trinidad, for example, the sizable Indian community considers itself the victim of attempts to promote a sense of national unity, since such efforts focus on cultural phenomena such as carnival or calypso music, both of which are predominantly associated with Afro-Caribbean culture. The selection of an Indian player to the West Indian national cricket team is likely to cause controversy throughout the cricket-playing parts of the region, since the game is still regarded as belonging to the Afro-Caribbean tradition.

Hindus attest that efforts to create a common Caribbean identity ignore the interests of the minority cultures and religions such as theirs. So long as this is the case, they argue, Caribbean societies will continue to be seriously fragmented through lack of cooperation.

Rites of passage (*above*) The eldest son sprinkles petals on the corpse before the body is cremated and the ashes sprinkled in sacred water, usually a river.

Religious revival (*right*) A devout Hindu prays at the temple. Caribbean Hinduism has been invigorated by better organization and political participation.

A place for worship (*below*) A new temple in Trinidad mixes Caribbean and Indian styles. Hindu temples are shrines for individual, rather than congregational, worship.

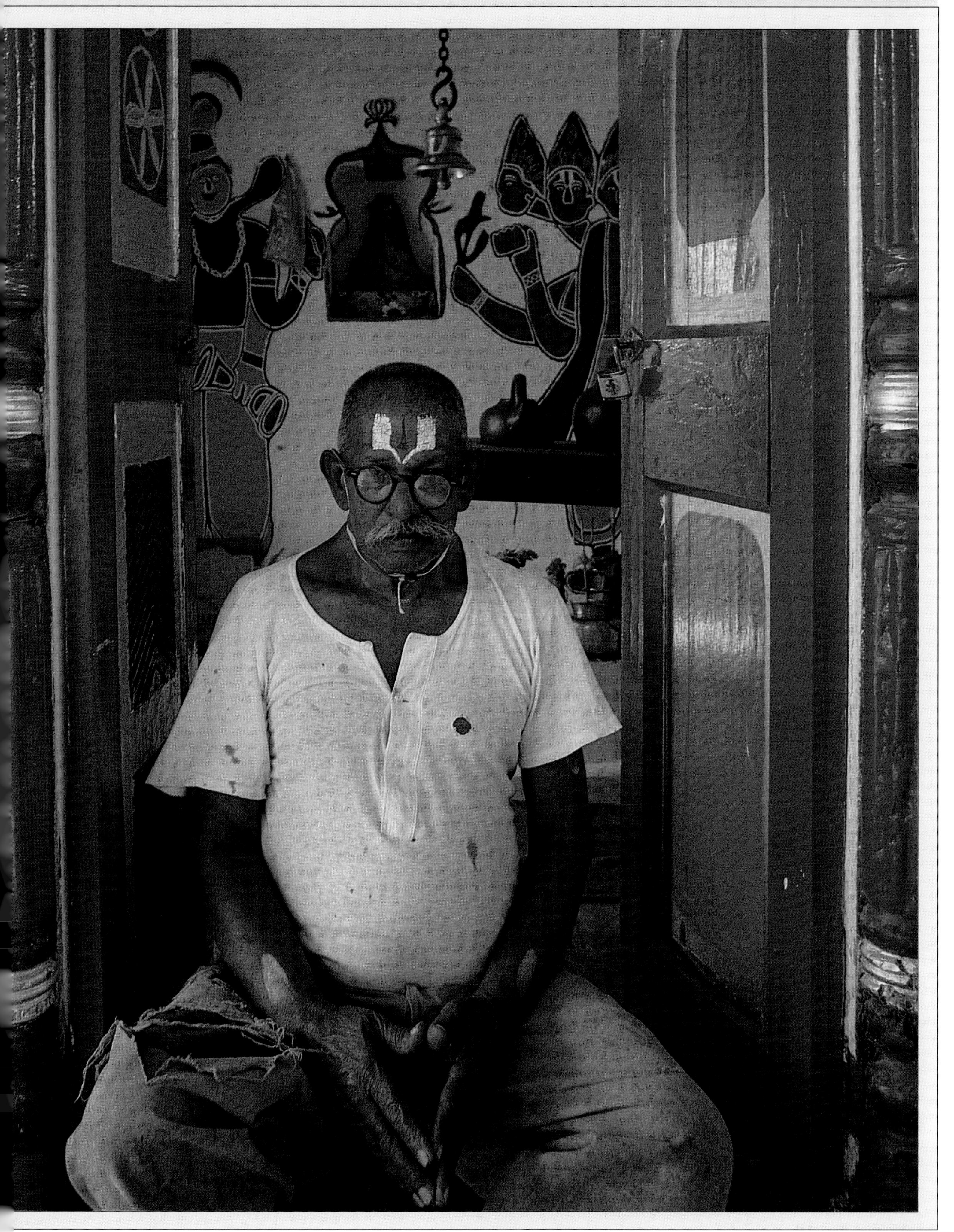

Carnival!

Carnival was traditionally celebrated in Christian Europe as a time of feasting and popular fun before the start of the 40-day period of abstinence known as Lent. It was observed in the Caribbean as early as the 18th century, and after the emancipation of slaves was taken over by black Africans, when it resumed its original character as a time of rule-breaking and role reversal. Men dressed as women, slaves became masters, night became day, people hid behind masks, and the streets were filled with music, revelry and drinking, banned at other times of the year.

Carnival had a variety of names throughout the Caribbean – Junkanoo (or John Canoe) in Jamaica, Belize and the Bahamas, and "mas" (short for masquerade) in Grenada. Some were held before Lent, others on New Year's Day or Twelfth Night, 12 days after Christmas. Its associations with disorderly behavior meant that it constantly came into conflict with the authorities and was frequently banned.

After independence, Carnival was made a day of national celebration in many of the islands. Cuba moved its Carnival to July to celebrate the revolution, while Bermuda and Grenada shifted theirs to August for the tourist season – Carnival has become a valuable source of revenue. The traditional mingling of costumed paraders and onlookers and the spontaneous routes taken by the parades were changed to create a more orderly spectacle. In doing so, much of the subversive character of the festivities has been lost, though in Trinidad – the greatest Carnival of them all – the calypso-singing competition still provides an occasion to mock politicians.

A swirl of color Every year, crowds of people fill the streets of Port of Spain in Trinidad to watch the costume parade, attend the dance competitions and listen to the calypsos, steel bands and reggae that make Carnival such a vibrant celebration.

PRISONER OF ITS PAST

A HISTORY OF CONQUESTS · INTERWEAVING CULTURES · A CHANGING SOCIETY

The earliest humans in South America, perhaps as many as 30,000 years ago, were prehistoric nomadic hunters. Their descendants became settled farmers and craftsmen about 8,000 years ago. The European expansion into the New World in the early 16th century rapidly transformed the social organization of the region. The Spanish and Portuguese, followed later by other nationalities, all added new languages, religions and customs to the cultural mix, as did millions of imported African slaves. Despite this ethnic diversity, South Americans are united by the Spanish language (understood if not actually spoken by most people outside Brazil, where Portuguese is the majority language), the Roman Catholic religion, and the twin popular obsessions of music and soccer.

COUNTRIES IN THE REGION

Argentina, Bolivia, Brazil, Chile, Colombia, Ecuador, Guyana, Paraguay, Peru, Surinam, Uruguay, Venezuela

POPULATION

Over 147 million Brazil

10 million–31 million Argentina, Chile, Colombia, Peru, Venezuela

1 million–10 million Bolivia, Ecuador, Paraguay, Uruguay

Under 1 million Guyana, Surinam

LANGUAGE

Countries with one official language (Dutch) Surinam; (English) Guyana; (Portuguese) Brazil;(Spanish) Argentina, Chile, Colombia, Ecuador, Paraguay, Uruguay, Venezuela

Country with two official languages (Quechua, Spanish) Peru

Country with three official languages (Aymara, Quechua, Spanish) Bolivia

Other languages spoken in the region include Arawak, Carib, Jivaro, Lengua, Mapuche, Sranang Tongo, Toba and numerous other indigenous languages

RELIGION

Countries with one major religion (RC) Argentina, Bolivia, Brazil, Chile, Colombia, Ecuador, Paraguay, Peru, Venezuela,

Countries with more than one major religion (A,N,P,RC) Uruguay; (H,I,M,P,RC) Guyana, Surinam

Key: A–Atheist, H–Hindu, I–Indigenous religions, M–Muslim, N–Nonreligious, P–Protestant, RC–Roman Catholic

A HISTORY OF CONQUESTS

When Spanish and Portuguese explorers reached South America in the 16th century they found a marked contrast between the Amerindian populations who lived in the highlands of the Andes and those who lived in the lowlands – and these differences still hold good today. The highlands were dominated by the Incas, who had rapidly expanded their empire by conquest in the previous century and controlled a territory that stretched from what is known today as Ecuador through the length of Peru and into central Chile; an extent of nearly 4,000 km (2,500 mi). Their southward expansion was halted by the Mapuche Indians who held back the Inca armies, as they later would the Spanish.

The Inca empire was highly centralized, with a complex social structure. Its villages and large urban centers contained well-constructed massive stone buildings, and were linked by an excellent network of mountain roads. In the absence of wheeled transportation, goods were carried on people's backs or on llamas. The Quechua language acted as the lingua franca of the Inca empire, uniting its linquistically diverse people. There was no writing, but records were kept by means of the *quipu*, an abacuslike device. In lowland areas – the savannas of central South America and the forested Amazon basin – Amerindian society was far more fragmented, with innumerable groups lacking a common language or culture. Communication was possible only by river. There were virtually no buildings made of stone, nor any settlements on the scale of the Inca cities.

European colonization

European settlement patterns enlarged these Amerindian differences. Rivalry between the colonizing powers of Spain and Portugal, who divided the continent between them, drove the Portuguese to push the frontiers of their empire in Brazil westward, deep into the South American lowland interior. This ensured that what was to become the largest, most populous and economically most powerful country on the continent was Portuguese- rather than Spanish-speaking; to this day Brazilians are regarded as a race apart by Spanish-speaking South America.

The immediate effect of colonization on the indigenous populations of South America was massive mortality, as many ethnic groups were wiped out by imported European diseases to which they had no resistance, notably influenza, measles and syphilis. Survivors were often organized into missions run by Christian orders (with the Jesuits and Dominicans competing with each other to set up mission stations), or simply enslaved as agricultural laborers to work the Andean silver mines and gather the forest products – vanilla, rubber, sasparilla and cocoa – valued by the Spanish and Portuguese. In Colombia, Venezuela, Brazil and Peru large mixed populations evolved, called variously *criollo* or *mestizo* in Spanish South America, and *caboclo* or *mestico* in Brazil.

European cultural influences, however, were not confined to the Iberian countries alone: Britain, France and the Netherlands all established footholds on the northeast coast in the 17th century. Since the 19th century, migrants from many other European countries have settled throughout South America. Between 1850 and 1940 some 4 million Italians migrated there, the majority attracted to the coffee plantations of southern Brazil and the burgeoning cities of Uruguay, Argentina and Chile. There are small, but economically important, communities of Anglo-Argentines and Anglo-Chileans – reminders of the decades after independence in the 19th century when Britain dominated South American trade – and significant German populations exist in Paraguay and southern Brazil.

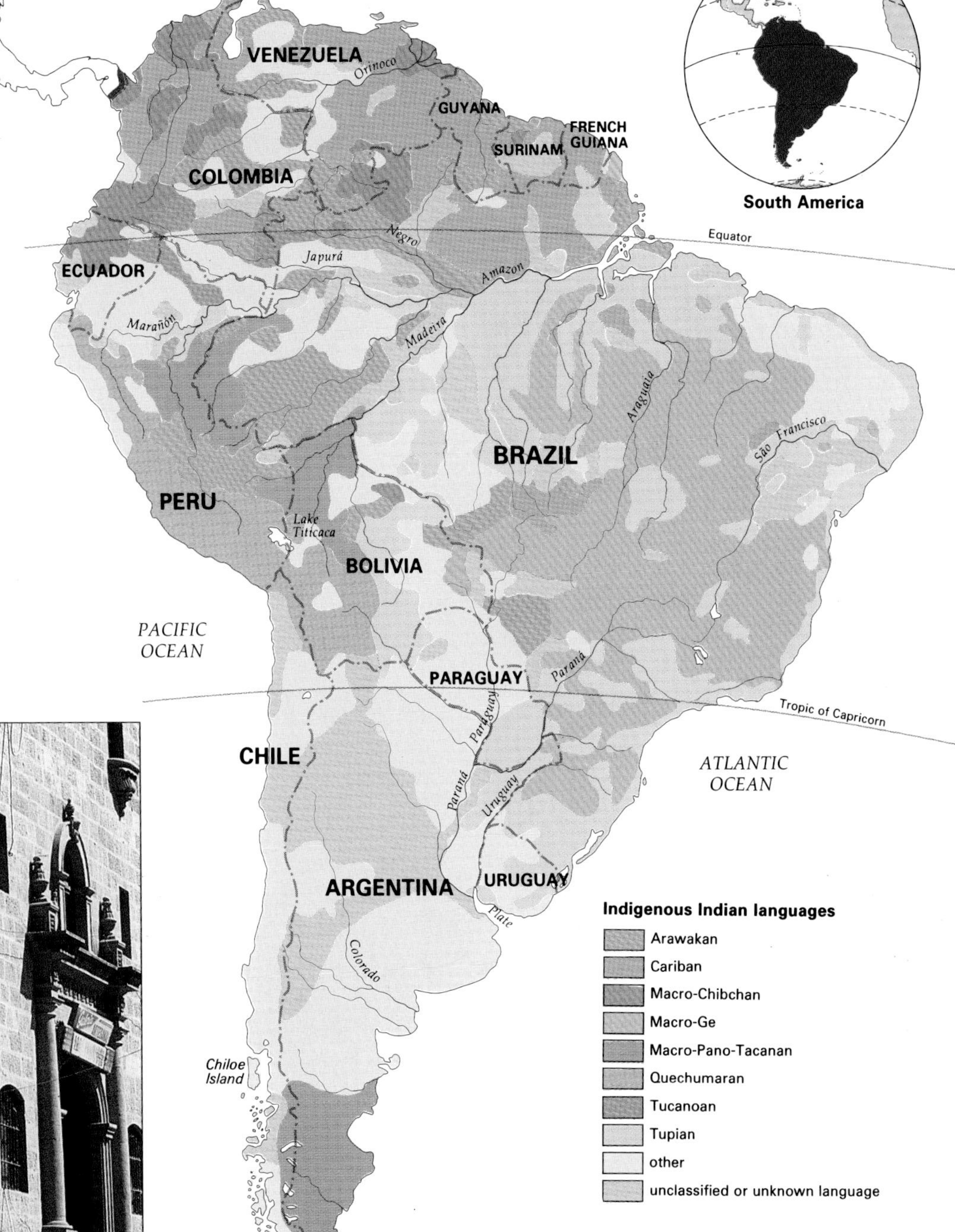

Amerindian languages (*above*) The mosaic of languages indicates the complexity of pre-Columbian society. Although many of them are vanishing, indigenous languages are still prevalent in the Amazon Basin. Quechumaran (designating both Quechua, spoken by the Incas, and Aymara) is still widely spoken by inhabitants of the Andes.

Colonial heritage (*left*) The magnificent architecture of a Roman Catholic church in La Paz, Bolivia, is a lasting reminder of Spanish colonial rule. Over half the country's population is pure Amerindian; most are practicing Roman Catholics, though some blend this with a continued belief in some of their traditional deities.

Other traditions

The number of African slaves transported to South America to provide labor on European-owned plantations was ten times higher than the number shipped to the United States, but higher death rates ensured that only about 12 percent of South Americans today are black, roughly the same proportion as in the United States. They are concentrated in the old plantation zones of northeastern Brazil and along the Caribbean coast. Their impact on popular South American culture has been highly visible.

Early in the 20th century, a quarter of a million Japanese, mostly small farmers, were attracted to Brazil's open spaces. Every South American country also hosts an Arab minority, whose trading skills underlie an economic importance out of all proportion to their size.

An exciting mixture of pagan and Christian influences, carnival is the supreme expression of the South American passion for dance. The lavish carnival celebrations in Rio de Janeiro, in Brazil, are unsurpassed throughout the world. Competing neighborhood "samba schools" – possibly consisting of thousands of people – spend months preparing for their carnival procession, which is usually based around a religious, social or political theme.

INTERWEAVING CULTURES

South American society is the product of the interweaving Amerindian, European and African traditions, but there are marked cultural variations between the different countries, and between the life of the cities and of the countryside. These differences have been molded over the centuries by geographic and economic factors that have shaped the history of settlement in the region.

Festivals and cults

The missionary efforts both of the Spanish and Portuguese conquerors ensured that Roman Catholicism became the dominant religion of the region. For example, in the week before Ash Wednesday, which marks the beginning of the Christian season of abstinence, most of the continent erupts into carnival, with spectacular parades, music and characteristic dances such as the samba. But Christianity is not the only religion. Most characteristic of South America is a form of religion known as *candomblé* in Brazil and *santaria* in Spanish-speaking South America, formed by the blending of West African religious beliefs with those of Roman Catholicism. The Virgin Mary, for example, is transformed into Iemanjá, goddess of the sea, who is represented in many cult houses by images of the Virgin.

Such Afro-American cults have millions of adherents, from all ethnic groups, most of whom remain practicing Roman Catholics as well. In the Andean countries of Colombia, Ecuador, Peru, Bolivia and Chile, gods that the Incas would have recognized are still worshipped in traditional ceremonies. In Peru and Bolivia, for example, miners still worship Tio, which literally translates as "Uncle", and represents both the Inca god of the Earth and the Christian devil. Equally ancient beliefs and myths, with their attendant ceremonies, survive in the animistic religions of the lowland Amerindians.

Questions of identity

South America is the most urbanized continent in the developing world, with some 60 percent of its population living in cities. Nowhere is the intriguing mix of South American diversity and homogeneity more evident than in its urban

FOOTBALL CRAZY

Undivided loyalty Argentinian emigrants to the United States retain links with their homeland through undiminished support for the national football team during the 1986 World Cup final.

Football (soccer) in South America has a hold on the popular imagination that is unrivaled anywhere in the world. The largest football stadiums, the most fanatical supporters in the world, and a very distinctive Latin American style of play combine to give the region a supremacy in the sport that is matched only in a few European countries.

The game was first imported to the region by British railroad engineers at the turn of the century: to this day the British influence is obvious in team names such as Newells Old Boys, Racing Club and Corinthians. By the 1920s regional football leagues had been established in Argentina, Brazil, Chile and Uruguay. Right from the beginning football was a mass spectator sport: over 100,000 people watched Uruguay defeat Argentina 4–2 in the first World Cup Final in Montevideo in 1930. Since then South American teams have won the World Cup – the sport's premier trophy, contested every four years – on seven occasions.

South American football has two main centers: the river Plate countries (Argentina and Uruguay) and Brazil. Spanish and Italian migrants dominated the early years of the game in the former countries, and players have developed a powerful, muscular style of play that is closely related to the European game, albeit with a level of skill and tactical subtlety unknown outside Italy. But Brazil, the winner of three World Cups, has produced some of the most memorable football of all, and in the player known worldwide just as Pelé, probably the most skillful player of the ball ever seen, it has a national hero and international star.

centers. The metropolis of São Paulo – the continent's largest city – has a population of some 15 million. It is an ethnic melting pot, in which many languages are present. In the Japanese *barrio* of Liberdade, for example, old people still speak Japanese among themselves, but their sons and daughters use the drawling Portuguese of native Brazilians, indistinguishable from that spoken by the descendants of Italian, Spanish and Portuguese migrants. In the teeming life of the city, other aspects of cultural identity, such as music, have become hybridized, to create a distinctively Brazilian identity.

In rural areas, by contrast, it has been relatively easier for many ethnic groups to preserve a separate cultural identity. Fiercely guarding their languages, rituals and common history, and sheltered by their geographical isolation and deep attachment to the land, they have resisted absorption into national cultures they perceive as alien or hostile. Many of these groups are Amerindians, both highland and lowland, but some are descendants of Africans and Europeans. There are, for example, many small rural black communities in the northern Amazon basin. These were originally established by African slaves who escaped into the interior from plantations along the coast and rivers. Many elements of African culture, particularly its religion, music, cuisine, and rich tradition of oral and written literature, have entered the popular culture of South America.

With the opening up of the savanna grasslands of South America to cattle during the 19th century a new way of life, celebrated in art and literature, came into being, that of the *gaucho*, or South American cowboy. The *gaucho* way of life still survives, more or less intact, on the huge cattle ranches of northern Argentina, Uruguay and southern Brazil. *Gauchos* are almost exclusively the descendants of European immigrants, with a sprinkling of Indian blood. A similar, cattle-based subculture is to be found in the extensive savanna areas of Venezuela, Colombia and northern Brazil, exemplified by the *llaneros*, the cowboys of the Venezuelan interior.

At the farthest extreme of rural life are the peoples of the Amazon basin. This vast area has been transformed since the 1960s by highway and dam construction for hydroelectric power schemes, leading to the destruction of vast tracts of rainforest. The indigenous Amazonians, Amerindian and non-Amerindian alike, are struggling to preserve their way of life – based on fishing and the extraction of forest products such as rubber – from large-scale exploitation of the rain forest.

A CHANGING SOCIETY

The rapid pace of socioeconomic change, most visible in the very large numbers that are migrating from the countryside to the city everywhere in the continent, has transformed South America in recent decades. Yet, in many important respects, it is still a prisoner of its past, grappling in the 1990s with issues familiar to generations of South Americans reaching back to colonial times: how to maintain minority rights and identities in national states that may be hostile to them; how to organize and control the ownership of land; and how to ensure that there is a reasonable division of the region's natural resources and wealth.

Mountain people *(above)* Aymara women use their colorful shawls to carry anything from babies to goods for market. Hats are a feature of Amerindian traditional costume; the shape varies from one area to another.

Playground of the privileged *(right)* The beach is a popular weekend destination for Brazilians with the leisure to enjoy it.

Forgotten children *(left)* Out of sight of the affluent consumers and business people from downtown, families struggle to exist in the slum areas that fringe South America's large cities.

Life in the cities

In all the cities of South America cultural and economic differences between the educated, Westernized middle class and the mass of poor people living in shanty towns or *favelas* are all too visible. Nevertheless, across the continent the urban poor have responded in a number of similar ways to defend their livelihoods and improve their lot. The first line of defense is the extended family, which is the basic unit of social organization and provides a network of related members who can offer mutual support. Then there is religion: Roman Catholic parishes, Afro-American cult houses and Protestant congregations offer their members a group identity and form a basis for community action.

Foremost in encouraging urban action have been the Roman Catholic "base communities". These arose out of the "liberation theology" movement that seeks to apply religious faith by aiding the poor and oppressed through involving them in community affairs. It is led at the local level by parish priests who organize residents' associations to pressure politicians to give them the facilities they need: health centers, street lighting and sewage disposal. Although South American slums seem shockingly poor to outsiders, the high level of community organization that is also a part of slum life means that a poor city dweller is nearer to schools, hospitals and jobs than most rural peasants.

Urban growth and the continuation of land tensions in the countryside have both provided fertile grounds for new social movements. Many are religious: evangelical Christian sects flourish, and new religions appear, such as the Santo Daime cult, based on Amerindian beliefs, which initially radiated out from western Amazonia to the great cities of southern Brazil. But other movements are political, and take the form of populism, stoked up by politicians who try to imitate the mystical language and imagery of popular religion in order to present themselves as the saviors of the people. So long as the social conditions that encourage it remain, populism will be South America's main form of politics, both in the cities and in rural areas.

Intergroup dissension

There are deep divisions between urban and rural populations, the most dramatic of which are perhaps to be found in the Andean countries. Elites of largely European descent cluster in the cities, where Spanish is the dominant language. In the surrounding rural areas the population are Quechua- and Aymara-speaking Amerindians, who live in small communities clustered around the silver and tin mines, or survive as landless subsistence farmers. Here the smoldering cultural divide that exists between city and countryside, nourished by ethnic divisions and economic inequality, often

SYMBOLS OF DOMESTICITY

Life for most women in South America is much more restricted than it is for those in the countries of Western Europe and the United States. They have still to achieve significant political and economic power, yet, ironically, property-owning women were voting in South American elections in the 19th century long before European women had been enfranchised. In this respect they were the beneficiaries of the political liberalism that fueled the independence movements of South America.

Working counter to these ideas, however, has been the narrow view of women that the Roman Catholic church imposes. The numerous cults of the Virgin present a sanctified image of motherhood and domesticity, and this makes it difficult for women to break out to create an independent role for themselves. Women tend to marry early by European and North American standards, and have more children: the church successfully manages to restrict the availability of family planning. South Americans are generally deeply conservative on social issues, whatever their political complexion, and most men believe that a woman's place is in the home.

Moreover, South American women have sometimes been able to exploit their positions as wives and mothers, by speaking up and fighting against political repression in ways not open to their menfolk. During Argentina's "dirty wars" (1976–83), the Mothers of the Plaza de Mayo in Buenos Aires, protesting against the disappearance of their sons, issued a challenge that the country's military rulers found it hard to ignore, coming from mothers.

flares up to produce guerrilla movements and aggressive peasant rebellions.

One such movement in Peru is called the Shining Path. Nominally communist, on one level the Shining Path is a highly politicized campaign by the economically disadvantaged, pressing for a larger share of national resources. On another, it is a social movement of rural, Quechua-speaking, highland Amerindians who violently reject domination by an urbanized, Spanish-speaking Roman Catholic state, centered in Lima. In this respect the insurrection, which began in highland Peru in the 1970s, can be seen as a modern counterpart to the periodic Inca rebellions against Spanish rule that took place in the 17th and 18th centuries.

In Surinam, "bush blacks", the descendants of escaped slaves, live in remote villages in the interior of the country, where they form a numerical majority. They are dependent on subsistence farming and fishing, and are struggling to preserve their independence from the Creole-dominated government of the coastal area. Ethnic and racial divisions are evident elsewhere in South America: in Bolivia, for example, tension is high between *kollas*, the highland Amerindians, and *cambas*, the descendants of European settlers in the lowlands.

Although open racial discrimination is illegal throughout the continent, both blacks and Amerindians are often limited to marginal participation in the economic and political life of their countries. Everywhere in the continent, compared to the national average, blacks earn less and die earlier, and the same is true of many Amerindian groups. However, the formation of political parties and pressure groups such as the Union of Indigenous Peoples in Brazil – a national Amerindian organization lobbying to defend land rights – gives cause to hope that the future for ethnic minorities in South America may improve.

The fate of the Amerindians

The indigenous peoples of South America suffered disastrously from their first contacts with Europeans. Although the Indian populations of the Andes have since recovered, the lowland Amerindians still number just 3 million – less than a fifth of precolonial levels. Land belonging to the Amerindians was appropriated by the European settlers, and systematic attempts were made to break down their cultures, often with great cruelty: some early settlers even argued that Amerindians had no souls and could therefore be treated as animals.

Nor has this process been restricted to the colonial era. In the Amazon basin, many ethnic groups are suffering the disastrous effects of intrusion into their traditional homelands, with increased mortality as a result of introduced diseases, and rapid destruction of the forests on which they depend for subsistence. Their traditional religion and cosmology are being undermined by Christian evangelists of all denominations. A particularly poignant modern example are the Yanomami Indians, some 20,000 of whom living in northern Brazil and southern Venezuela make up the largest population of lowland Amerindians in South America. Their lands have been invaded by gold miners and missionaries, some 10 percent of the population has died of infectious diseases since 1985, and only in recent years have some efforts been made by governments and relief agencies to help.

Nevertheless, Amerindian cultures, languages and religions have survived, and have left an imprint on modern South American culture out of all proportion to the size of the contemporary Amerindian population. Critical to this process was the independence movement from Spain and Portugal in the 19th century. Resistance to European rule led South Americans to want to create a separate cultural identity from their European masters, and Amerindian culture was consequently invested with a new symbolic importance: names, folk heroes and imagined aspects of Amerindian culture were all annexed by the movement.

Surge of creativity

Non-Amerindians celebrated their discovery of an indigenous culture in a surge of novels, plays and operas with Amerindian heroes. The Brazilian José de Alencar (1829–77) wrote the novel *Iracema* in 1865

and named it after its Amerindian heroine, who became a symbol of selfless love and harmony with nature. Another novel *The Guarani* (1857), a paean to the first inhabitants of the pampas, subsequently provided the libretto for a popular opera that is still regularly performed in the theaters of Asunción in Paraguay and Montevideo in Uruguay.

In towns and cities across the continent, streets are named after Amerindian ethnic groups and squares are dominated by statues of Amerindians, always portrayed as highly idealized representations of the "Noble Savage". Even today, social movements are linked to Amerindian history and culture. A Peruvian guerrilla movement calls itself *Tupac Amaru* after an Inca king who led resistance to the Spanish. The flute and drum music of the Andean Indians is now reaching a far wider audience, as South Americans turn to what they see as pure forms of Amerindian expression in reaction against the culture of the West.

The appropriation of Amerindian symbols, however, did little to protect the Amerindian people themselves from the continued seizure of their land and erosion of their culture. Many of the ethnic groups who gave their names to streets and squares are now extinct. Practical concessions have been few; for example, the teaching of reading and writing in Amerindian languages, rather than the enforced learning of Spanish or Portuguese, has only recently become the norm in lowland South America.

In many parts of the region the idealization of the Amerindian as the "Noble Savage" has restricted rather than liberated. The accepted image of Amerindian culture paints it as quintessentially rural, with Amerindians living in harmony with their natural surroundings. This makes it particularly difficult for the many urbanized Amerindian groups to maintain a distinctive identity. Yet in the still frequent land conflicts in the Amazon basin, the even older European view of the Amerindian as "Ignoble Savage", ignorant, non-Christian, and an obstacle to economic development, resurfaces in cultural clashes in which the Amerindian is always the loser.

Music of the high Andes The haunting, spacious sound of Andean music is based on the traditional instruments of pan flutes and drums; harps, violins and close vocal harmonies were adapted from the music of the European colonists.

NORTHERN PEOPLES

A PIONEERING SPIRIT · CULTURAL UNITY · PROBLEMS OF AFFLUENCE

The peoples of the Nordic Countries show great uniformity in their ethnic and cultural composition, reflecting their common historical roots and the extent to which migration has taken place within the region. With the exception of Finnish and the Sami (Lapp) languages, they all speak closely related languages; Lutheranism, a division of the Protestant Christian church, is the predominant religion. The Nordic people have always possessed a great spirit of adventure, as evidenced by their Viking ancestors and more recently by the great polar explorers, Roald Amundsen (1872–1928) and Fridtjof Nansen (1861–1930). Shortage of land, which fed this restlessness, accounted for the emigration of many Scandinavians in the 19th and 20th centuries. Today affluence shapes the prevailing lifestyles of the region.

COUNTRIES IN THE REGION

Denmark, Finland, Iceland, Norway, Sweden

POPULATION	
Sweden	8.5 million
Denmark	5.13 million
Finland	4.95 million
Norway	4.2 million
Iceland	253,482

LANGUAGE

Countries with one official language (Danish) Denmark; (Icelandic) Iceland; (Norwegian) Norway; (Swedish) Sweden

Country with two official languages (Finnish, Swedish) Finland

Faeroese is recognized with Danish as an official language in the Faeroe Islands

RELIGION

Denmark Protestant (95%), nonreligious and atheist (3%)

Finland Protestant (92%), nonreligious and atheist (5%), Eastern Orthodox (1%)

Iceland Protestant (96%), nonreligious and atheist (2%), Roman Catholic (1%)

Norway Protestant (98%), nonreligious (1%)

Sweden Protestant (68%), nonreligious and atheist (28%), Eastern Orthodox (1%)

A PIONEERING SPIRIT

The first inhabitants of the Nordic Countries were hunters who gradually spread northward from Central Europe when the ice sheets retreated between 10,000 and 13,000 BC. They lived primarily on reindeer and fish, and supplemented their diet with aurochs and moose. During Neolithic times, from 3000 to 1500 BC, agriculture and animal husbandry were introduced into southern Scandinavia.

Between 1500 BC and 500 BC Finnish peoples, originally from the Urals, moved into what is now southeastern Finland. The origins of the Sami (Lapps) of Finland, northern Norway and Sweden are obscure, but they appear to have inhabited those areas (and parts of the Soviet Union) for at least 2,000 years. Like the Finns, their languages belong to the Uralic family of languages. There is considerable evidence to indicate that the climate of the region worsened about 500 BC, putting pressure on the human population. The scarcity of archaeological remains suggests that numbers dropped, and the people that stayed moved their livestock into permanent shelters to help withstand the harsh winters.

The Vikings

The ending of these inhospitable conditions coincided with the Viking age (800–1050 AD) – a term that covers the period of intense seafaring activity when adventurers and raiders from Denmark, Sweden and Norway (commonly called Vikings or Norsemen) sailed their narrow ships along the coasts of Europe, raiding as far south as the Mediterranean. In the east they followed the rivers of Europe to penetrate into the heart of Russia, and they had sailed west to settle the Faeroe Islands (still belonging to Denmark today) and Iceland by the 9th century. From here colonies were established in Greenland, and even attempted in North America.

The Viking age brought tremendous change to the Nordic Countries. Christianity, which has since remained the dominant religion in the region, was introduced by the French missionary, Ansgar (801–865), first to Denmark and then to Sweden. At the same time, the earliest divisions in the common Scandinavian language – known as "the Danish tongue" – began to appear. The languages spoken in Norway, Sweden, Denmark and Iceland today all belong to the Germanic branch of the Indo-European language family. They slowly developed into distinct languages during the early medieval period as separate national identities began to take shape and became established among the Scandinavian peoples of the region.

By the mid 13th century there were Norse-speakers in present-day Norway and Iceland, Swedish-speakers in both Sweden and the colonized parts of Finland, and Danish-speakers in Denmark and parts of southern Sweden, as well as northern Germany. Subsequently, Norse split into two separate languages – Icelandic and Norwegian. As the Norwegians came under Danish domination for several centuries, their languages grew closer again. Consequently, Danes, Norwegians and Swedes are still able to understand each other fairly easily, whereas Icelanders – and to some extent Faeroese – whose languages have remained closer to their medieval roots, are unintelligible to other Scandinavians.

Between the 16th and 19th centuries, during the period of Norwegian rule from Denmark, Danish became the official written language in Norway. However, over the years Norwegian words were absorbed into this written language to produce a form known as Bokmål. In the mid 19th century a language called Nynorsk (New Norse) was created, which drew on the dialects still spoken in rural areas of western Norway, in an attempt to carry on the tradition of Old Norse. In 1907, following Norway's independence from Denmark, both types of Norse were accepted as national languages. Today Bokmål is taught to about four-fifths of Norwegian schoolchildren; the remainder – the majority of whom live in rural areas – learn Nynorsk.

A tradition of migration

There has always been considerable movement of peoples both within the region, as well as to and from it. During the 19th and early 20th centuries the twin pressures of population growth and the worsening economic conditions encouraged those brave enough to do so to emigrate, principally to the United States, Canada, Argentina and Australia: there are today more Norwegian-Americans than there are Norwegians. Scandinavian emigrants frequently settled in distinct communities in their new countries

Winter wonderland A Swedish family enjoys the thrills of cross-country skiing. This sport, which is increasing in popularity in the Nordic Countries, originated in the region as a practical means of traveling from place to place during the winter months across snow-covered, hilly terrain.

and married within them, thereby establishing the Scandinavian languages and traditions overseas, as well as keeping alive a sense of national feeling toward their country of origin. For example, even today about 67 percent of the population of Solvang, a town in California, USA, founded in 1910 by Danish immigrants, remain fluent Danish speakers.

Over the centuries there has been a considerable influx of people from other European countries, such as Scotland and Germany, into the region. In the second half of the 20th century there have been increasing numbers immigrating from Asia, and the Nordic Countries gained a reputation for receiving political refugees from all over the world, but especially the Middle East. These newcomers tended to settle in their own distinct communities on the outskirts of towns and cities; the only future for many refugees is as social welfare clients in the small communities to which they have been sent. In the late 1980s the former consensus on immigration and refugee policy came under increasing pressure from populist political parties in Sweden, Norway and Denmark – the Swedish government, for example, refused entry to 5,000 Bulgarian Turks in 1989, in response to public outcry – despite the comparatively low numbers of immigrants in all three countries.

A stave church at Heddal in Norway, built in 1250. These wooden medieval churches sprung up all over Norway with the spread of Christianity. The boxlike frame is built around a number of "masts", each supporting up to six tiers of double-sloped roofs.

CULTURAL UNITY

The four major Scandinavian languages of the region (Danish, Norwegian, Swedish and Icelandic) have played a central role in shaping and preserving the national identities of the respective countries in which they evolved, and today are spoken by more than 90 percent of their populations. None has acquired regional dominance, so English is the universal secondary language. It is taught in schools from the age of 10, and is frequently used when communication between speakers of different Scandinavian languages fails. German is spoken by some 20,000 people in Denmark living just north of the border with Germany. The Danish government provides separate schools for this minority group.

Only in Finland is there significant linguistic diversity. For a long period of its history (between the 14th and the 18th centuries) Finland formed part of the territory of Sweden. During this time all official business was conducted in Swedish, which was also the language of literature. Today Swedish is spoken by some 6 percent of the population of 5 million, and it is recognized with Finnish as being an official language. Even after Finland was ceded to Russia in 1809, Finnish had no official status – it did not achieve this until 1863. During the 19th century the language played a vital part in fostering Finnish nationalist aspirations, particularly with the publication in 1835 of the *Kalevala*, an epic poem based on themes from Finnish folklore.

In the north of the region three distinct forms of Sami, or Lapp, related to Finnish, are spoken. However, nearly all the 40,000 Sami of Norway, Sweden and Finland are now bilingual, speaking as their second language the official language of the country in which they live, and many no longer speak their native language at all. In the face of this evidence of cultural assimilation, the Sami continue to assert their right to remain an ethnic minority inside the Nordic nation-states, and recent struggles have consequently centered on the campaign to teach the Sami languages in schools.

Religious consensus

By and large, the Nordic Countries are culturally homogeneous and minorities of any kind are rare. One of the ties that links the peoples of the region together – apart from the bonds of a shared history – is a common religion. More than 90 percent of the population are members of the Lutheran church – the oldest and the largest of the Protestant Christian churches. This follows the teachings of Martin Luther (1483–1546), a monk from Germany who spoke out against the influence and corruption of the medieval Roman Catholic clergy.

Lutheranism spread quickly from Germany to the Nordic Countries in the 16th century, and was established by law in each of them as the state religion, thereby avoiding the bitterness of religious controversy that divided the rest of northern Europe in the 17th century. The intervention of the Swedish king, Gustav II (1594–1632) – better known as Gustavus Adolphus – on behalf of the Protestant cause during the Thirty Years' War (1618–48) in Germany was a powerful factor in ensuring its survival against Roman Catholic hostility.

THE LANGUAGE OF THE SAGAS

Norwegian Vikings first reached the uninhabited island they called Iceland late in the 9th century – tradition says that they were fleeing the tyranny of King Harald Fairhair (c.860–940), the first king to claim sovereignty over the whole of Norway – and by 930 some 40,000 people had settled there. The Icelandic sagas, written mostly in the 12th century, record the first 100 years of the Norse settlement in Iceland, starting with the establishment in 930 of the *Althing*, said to be the world's oldest parliament.

The sagas recount in detail the expeditions, family feuds and legal disputes of these island people, and in so doing provide a unique record of a changing society: because of this they find a place in the literature of the world. Even more important in cultural terms, however, is the role the sagas played in defining and preserving the Icelanders' identity and heritage over the centuries of their political domination, first by Norway, from 1264 to 1380, and then by Denmark until 1944.

During these years the Icelandic language remained very stable – not even Danish has had any major influence on it – so the language spoken today is very close to that of the sagas. Loan words are rare, and are only accepted if they can be both grammatically and phonetically adapted to the Icelandic language. An old Norse peculiarity preserved in the Icelandic language is that family names are not commonly used. Instead *son* (son) or *dottir* (daughter) is added after the father's forename. The name Einarsdottir, then, simply means the daughter of Einar.

The clean bare look of Scandinavian interior design had a huge influence on the rest of Europe during the 1960s. Floors and furniture are in natural wood against a pale background, and fabrics are woven in neutral colors decorated with traditional motifs.

Lutheranism remains the official religion in each of the Nordic Countries, and religious practice is strong by comparison with other European countries. A number of religious sects – Mormons, Pentecostalists and Jehovah's Witnesses – exist in small, isolated communities. As a result of immigration from outside the region some cities contain groups of Muslims, Hindus and Buddhists.

A rich cultural tradition

Although in the past, they were isolated to some extent from the mainstream of European artistic and cultural activity by distance and language, each of these small countries has nevertheless produced major figures of worldwide significance in a number of different fields. In philosophy, the controversial ideas of the Dane Søren Kierkegaard (1813–55) have influenced many later thinkers; in art, the Norwegian painter Edvard Munch (1863–1944) was one of the forerunners of the Expressionist movement; in literature, the many plays of the Norwegian dramatist Henrik Ibsen (1828–1906), such as *Peer Gynt*, *A Doll's House*, and *Hedda Gabler*, revolutionized European theater, and those of his Swedish contemporary August Strindberg (1849–1912) also achieved world fame. The stories of the Danish writer Hans Christian Andersen (1805–75) have enchanted generations of children.

However, perhaps the greatest achievement in the arts has been in music. Both the Norwegian composer Edvard Grieg (1843–1907) and the Dane Carl Nielsen (1865–1931) found the inspiration for their symphonic compositions and songs in the landscape and folk music of their native countries. In Finland the music of Jean Sibelius (1865–1957) has come to symbolize the nationalist aspirations of his countrymen and even Finnish nationhood itself. He drew particular inspiration from the myths of the *Kalevala*.

The development of Finland's distinctive 20th-century school of architecture also found its inspiration in the nationalist movement and the need to create a style completely independent of Russian influences. Foremost among its members were Eliel Saarinen (1873–1950), who designed the National Museum and the Helsinki railway station, and Alvar Aalto (1898–1976), architect of the Finlandia Hall in Helsinki.

Folk art motifs and traditional handicrafts have been a major influence on the development of contemporary textile, furniture, ceramic and glass designs. The Nordic Countries have become particularly well known for these designs in the 20th century, and many are exported around the world. In spite of this, traditional crafts have disappeared in many urban parts of the region; they have been preserved longest in remote rural areas such as western and northern Norway. Books and magazines, as well as television, play a major role in spreading popular culture – Norway, for example, publishes more books per capita than almost any other country in the world.

Among young people, rock music festivals are becoming increasingly popular. Large open-air concerts, lasting several days, are held throughout Scandinavia every summer, and are likely to attract audiences of more than 100,000 people. Internationally known Western rock bands and also, increasingly, bands from Eastern Europe are regular performers at these events.

PROBLEMS OF AFFLUENCE

Unlike many of their European counterparts, most people in the Nordic Countries have retained close links with their rural origins, even though the majority now live and work in towns and cities. Part of the appeal of the unspoilt natural areas of the countryside lies in the escape route it gives from an all-encompassing social system.

In material terms, the Nordic Countries are extremely affluent. They are often represented as model examples of the 20th-century welfare state, and few countries have succeeded in creating such elaborate social security systems for all their citizens. From nursery education and disability allowances to old people's homes and the free use of hospitals, a wide range of services is provided. Unemployment benefit is superior to that found in most other countries; basic education is free and young people are able to obtain higher education by combining state grants with loans from banks or the state.

The negative side to such superb benefits is that it may encourage some people to drop out of society, turning instead to the drug culture or "alternative" lifestyles, while others find their escape by going in search of the simple life. Modern technology makes "going back to nature" a relatively relaxed and comfortable business. Transportation is readily available, food is assured, and medical services are rarely far away.

Leisure-time cultures

At weekends, many people retreat to their country cabins in the mountains, by the coast or in the forest. These cabins vary from the most simple, with little insulation, to those boasting modern facilities such as washing machines and double glazing. The renting of country cabins is also widespread, so the number using them for rural retreats is even greater than the quantity of cabins suggests. They are frequently let to foreigners, especially Germans: in summer parts of the west coast of Jutland in Denmark are crowded with German visitors, and German is the most frequently heard language in shops and restaurants.

Membership of a sports club is a vital part of most Scandinavians' daily lives; it provides them with physical exercise as well as a focal point for social and cultural contacts. Every year Nordic championships in a variety of sports such as weightlifting, athletics, speedway racing, swimming, and billiards are held in one or other of the countries in the region, and these attract a large number of participants. Soccer, too, is popular both as a participatory and a spectator sport. Those living near the mountains in Norway and Sweden enjoy regular skiing in winter and even, in some places, in summer. Crosscountry skiing is popular on the more level surfaces of Finland and southern Sweden.

The increased use of the countryside for recreation and escape from the varied demands of everyday life has put Nordic governments under pressure from their voters. During the 20th century nature has been culturally redefined. Instead of being pure wilderness, the countryside is now perceived as an essential part of Nordic leisuretime pursuits. It is regarded as a heritage that should be protected in its own right, even at the expense of shortterm profit.

Perhaps the most visible sign of this new attitude to nature is the establishment of the region's large national parks, whose environment and uses are protected by state legislation. Although certain activities, such as the use of automobiles and the lighting of open fires, are banned, camping is permitted – indeed, in Norway and Sweden it is legal

The pleasures of affluence Many people in the Nordic Countries now enjoy sailing as a leisure activity rather than the means of survival it was in the past. Most of these buildings along the waterfront are holiday homes, often rented to other families and foreign tourists.

A relic from the early 1970s *(above)*, this alternative lifestyle commune was set up in a former barracks in Copenhagen, Denmark. Some 20 years on, the authorities felt that its time had passed and proposed in 1991 that it be turned into a children's playground.

A Sami reindeer herder *(above)* in northern Finland. Over modern weatherproofed clothing he is wearing a traditional cape in the bright colors that are typical of Sami dress. The unique Sami culture and languages are under increasing threat.

pitch a tent anywhere except on cultivated land and close to private houses.

Rise of the mass media

The mass media play an increasingly important part in the lives of most Nordic people. During the 1970s, for example, the average Finn devoted nearly five hours a day to the mass media, mainly watching television and listening to the radio. In the 1990s, with the advent of satellite and cable broadcasting, there is a greater choice of programs, including many from the United States. Cable television in Denmark offers its viewers 18 channels to choose from.

Despite frequently expressed fears that the popular success of this international multimedia culture will suppress the indigenous cultures of the individual countries in the region, such effects are difficult to measure. However, there is evidence to show that in Finland, far from bringing about a decline in interest in the traditional arts, they are now enjoying greater popularity than ever.

SAMI SURVIVAL

Many of the Sami, or Lapps, living in scattered groups in the far north of the region, have become thoroughly assimilated within the prevailing culture of the region. This is particularly true of those living in Norway, where their numbers are greatest. Yet the Sami possess a strong sense of their separate ethnic identity, and this has fostered the determination to preserve their traditions and language against further erosion by modernizing pressures. Today the Sami face the dilemma of having to find a means of economic survival that will enable them to remain as they are, without turning them into living museum exhibits or forcing them to live in reservations.

For more than 300 years the Sami have supported themselves by herding reindeer – a way of life they turned to from a hunting–gathering existence when their natural food supply began to be depleted through overhunting. Today, however, reindeer husbandry is no longer viable, except for a very few, and many have turned instead to farming and fishing.

Yet much of their traditional culture is closely linked to this nomadic way of life, and many Sami are quick to defend their historical grazing and migratory rights – along with their language and traditional forms of dress – as a way of ensuring their cultural survival. Sami ethnic consciousness transcends national boundaries, and in recent years Sami living in the Soviet Union have begun to attend reunions ,in Scandinavia.This would have been impossible a few years ago, when the Soviet–Finnish and Soviet–Norwegian borders were firmly closed to the Sami.

Fishing traditions

For millennia, the sea has provided the people of the Scandinavian peninsula and islands with an easy means of communication along the coast with other parts of the region and beyond it, and a ready source of food. Fishing has been a reliable and constant resource, providing a basis for exchange: in the Middle Ages, traders from the Hanseatic ports of the Baltic traveled annually to the northernmost part of Norway to exchange grain, rope and luxury goods for furs and dried fish. Both commodities were greatly in demand among the people of central Europe – the first as an item of dress that conferred status upon the wearer, the second as a food that could be consumed on fast days when the Roman Catholic church forbade the eating of meat.

Different local conditions have given rise to a variety of fishing techniques within the region, though most traditional fishermen rely on nets and long lines. The calm waters of the Swedish and Finnish archipelagos and of the Danish and Norwegian fiords are particularly suited to small, open fishing boats, and here the traditional Faeroese boat excels. This elegant, seaworthy vessel closely resembles the Viking ships of the past, being slender and tapered at both ends with a high bow and stern. Traditionally, the boats were owned by the more prosperous fishermen themselves, and any additional crew members were paid from a shareout of the catch at the end of a fishing trip. Few fishermen could afford to depend solely on the sea to provide them with their livelihood.

This Viking longship, preserved in the Roskilde Museum, Denmark, is typical of those used by colonizing warriors; it is 1,000 years old and probably held about 80 men. The sleek shape of its high prow is still common on fishing vessels in use today.

Although women rarely participated in fishing expeditions, they played an important role in the fishing communities strung out along the Scandinavian coastline. Often assisted by their children, they would mend the fishing nets, bait the long fishing lines, and dry the fish for marketing. The development of modern technology in the fishing industry means that these activities are now carried out by machinery, but many women work instead in local fish processing plants.

A fisherman's "luck"

Although the prevailing calm conditions of the Scandinavian coastal waters make inshore fishing a relatively safe business, storms may suddenly occur without any warning. Over the generations, many

Northern waters A fishing trawler from Tromsø, in the far north of Norway, recalls the Nordic peoples' long links with the sea. Fishing communities are strung out along the length of Norway's rugged, indented coastline.

families have lost fathers, husbands and brothers to the sea. The unpredictable nature of fishing has inspired a variety of beliefs and legends among the fishing communities, which enjoyed currency well into the 20th century.

In the Faeroe Islands, for example, it was believed that if a seal appeared in front of a boat it was a sign of misfortune, but if it appeared behind the boat it was an omen of good luck. Likewise, certain women signified misfortune. Should a fisherman meet a red-haired woman on the way to his fishing boat, he would be well advised to stay at home. Similar advice would apply if he met a woman carrying ashes from the fireplace.

It was also believed by the Faeroese that a boat's future could be foretold by the boat builder if he took a curly wood shaving from the boat and threw it on the floor. Should the shaving on landing resemble the shape of a small boat, then the omens were favorable. It was held to spell bad luck if a new boat was to touch the ground on being launched for the first time – a boat that did so was destined to run aground. To avoid such misfortune, elaborate precautions were taken to safeguard the passage of the boat from the building yard to the water.

Particular ceremonies, such as the ritualistic rubbing of fishing boats with seaweed to symbolize their partnership with the sea, were observed at the start of each fishing season. Today, with the development of sophisticated fishing technology, almost all these fishing traditions have disappeared, and many of the small fishing communities themselves are likely to become a thing of the past, their way of life ended by the fleets of industrial trawlers, owned by large companies. With their huge catches, commercial fleets have depleted the fish stocks of the North Atlantic, making it harder and harder for individual fishing boats to make a living from the sea.

Environmentalist organizations such as Greenpeace, which seek to restrict the size of catches, are also a source of friction. In particular, many fishermen in the Faeroes and Norway see the international ban on whaling as a direct threat to their livelihood. For generations whaling has been a mainstay of their fishing communities and many fear they will be unable to survive without it.

THE PEOPLES OF THE ISLES

UNITY AND DISSENT · SCEPTERED ISLE, EMERALD ISLE · OLD NATIONS AND NEW INFLUENCES

Evidence of centuries of internal conflict and migration, during which the English of the lowland south and east established ascendancy over the rest of the British Isles, is to be found in the distribution of placenames and dialects, and in the myths from which the English, Scots, Welsh and Irish forged their national identities. From the 16th century British emigrants started new colonies around the world, leading to the global spread of the English language and of British cultural ideas and pastimes. The growth of nationalism that eventually caused the British empire's breakup developed early in Ireland, where deep differences divided the rulers from the ruled. Preservation of their distinct national culture has been a major preoccupation of the Irish since they gained independence.

COUNTRIES IN THE REGION

Ireland, United Kingdom

POPULATION

United Kingdom	57 million
Ireland	3.54 million

LANGAUGE

Country with one official language (English) UK

Country with two official languages (English, Irish) Ireland

Local minority languages are Gaelic, Irish and Welsh. Significant immigrant languages include Bengali, Chinese, Greek, Gujarati, Italian, Polish and Punjabi

RELIGION

Ireland Roman Catholic (93.1%), Anglican (2.8%), Protestant (0.4%), other (3.7%)

United Kingdom Anglican (56.8%), Roman Catholic (13.1%), Protestant (12.7%), nonreligious (8.8 %), Muslim (1.4%), Jewish (0.8%), Hindu (0.7%), Sikh (0.4%)

UNITY AND DISSENT

The first peoples to inhabit the British Isles after the retreat of the ice sheets crossed by a land bridge from mainland Europe about 7000 BC. Once this was ruptured, migration became more difficult, but over the millennia successive waves of people brought new knowledge and skills, including the use of bronze. Among them were the people who built the great circle of standing stones at Stonehenge, between 1800 and 1400 BC. About 450 BC Celtic-speaking peoples, originally from Central Europe, brought a common iron-using culture to most of the islands, including Ireland.

The Roman conquest and colonization of mainland Britain in the 1st century AD brought sweeping changes in the agriculture, social organization and urban development of the lowlands (the Romans did not penetrate the hilly regions of the north and west). Latin was the language of the ruling elite, though Celtic continued to be spoken by the majority; oriental religious cults, including Christianity, were introduced by the empire's soldiers and traders.

As this civilization broke up during the 5th century AD, Anglo-Saxons and Jutes – Germanic-speaking peoples from northern Europe – began to move into the region. The medieval legends of King Arthur, a mythical warrior-king, may have their origins in battles fought between these invaders and the Celtic-speaking Britons, who were gradually driven into the southwestern peninsula of Cornwall and into Wales in the west.

The consolidation of Anglo-Saxon power saw a great flowering of art and literature, especially after the reintroduction of Christianity by missionaries from Rome in the 7th century and by Celtic monks from Wales, Ireland and Scotland who had kept alive a separate Christian tradition and culture since Roman times. From the 8th century onward Viking raids from Norway and Denmark inflicted increasing damage and disruption, and for nearly 100 years the Danes controlled a large area of eastern and northern England, known as the Danelaw. Placenames ending in "by", such as Grimsby, are evidence of this occupation.

A single English kingdom was established for the first time in the 10th century (Scotland was unified at about the same time), but power continued to be disputed with the Danes until 1066 when the Normans, a French-speaking people of Viking descent, invaded from northern France. They quickly extended military and administrative control over most of the mainland, except parts of Wales and all of Scotland, introducing new systems of laws and landholding. French became the language of the ruling elite; Latin was the written language. By the 14th century, however, English began to replace both for most purposes. The language still contains numerous French and Latin borrowings from these centuries.

The English ascendancy

Government and power was centralized in London in the southeast and gradually began to extend into the peripheral parts of the region: during the 16th and 17th centuries Wales was formally annexed, the crowns of England and Scotland were united, and Ireland was colonized by English and Scottish settlers. Political union, however, was accompanied by religious division, which almost at once assumed a regional dimension.

The Church of England, founded by Henry VIII (1491–1547) in opposition to the Roman Catholic church, in time overcame the challenge from a number of dissenting Protestant sects, influenced by the ideas of 16th-century European reformers, to become the established church of the ruling English land-owning class. Protestantism had greatest hold in Scotland, from where it was exported by settlers to northern Ireland: in southern Ireland an Anglican minority dominated the indigenous Roman Catholic majority. Religious dissent was strong in Wales, too, and Methodism, an 18th-century nonconformist sect, took deep root here.

An industrialized people

The peoples of the British Isles were the first to experience an industrial revolution. During the 18th and 19th centuries, the transformation of a rural, agricultural society into an urban, industrial one reemphasized regional differences. The new industrial centers lay in the north (Liverpool, Manchester), Scotland (Glasgow), Wales (Cardiff) and northern Ireland (Belfast), but political and economic power remained in the southeast. Migrants were attracted to these cities from all parts of the British Isles and settled with others sharing their religious or regional

A crofter's cottage (*above*) on the Orkney Islands off Scotland's north coast. The islands were settled by Vikings in the 8th century and were subsequently ruled by Norway and Denmark until the 15th century. Crofters are smallscale tenant farmers, but their traditional way of life is now in decline.

Festive occasion (*right*) Two strands in London's diverse culture come together at this exuberant meeting of Afro-Caribbean dancer and Cockney "Pearly King" at the Notting Hill carnival. This annual event has been the cause of friction between the police and the black community.

loyalties in closeknit communities. This still influences local rivalries: support for Glasgow's two major soccer clubs, for example, is traditionally based on Roman Catholic or Protestant affiliation.

During the 19th century, agricultural depression, the clearing of land for sheep farming in Scotland, and famine in Ireland, sent thousands of migrants from the region to North and South America, Australia and New Zealand, and South Africa. The growth of the British empire and a newly popular monarchy under Queen Victoria (1819–1901) presented a unifying focus for British national identity. Only in Ireland were the forces of separatism strong, eventually leading to independence in 1922.

SCEPTERED ISLE, EMERALD ISLE

The landed aristocracy, stately country houses, ceremonial occasions such as the annual Trooping of the Color or the State Opening of Parliament, the robes and rituals of the legal profession or the ancient universities, lavish sporting and society events such as the Royal Ascot horse races, or the Henley rowing regatta – these all contribute to the image of Britain that draws millions of tourists there every year and sells British products abroad. Yet it is one that confuses an essentially English, narrowly upper-class culture with that of Britain as a whole.

Culture and class

Although some parts of the region – Scotland, Wales, northern England and the southwest peninsula – suffer a greater level of social and economic disadvantage than the south, Britain today is a predominantly affluent, urbanized society. As a result, many of the traditional divisions between the classes have been eroded; yet Britain remains a class-conscious society. Not only wealth and professional status, but also values, religion, leisure pursuits and lifestyle, are indicators of social position.

In some respects, British education perpetuates the established social hierarchy. Although entrants from state schools outnumber students from private schools at most British universities and other higher-level educational institutions, the "old boy" network – an informal system of patronage based on attendance at elite private schools and the universities of Oxford and Cambridge – is still believed by many to open the door to career advancement in certain fields such as politics, the civil service and business.

Accent remains a useful key to defining family or regional background. There is evidence to suggest that a strong regional accent in the past was not the social drawback that some consider it to be today: both the Victorian poet Alfred Tennyson (1808–92) and the prime minister William Gladstone (1809–98) are supposed to have retained their boyhood regional accents into adult life. It was the growth of the English private school system, and later the promotion of "Received Pronunciation" by the British Broadcasting Corporation (BBC) that fostered the British preoccupation with having the "right" accent.

Nevertheless, whatever their class and allowing for some regional variations, the British are broadly united by a common cultural outlook. The most noticeable example of this, to Britain's more gregarious European neighbors, is the desire for privacy. Houses are preferred to apartments, and house-ownership is a major goal for most Britons. Home-based activities, such as gardening, watching television, or home improvements, are among the most popular leisure pursuits. Supporting or playing team sports such as soccer, rugby or cricket (all of which originated in Britain) are also popular, along with other outdoor pursuits from fishing to walking or horseback riding. Gambling on the outcome of sporting and other events also cuts across all classes.

The two Irelands

It is in Ireland that the cultural dominance of the English upper class has been most strongly rejected. Irish nationalism in the 19th and early 20th centuries found its inspiration in the island's Celtic past, including the Irish language, and in the Roman Catholicism of its predominantly

English eccentric *(above)* Dressed as an elephant a competitor in the London marathon – which is open to professionals and amateurs alike – passes over Tower Bridge. Sporting events such as this, where the emphasis is on everyone taking part, transcend social barriers, and are growing in popularity.

Dublin pub *(left)* The pub (public house) became popular as a working-class institution in Victorian days. Pubs in Ireland are generally more traditional than their British counterparts – women, for example, would still feel out of place in this public bar.

Privileged youth *(below)* Students at Eton College, Britain's largest public school, founded by Henry VI in 1440–41; it is traditionally a breeding ground for future prime ministers. The students' distinctive uniform of top hat and tail coat dates from the 19th century.

A POPULAR INFATUATION

The British royal family represents the most powerful symbol of unity in the islands. Many European countries have monarchies, but no other is so popular or profusely publicized. It is said that a third of the population have dreams about the various members of the royal family, and opinion polls show an 80 percent level of support. This is reported to be highest among women.

The popularity of the monarchy rose sharply after the new technologies of radio and television brought ceremonial occasions of state into the country's homes. The coronation of Queen Elizabeth II in 1953 was the first such occasion to be televised, and thousands of Britons purchased their first television sets in order to watch it. The reverential commentaries created a sense of national greatness.

A more recent trend has been the transformation of the royal family into an object of popular entertainment. In part, this arose from conscious efforts in the 1960s to remove the awe and formality that surrounded the royal family and bring it closer to the public, imitating the more relaxed style of the Dutch or Danish monarchies. Hardly a day passes without a story about the royal family appearing in the popular press, and its individual members are scolded or held up for approval according to the whim of the newspaper editors. Perhaps the true appeal of the royal family lies in a peculiar mixture of deference and curiosity: they are at once popular and privileged, both ordinary people and superheroes.

rural population. The furthering of a distinct Irish culture has been a persistent aim since independence.

Religious belief and the authority of the Roman Catholic church have consequently been protected by state censorship. Today 94 percent of the population are Roman Catholic, and 90 percent regularly attend mass. By contrast, less than a fifth of the United Kingdom's population are regular church-goers, though two-thirds profess a belief in God.

Such policies create pressures to conform. This, and the need to find work, have kept levels of emigration high, especially from the west of Ireland where the Irish language was traditionally strongest. Today only 1 percent of the population has Irish as its first language, though one-third can speak it. Even in the 1980s over 180,000 people left Ireland, mostly for Britain where the Irish need no passport and are entitled to vote in elections. Only one of Europe's highest birth rates has prevented drastic depopulation.

By the 1990s there were signs that Ireland was moving away from cultural introversion to find a self-confident identity as a member of the European community of nations. The church was no longer able to count on the unquestioning support it formerly enjoyed – only half of the population bothered to vote in a referendum on abortion. In 1990 the country elected its first woman and socialist president, Mary Robinson.

Ireland and Britain meet in the province of Northern Ireland, where the majority Protestant community, strongly linked through migration and religion with Scotland, gained economic power through the industrialization of the Belfast region. One-third of the population, however, is Roman Catholic. With largely separate schools, clubs and newspapers, the division between the two communities is deep, and both find an emotional appeal in the quarrels of the past. The Orange Order, for example, a Protestant organization founded in 1795, annually celebrates the defeat of Roman Catholic forces at the Battle of the Boyne in 1690, while Roman Catholics dwell on the remembrance of past discrimination.

The sectarian violence that receives international media attention is mainly the lot of Belfast's working classes. Middle-class neighborhoods are far from segregated, and the province has the lowest crime rate in the United Kingdom.

OLD NATIONS AND NEW INFLUENCES

England, Scotland, Wales and Northern Ireland have all adjusted differently to their dual identities as separate nation and integral part of the United Kingdom. Internal migration and the almost universal use of the English language means that cultural differences have been eroded: old rivalries are mostly subsumed into sporting occasions, but nevertheless, separatist movements do exist in Scotland and Wales, and the political status of Northern Ireland is a continuing source of conflict and debate both within the province and within the United Kingdom and the Irish Republic. Even in England, the sharp economic divide between north and south has reinforced perceptions of regional distinctiveness.

A sense of difference

Scotland has, perhaps, a stronger feeling of independent nationhood than Wales. It retained separate education and legal systems after the Act of Union with England in 1707, whereas those of Wales were fully integrated with England. The urbanized and industrialized lowlands are strikingly different from the sparsely populated highlands and islands. Here, the traditional way of life of fishing and smallscale farming is in decline, though years of depopulation are being partly reversed by an influx of outsiders (many from England) seeking an alternative to urban lifestyles. Resentments may arise when the incomers' liberal attitudes clash with the Protestant conservatism of the highlanders. Many of lowland Scotland's heavy industries have declined, and unemployment fuels the discontent with government from England.

Although Scottish nationalism has greater political support than Welsh nationalism, in the important aspect of language the Welsh have good cause to feel more distant than the Scots from the English. One-fifth of the population speak Welsh, a Celtic language closely related to Irish and the Gaelic of Scotland, as well as to the Breton of northwest France, while only about 80,000 Scots speak Gaelic. This reveals a conflict at the heart of Welsh identity – the more rural north has a tradition of strict Methodism, including observance of the sabbath and opposition to alcohol; the industrial south, on the other hand, has a more recognizably "English" working-class culture based on the local pub, the national game of rugby and a tradition of radical politics.

English influence in the south has restricted support for Plaid Cymru, the Welsh nationalist party, to the Welsh-speaking areas of the north and west. Even here Welsh identity is under increasing pressure: very few pubs now close on Sundays, and English incomers are snapping up cheap property as second homes. Nevertheless, as a consequence of persistent campaigning, there has been a separate Welsh-language television station covering the whole of Wales since 1982, and Welsh is a compulsory part of the school curriculum. After decades of decline, an increasing number of adults are now learning Welsh.

Although northern England lacks the separate institutions of Scotland and Wales, it has its own distinct personality, which in part has its roots in a history of resistance to the centralizing power of the south. The great industrial cities of the north possess a much stronger sense of

Local loyalties Soccer is a major passion in the northern city of Liverpool. When nearly 100 soccer fans were killed in a stadium disaster in 1988, a spontaneous display of public grief swept the entire community. The club's slogan "You'll never walk alone" captures the spirit of local feeling.

THE SATANIC VERSES AFFAIR

The publication in 1989 of *The Satanic Verses* by the British-based, Indian-born writer Salman Rushdie sparked off a bitter controversy that revealed some of the tensions in Britain's multi-ethnic society. The book, its detractors alleged, blasphemed against Islam by treating the life of the Prophet in a fictional and satirical manner. Ayatollah Khomeini, Iran's fundamentalist religious and political leader, proclaimed a death sentence against the author, forcing him to go into hiding with police protection. British Muslims demanded that the book be banned; copies were burned at protest marches, and the publishers, bookstores and their employees were threatened.

The controversy united the various factions and organizations of the 800,000-strong Muslim population, giving them the chance to voice a set of wider demands that had previously gone largely unheard. These included the extension of the blasphemy laws to cover religions other than Christianity, recognition of Muslim family laws on marriage, divorce and inheritance, and the right of Muslim-run schools to receive public funding.

The means by which Muslim outrage at *The Satanic Verses* was expressed provoked a groundswell of support for Salman Rushdie from the intellectual establishment and a reserved response from the government. While the Muslim protest sprang from deeply held beliefs, it attacked an equally deep tenet of British society: the right to free speech. The affair highlighted the sometimes uneasy relations between Britain's different cultures, raising questions of whether equal status should be given to all religions, and to what extent censorship is justified in protecting religious sensibilities.

local identity than their counterparts in the south of the region. This reveals itself at the simplest level in particular food preferences and also in distinct regional dialects: "scouse" coming from Liverpool or "Geordie" from Newcastle.

The southwest Cornish peninsula of England, at as great a distance from the center of power in London as the north, shares some of the same resentments at the perceived neglect of its interests. A granite outcrop, its physical geography is very similar to Wales and its people are of the same Celtic descent – use of the indigenous Cornish language, however, died out in the 18th century. The area shares with Wales the same tradition of Methodist nonconformity.

Other cultures, other traditions

Until the 1930s, Britain exported many more people than it imported. Millions of migrants left to settle in British colonies all round the world or to start new lives in lands of opportunity such as Australia or the United States. After World War II, however, the flow reversed, and for almost 40 years large numbers of immigrants arrived in the British Isles. Most of these new arrivals were from the former empire, especially the Indian subcontinent and the Caribbean, and were already in possession of British citizenship.

The immigrants filled gaps in the labor force of the large industrial cities and northern textile towns. At first welcomed, as the postwar economic boom gave way to stagnation, they became the target for resentment (a role that had earlier been held by Jewish immigrants from Eastern Europe, and by the Irish), and were increasingly regarded as a threat to the integrity of white British culture.

Racial tension grew throughout the 1950s and 1960s as Caribbean and Asian immigrants closed ranks to form their own communities in the poorer districts of the cities. Individuals were frequently subjected to racial discrimination, abuse and sometimes violence. British governments reacted by imposing new immigration laws and nationality laws that removed citizenship from thousands of people in former British colonies.

Some 40 percent of Britain's black and Asian population (accounting for 4 percent of the total population) are British born. Their cultures are now an indelible part of British life. Hindi and Urdu have become significant second languages, mosques and temples have proliferated, and Asian-run grocery stores are familiar features of city life. The high visibility of Afro-Caribbeans in sports, popular music and fashion attest to the contribution they have made to British culture. Nevertheless there is very real prejudice against them, which expresses itself in violent attacks from gangs of youths at one extreme, and in job restriction and lack of opportunity at the other.

A further challenge to traditional concepts of British identity arose from Britain's forging of closer links with its European neighbors after its entry into the European Community (EC). Fears were expressed about the loss of parliamentary sovereignty. In 1990 half of Britons were opposed to the building of the Channel Tunnel with France, then already under construction. These attitudes showed that British insularity and fear of absorption by Europe were by no means a thing of the past.

A Sikh wedding at Chapeltown, Leeds, an inner city area that has a large Asian community. The Sikh religion requires men to wear a turban, and this has caused problems for immigrants. Sikhs fought for a long time to alter the law that makes the wearing of safety helmets obligatory for motorcyclists .

Selling history

Rural nostalgia *(above)* Gold Hill in Shaftesbury, Dorset, is the epitome of the classic English village. The novelist Thomas Hardy (1840–1928) described it in his Wessex novels. More recently it has featured in a television commercial to sell a brand of bread – evoking the rural past to promote a healthy image.

Peoples and nations define themselves through a shared history; every country tries to conserve old objects, buildings and landscapes as a way of understanding and representing its own past. In Britain, this process began early and is so widespread that it has now become a major industry and pastime.

For over a century Acts of Parliament and civic organizations have sought to preserve and protect the British national heritage. The Ancient Monuments Act of 1882 listed 68 sites; now over 12,000 are protected. Museums abound: the British Museum in London, opened in 1753, was the world's first state collection and public museum; now there are over 2,000 museums and more are being opened, catering for every specialist interest.

The National Trust, founded in 1895 to preserve threatened landscapes, is the largest civic body in England and Wales and Britain's largest landowner, after the government. Its membership more than doubled between 1975 and 1990, and over 4 million people visited its country houses in 1986. Visiting historic buildings is more popular among the British public than the live arts, going to movies, or watching soccer and other sporting events. Foreign visitors, too, flock to Britain's historic centers. The United Kingdom is the world's fifth most important destination for tourists, attracted by its colorful pageantry and history, rather than by its climate.

The past seems to be getting closer. Early conservationists were concerned with the protection of rural landscapes and ancient monuments such as castles and abbeys. They turned later to country houses, and then to city buildings. Now there is increasing emphasis on preserving the country's comparatively recent industrial heritage. Coal mines, docks,

Reliving history *(above)* Enthusiasm for the past leads many people to recreate it. These members of an organization known as the Company of the Sealed Knot meet several times a year to refight the battles of the English civil war, playing out once again the encounters between Parliamentarians and Royalists. They have acquired a close knowledge of historic warfare, and their costumes and weapons – even their language – are correct in every detail.

textile mills and even 1930s' power stations all receive protection.

A rash of heritage centers around the country shows how far the presentation of the past has moved from the display of discreetly labeled objects in museum showcases, for passive contemplation and enlightenment, toward active entertainment. At the Jorvik Viking Center in York, visitors can experience tableaux of Viking life, including its smells, on a carousel ride. Other centers offer visitors the opportunity to converse with authentically costumed actors impersonating people from the past, or watch former coal miners reenact the work they were doing every day until their pit was closed and turned into a museum.

Sightseers can combine a love of landscape with that of literature by taking conducted tours through areas of country associated with particular writers. The most popular of these is still to "Shakespeare Country", visiting the sites of William Shakespeare's (1564–1616) birthplace, supposed boyhood exploits and later married life in Stratford-upon-Avon. Thousands of visitors are attracted to the parsonage in the small Yorkshire mill town of Haworth where the Brontë sisters lived and wrote their world-famous novels. Even present-day authors may be marketed in this way. Trips can be made to "Cookson Country", the area around Tyneside in the northeast of England described in the popular romantic novels of Catherine Cookson.

Education or economics?

The boom in the heritage industry has been the subject of some controversy. Critics argue that the marketing of the past as entertainment can result in passing on inaccurate or incomplete history. The Robin Hood Center in Nottingham, for example, rests on legend: the story of Robin Hood, a 12th-century outlaw who robbed the rich to help the poor, is probably more fiction than fact. To some observers, the British obsession with the past arises from a sense of national decline, as the world's former leading manufacturing power closes its factories, and the empire fades into memory. However, the British are not alone in this obsession: the United States' much briefer documented history, for example, is also packaged and sold in numerous small museums, history trails and heritage centers.

Support for the industry is not just restricted to those who benefit from it financially. Its advocates stress its educational value: by interpreting the past and presenting it in an accessible way, visitors to heritage centers can understand more about historical processes than they might gain simply by viewing historic sites or monuments in isolation. The recent interest in Britain's industrial past adjusts the historical record. In centers such as the museum at Ironbridge in Shropshire, the birthplace of the Industrial Revolution, the lives of ordinary people, and not just the aristocrats of the great country estates, are brought into focus giving visitors an insight into the way their parents and grandparents lived.

THE FRENCH PEOPLE

A CHANGING PICTURE · THE FRENCH WAY OF LIFE · TOWARD A NEW FRANCE

The French combine the exuberance and vitality of southern Europeans with the practicality and tolerance that are characteristic of the north. More than most countries, France is marked by a strong spirit of unity; French men and women derive a real sense of common identity from their use of the French language, their love of good food and wine, their enthusiasm for intellectual ideas – even from their passion for national sporting events such as the Tour de France – which marks them out from other nations. Although it is also a highly centralized state, with Paris lying at the heart of the country's political, economic and cultural activity, regional differences – reflected in contrasting architectural styles, in food preferences and dialect – still contribute a particular flavor to the contemporary scene.

COUNTRIES IN THE REGION

Andorra, France, Monaco

POPULATION

France	56.18 million
Monaco	26,000
Andorra	25,000

LANGUAGE

Countries with one official language (Catalan) Andorra; (French) France, Monaco

Local minority languages in France are Basque, Breton, Catalan, Corsican, German (Alsatian) and Occitan. Significant immigrant languages include Arabic, Italian, Polish, Portuguese, Spanish and Turkish. English, Italian and Monegasque are spoken in Monaco; Spanish in Andorra

RELIGION

Andorra Roman Catholic (99%)

France Roman Catholic (76%), Muslim (3%), nonreligious and atheist (3%), Protestant (2%), Jewish (1%)

Monaco Roman Catholic (91%), Anglican (1%), Eastern Orthodox (1%), Protestant (1%)

A CHANGING PICTURE

Although the French regard themselves as a single nation, they are actually an amalgam of a succession of peoples who moved into the region from farther east and settled there because they were unable to migrate any further. The Celts, a people that originated in Central Europe, spread here from across the Rhine between 800 and 400 BC; they were known to the Romans as Gauls. Both the Greeks and Romans established trading colonies on the Mediterranean coast, and from their base in the south the Romans began to expand northward in the 1st century BC; by 54 BC the emperor Julius Caesar (c.102–44 BC) had completed the Roman conquest of the Gauls. The Roman occupation, which lasted nearly 500 years, had enduring effects; the Gauls adopted the Latin tongue of their conquerors, which gradually evolved to become modern French, and Roman law remains the basis of French law.

As the Roman empire began to disintegrate in the 5th century AD, the Franks, a Germanic people, moved into northeastern France and gradually extended their rule southward; only in Brittany, in the far northwestern peninsula, did the Celtic language survive. Charlemagne (768–814) made France the center of a great Frankish empire that extended into Germany and northern Italy; it did not long survive his death, fragmenting into numerous feudal lordships. Viking invaders from Scandinavia settled in Normandy, in northern France, from where they launched their invasion of England in 1066.

The making of modern France

All these peoples have left their mark on the French nation. The modern French state first began to emerge as a small feudal power in the Ile de France in the center of the Paris basin in the 10th century. From this heartland, with many advances and retreats, French power was progressively extended over neighboring duchies and kingdoms by a process of conquest, dynastic marriage and even outright purchase.

By the 18th century the boundaries of France had been established substantially as they are today. The only important subsequent changes were the acquisition of Nice, on the Mediterranean coast, and Savoy, in the western Alps, from Italy in 1860, and the temporary loss of Alsace and Lorraine, on the west bank of the Rhine, to Germany between 1871 and 1918, and from 1940 to 1944.

French society underwent many fundamental changes after the Revolution that overthrew the monarchy in 1789. Instead of the old order, founded on the twin pillars of the monarchy and the Roman Catholic church, attempts were made to forge a secular national identity that embodied the egalitarian ideals of the Revolution. Despite the interruption of the empire (1804–15) created by Napoleon Bonaparte I (1769–1821), and subsequent episodes of monarchy and empire (finally ended by defeat in the Franco–Prussian War of 1870–71), Republicanism had triumphed by the end of the 19th century. A universal French "tradition" had been institutionalized in the use of such symbols as the Tricolor, the national flag of vertical blue, white and red stripes, and the female figure of Marianne to represent the French Republic. A spirit of unity was also fostered through the status given to French as the offical national language; it was protected from misuse by government edict, and spellings were standardized.

From the 17th century onward France acquired a number of overseas colonies,

National icon *(above)* Eugène Delacroix's painting *Liberty Leading the People*, displayed in the Louvre in Paris, commemorates the revolution of 1830. The French derive a strong sense of national identity from symbols of their revolutionary and republican tradition.

The old face of France *(left)* Peasant farmers returning from market with their loaves of bread represent a way of life that is changing with the increasing affluence and urbanization of many areas.

This policy of colonization intensified during the 19th century, and the consolidation of its empire, particularly in Northern and sub-Saharan Africa and in Indochina, was a contributory factor to France's growing awareness of national identity and destiny.

Rapid population growth

Compared with other European countries, the French population grew very slowly in the 150 years after 1800; it consequently remained a predominantly rural country for longer than its neighbors. Perhaps for this reason, images of rural France – of carefully tended fields and vineyards, and of gray stone houses huddled around the village church – still retain a powerful hold on the popular imagination, though France's population is now highly urbanized.

This rapid transformation took place after 1945. With the ending of World War II and the trauma of occupation by the Germans, population levels began to rise again, encouraged by social legislation. Immigration from other European countries, particularly Italy, Spain and Belgium, and from former French colonies in Africa added another 28 percent by the 1980s. There were also significant inflows from Indochina, and from remaining overseas territories, such as Martinique. Most of these newcomers were concentrated in the big cities, especially Paris, Marseille and Lyon.

THE FRENCH WAY OF LIFE

The French are justly proud of their rich cultural and intellectual heritage. From medieval times they have been at the forefront of European art and civilization, contributing to all aspects of cultural life, from architecture to music, literature, philosophy, painting, sculpture and the theater, and more recently to cinema. Literature remains an important element of contemporary French life, despite the growing attraction of television. Not only is the announcement of the Prix Goncourt, the annual literary prize for fiction, an eagerly awaited national event, but the works of past novelists such as Emile Zola (1840–1902) and Marcel Proust (1871–1922) are still widely read by men and women of all classes and occupations.

A wealth of variety

These generalizations should not disguise the fact that beneath France's apparent national unity lies a surprising degree of diversity. Although French is the universal language of everyday communication and of higher education, other languages prevail in some areas. Breton, a Celtic language, is widely spoken in Brittany; approximately 1 million people speak a German dialect in Alsace and Lorraine. In the Pyrenees, Basque (the oldest of Europe's extant languages) and Occitan (related to Catalan, spoken in eastern Spain) are spoken by separate groups; Italian dialects survive in the French Alps and on the island of Corsica (Italian until 1769). In addition, regional dialects persist among many people in rural areas.

The survival of these linguistic differences testifies to the regional variety that is still found in France, despite the leveling influences of modern mass media and communications. The greatest cultural divide is between the north and the south. South of the Massif Central, the people of the Mediterranean region of Provence have a reputation for warmth and exuberance. Their houses are typically roofed with curved, red "Roman" tiles; their landscapes are dominated by the vine and the olive. In the north lie the great agricultural heartlands of France; the people here are more dour and more cautious; beer drinking predominates over wine.

Other regional differences are rooted in the separate historical identities that the provinces had before their absorption into the French state, and are manifested in distinctive dress, customs and folklore. The Bretons, for example, are typified by a passionate link with the sea, and by their traditional religious processions, which are the genuine expression of both their piety and individuality. Germanic influences in Alsace range from food preferences such as *choucroute* (pickled cabbage) to a strong sense of civic pride, reflected in their clean and tidy, flower-decorated cobbled streets.

Perhaps the most fundamental regional difference, however, is between the people of Paris and those of the rest of France. More than any other European capital, Paris dominates national life, reflecting the way France has grown from the inside out. Nearly every official national organization and commercial corporation has its head offices in the Paris region, and its eight universities and specialist institutes of higher education (the *grandes écoles*) open the way to successful careers in government and business. As a result it attracts a steady stream of amibitious, upwardly mobile young people from all over the country.

Many of the new urban residents retain their rural roots – the former family farm often becomes a vacation home or is bought as a second home by other town dwellers. The overwhelming majority of Parisians leave the city during the annual August vacation, and accidents on the *autoroutes* (expressways) are notoriously high at the peak weekends at either end of the season. Retirement at the end of a couple's working life often takes the form of a *retour du pays*, a return to the countryside of their origin.

Changing attitudes

As in other Western countries, the extended family has diminished in importance as more and more people have moved to the cities. The nuclear family generally has two or three children, who frequently leave the family home in their late teens. The postwar population boom means that there is a high proportion of young people who play a special part in French culture, with their own fashions, lifestyles and speech patterns.

Although women have complete legal equality with men, and pay differentials between women and men are lower than in other Western countries, many French women have continued to accept the

AN APPETITE FOR FOOD

One passion that unites French men and women is their love of good food. Whether in the home or outside it, French cuisine is based on the discriminating use of good ingredients. It has a marked regional character that reflects the produce of a particular area. The orchards and pastures of Normandy, for example, have given rise to characteristic dishes that are cooked with cream and apples; Provence is famous for its liberal use of olive oil, garlic, tomatoes and peppers; the Loire region of central France specializes in stews of local eel and pike.

Today, however, much is changing in the French attitude to food, particularly marketing. The mushroom growth in popularity of enormous *hypermarchés*, or supermarkets, on the edge of towns and cities throughout the country means that shopping for food has become a weekly event, particularly when both husband and wife work, with the whole family helping to push the trolley and load up the car. The widespread use of convenience foods is supplanting the need for a daily visit to the local market for fresh produce, and fast foods such as pizzas and hamburgers are commonly served in cafés.

All these things are eroding the regional basis of French cooking. Some things, however, never change – the crusty long loaves of freshly baked bread that no French person would sit down to a meal without still need to be collected daily, or even twice a day, from the local *boulangerie*, or baker, as they quickly become stale.

A Breton *pardon* (*above*) The women wear elaborate lace caps at this annual pilgrimage to a local church. The people of Brittany in the northwest retain many of their distinctive local customs.

A sense of style (*left*) The French are renowned for their love of good clothes. A world center for dress design, the elegant avenues and boulevards of Paris – such as the Champs Elysées – are lined with exclusive fashion houses.

The pleasures of the table (*below*) A group of men enjoy a hearty meal together at the end of a day's work. The distinctive styles of French provincial cooking are based on seasonal use of locally produced ingredients.

conventional role of mother and housewife longer than in some other Western societies. Nevertheless, female employment is increasing. More and more women are taking up lower-level jobs in the expanding service industry sector and in some of the professions, notably teaching and medicine. It is likely that the changing attitudes of young people will further erode the distinctions between the sexes at home and at work.

Religion – once the most divisive issue in national life – is today of lesser importance. The majority of people belong to the Roman Catholic church, but only a minority participate in regular religious worship. There are about 1 million Protestants, with an important concentration in Alsace, reflecting its German traditions. Before World War II France had a substantial Jewish population, many of them emigrants from Russia and Eastern Europe. Large numbers were killed by the Nazis, but several hundred thousand Jews remain (mainly in Paris, Marseille and the large eastern cities), many of them repatriates from former French territories in Northern Africa.

Recent immigration from Northern Africa has led to a considerable increase in the number of French Muslims. Today there are about 2 million of them – more than in any other West European country. They outnumber Protestants and Jews together, and form the second largest religious group in France. There are over 400 mosques in France. The first was built by the government in Paris after World War I as a mark of gratitude to the Algerians who died fighting for France.

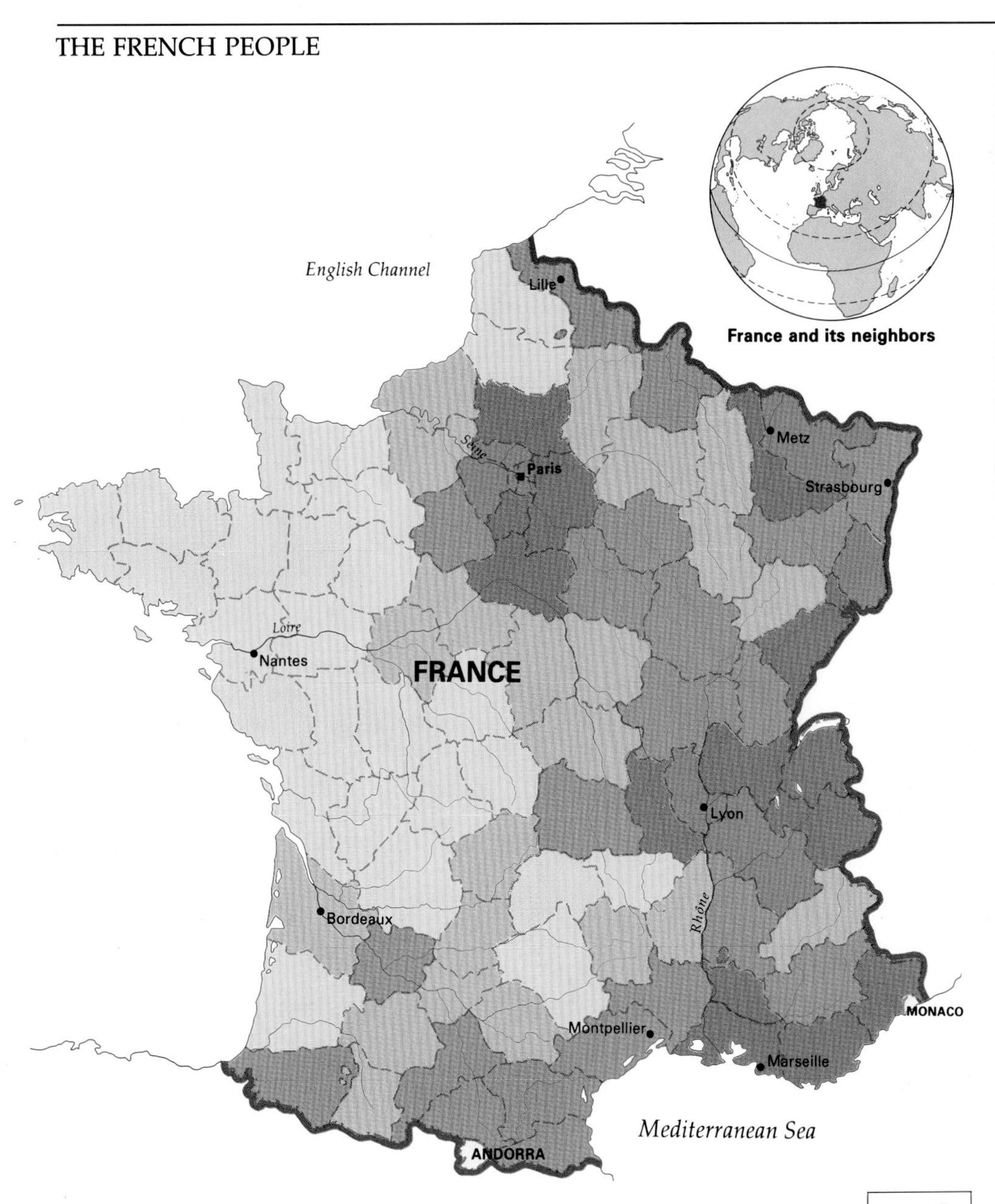

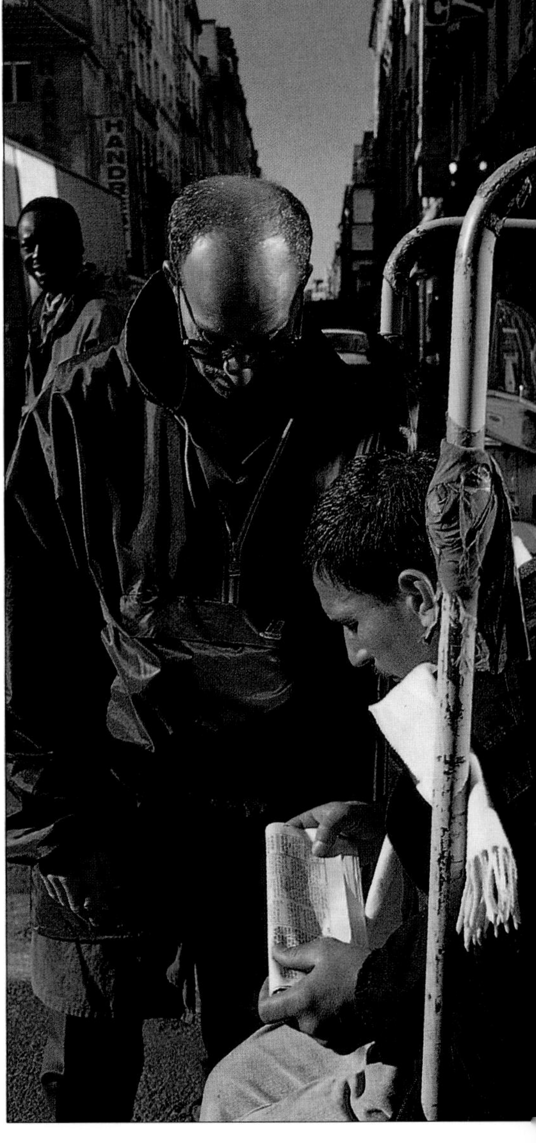

On the edge of society *(above)* Migrant workers in Paris discuss the day's news in an Arabic newspaper. The postwar influx of foreign workers – particularly Arabs from former African colonies – has led to increasing racial tension in the big cities.

The pattern of immigration *(left)* The map shows clearly that France's immigrant populations are greatest in the industrialized north and east of the country, with strong concentrations around the conurbations of Paris and Lyon.

Immigrants as % of population by department, 1986

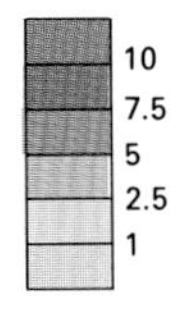

TOWARD A NEW FRANCE

The creation of the French state has left a history of resentments between the different regions, and particularly between the regions and Paris. These began to be publicly voiced in the 1960s. Regions on the periphery of the country, such as Flanders, Brittany, Alsace, Languedoc and Corsica, felt that Paris was not only neglecting their economic development, but also suppressing their culture, particularly their languages. The time was not far past when children were punished for speaking regional languages in state schools. Reforms in the 1980s increased regional self-rule; for example, the office of prefect – which was originally created by Napoleon – was abolished. These regional heads of government had previously been appointed from Paris.

Culture in the service of the state

In characteristic French fashion, high culture, especially the arts and architecture – long accustomed to being called upon to serve the purposes of the state – was mobilized in the effort to redress the balance between Paris and the regions. Provincial universities were expanded and local cultural facilities established through *maisons de la culture*. Devised by the novelist and critic André Malraux (1901–76), these multipurpose arts centers were designed to spread culture to the provinces, elevate the country's international cultural profile, and extend enjoyment of the arts beyond the middle class. They did not win wide acceptance. Regarded locally as yet another example of central imposition, they were very expensive, and only 15 were established.

It is likely to be economics rather than culture that restores geographical balance to the country. The Midi, or "sun belt", long neglected and poor, has emerged as the fastest growing region of France. High-technology towns are now drawing young professionals from all over France. At the same time, Mediterranean resorts from Menton to St Tropez, long favored as places for retirement and as fashionable vacation venues for the affluent classes of Paris, are being supplemented by the new state-led tourist development west of Marseille to Perpignan. In addition, more and more houses are being bought as holiday homes by foreigners, particularly the Dutch, the Belgians and the British. Although this provides a welcome source of revenue in depopulated villages, it also means that their houses stand shuttered

A MIXING PLACE FOR CULTURAL INFLUENCES

In great contrast to French attitudes toward recent immigrants is the warm welcome that has been given to African art and artists. One of the earliest examples of the impact of black culture came at the beginning of the 20th century when the painter Pablo Picasso (1881-1973), then working in Paris, was influenced by both West African and Polynesian masks in developing the Cubist style of painting.

There exists an extensive African literature in French. One of the major influences on this was the poet Léopold Senghor. Born in Senegal in West Africa in 1906, he was educated and worked as a teacher in France, where he began to write in the 1930s in protest against the soullessness of French culture. With a number of other African and Caribbean French writers he set himself the task of spreading black cultural values through his writings. In 1960 he became the first president of the newly independent state of Senegal, and in 1984 he was elected the first black member of the French Academy, the French literary society founded in 1652, whose members are limited to 40 at any one time.

African music and dance have also been very influential, bringing new rhythms and energy to the staider measures of the West. Initially, the impact came from African–American music, particularly jazz. The explosion began in the 1920s with an influx of black musicians who welcomed the freedom and apparent lack of racial prejudice that France offered. Some artists came to stay: perhaps the best-known was the flamboyant dancer and singer Josephine Baker (1906–75).

More recently, Paris has become the European center for a new wave of African music. From the late 1970s an ever increasing number of African musicians came to Paris, attracted by the superior recording facilities, more lucrative contracts and a receptive audience for new sounds in popular music. Many of these artists and bands made international reputations from their Parisian base, including Sunny Ade from Nigeria, Manu Dibango from Cameroon, Mory Kante from Mali and a large number of musicians from Zaire such as Kanda Bongo Man, Quatres Etoiles and Tabu Ley. Paris also proved to be fertile ground for mixing international styles of music, Western and African, to produce new sounds.

and silent for most of the year. In the summer tourist season small towns and villages are overwhelmed by the influx of visitors, who very often make no attempt to learn French and crowd out residents in shops and restaurants.

A succession of presidents in the postwar period have continued the tradition, first established by the Bourbon monarchy, of leaving their mark on Paris. Georges Pompidou (1911–74), president from 1969–74, left as his legacy the controversial Beaubourg Center, now the most visited landmark in France, and Valéry Giscard d'Estaing (president 1974–81), the shopping center at Les Halles. Neither of them matched the grandiose schemes of President Francois Mitterrand (president 1981–). The stunning Arc de Defense; the very unusual transparent pyramid in the forecourt of the Louvre; a new Opera house inaugurated in 1989 to celebrate the bicentenary of the French Revolution, and the ambitious new National Library, attest to the significance placed on architectural monumentalism under state patronage.

Such cultural ambition can run into trouble, however. The EuroDisneyland theme park 32 km (20 mi) east of Paris, planned and built in the early 1990s, was Europe's second largest construction project (after another French dream – the Channel Tunnel), partly financed by US Disney and designed to be a major international attraction. However, many people both locally and throughout France were enraged by this example of cultural imperialism from the United States. In few countries do such things as theme parks (or the importation of foreign words such as *hamburger*) excite such impassioned public debate.

New citizens, old resentments

France's postwar chronic labor shortage was initially filled by Spanish and Italian migrant laborers, but since the revival of their respective national economies, new sources of workers from farther afield have been used. There are about 5 million non-French nationals in the country, mainly from Portugal and Algeria, but increasingly from other former colonies in Africa. As many as one in six Parisians is a foreign national.

The Algerians came from France's largest and richest colony, one lost after a violent war that divided the French people. They came originally as temporary workers, but once the men had found jobs and accommodation they sent for their wives and families. Most were recruited into low-level jobs spurned by the increasingly affluent French: construction work, unskilled jobs in manufacturing, and lowgrade service jobs such as refuse-collecting and road-sweeping.

Until recently the French could, with some reason, pride themselves on their willingness to accept cultural incomers. This has its roots in the ideals of the Revolution: applied to the colonies, this meant that they were treated as extensions of mainland France. Among the colonial nations, France was unique in according full rights of citizenship to its overseas population. This was acceptable so long as those who emigrated to France were few and highly qualified, but has come increasingly into question as the numbers of incomers have risen. Since 1981 the right to French citizenship has no longer been automatically given to everyone born there.

Like immigrants elsewhere, France's new citizens settled in ghettoes in run-down quarters of the inner cities such as La Goutte d'Or on the edge of Montmartre in Paris. Here the streets are lined with Arab cafés, hotels and shops. However, in contrast to other countries, immigrants also moved into the suburban high-rise housing projects built to solve the postwar national housing crises, but since neglected. The 1990s saw riots in such areas in Lyon and Paris, often sparked by allegations of police brutality.

The rapid rise of extreme right-wing political groups, notably the National Front, intensified these conflicts. Combining demands for repatriation with a renewed anti-Semitism (the 1980s saw a whole spate of desecrations in Jewish cemeteries), groups like these are being countered by organizations such as SOS-Racisme, which seek to defend the immigrants, particularly the *beurs* – those of the second generation. Racism is most prevalent in those areas with large immigrant populations. However, it is also common in the south, especially in Marseille and Corsica, where the *pieds-noirs* – French settlers returned from the former colonies of Northern Africa – remember the bitterness of the colonial wars.

Isle of beauty, isle of terror

Signs of neglect Tourists may find the narrow streets picturesque, but the dilapidated housing reveals poverty. Islanders blame central government for policies that have benefited wealthy farmers or development companies more than ordinary people.

Of all the regions of France, the Mediterranean island of Corsica stands out as a special case. Here alone in recent times has a separatist movement erupted into violence. This "mountain in the sea" possesses great natural beauty, with more potential for tourism than agriculture. It is also known as the "scented isle" – a name it gains from the aromatic maquis scrub that covers the lower slopes of its central mountain ridge. Higher up there are forests of chestnuts, whose fruit formerly provided the islanders' staple diet – fritters of chestnut flour are characteristic of Corsican gastronomy. The pine forests still farther up contain herds of semiwild pigs, the basis of the island's brand of *charcuterie* – cooked or smoked pork products of every imaginable variety.

Lying closer to Italy than to France, Corsica has a long history of invaders and occupiers. Between the 12th and 18th centuries it was ruled from the Italian mainland, successively by Pisa and then Genoa. There were local uprisings in the 18th century, but on 15 August 1769 – the very day on which Corsica's greatest son, Napoleon Bonaparte, was born – the island was declared part of France.

A separate entity within France

Not surprisingly, Corsicans consider themselves to be very different from the French; their local language is, after all, closer to the Tuscan version of Italian. As on Sicily and some other Mediterranean islands, clan loyalties are very strong, and vendettas – prolonged blood feuds between families – are characteristic of the culture. As in Sicily, too, the state machine has proved powerless to prevent the settling of accounts with a gun.

Like other marginal areas in France, Corsicans believe that the island was economically ignored for years by Paris, with the result that many left it to earn a living as civil servants, soldiers, policemen or even gangsters on the mainland, especially in Marseille. There was a corresponding influx of people from other areas of France. Algerian independence in 1963 brought several thousand *pieds-noirs* to Corsica, and French government officials also helped to swell the number of outsiders (the state accounts for one-third of all jobs on the island). By 1978 the non-Corsican population amounted to 48 percent of the island's total.

Well-meaning attempts by the French government to regenerate the island's economy in the postwar years often provoked local resistance. The island's eastern plains were drained for agriculture, but those who benefited most were resettled *pieds-noirs*. The financial profits from the expansion of tourism along Corsica's incomparable coastline were all too frequently reaped by multinational development companies.

Although the French government in the 1970s met Corsican aspirations for independence with a greater amount of local control than has been offered to other French regions – thereby recognizing the claims of the Corsican people to be treated as "a separate entity within France" – antagonism continues. Not all the French Corsicans favored greater autonomy, and many resisted moves to increase it still further in 1990.

Throughout the island, therefore, the campaign for separate identity goes on. Painted slogans demand "Freedom for Corsica" and "Death to the French". Tourists are advised to hire cars with local registrations, since those with foreign (including French) license plates are liable to be attacked. French or foreign-owned holiday complexes are bombed from time to time. All symbols of French authority are particularly vulnerable to attack, including police officers, many of whom have been shot.

In the shadow of the mountains The historic port of Calvi is a major tourist attraction, though visitors might not relish the thought that bandits still operate from the shelter of remote hideouts situated high up in the island's mountains.

PDM
R.M.O
89
155

The Tour de France

The Tour de France cycling race takes place every year for three weeks in June and July. It is run over 4,000 km (2,500 mi), watched by up to a third of the French population and contested by between 120 and 140 professional cyclists, accompanied by a mobile army of followers, officials and police. Apart from the finish, which is always in Paris, it has no fixed course. This means that year after year, the race passes through a wide range of villages, towns and cities, making it a truly national spectacle. Sometimes spectators line the route for hundreds of kilometers, while many small towns hold fairs and close their schools for the day when the Tour passes through.

The event began in the early 1900s as a ploy in a circulation battle between two rival journals, *Velo* and *L'Auto*. The latter dreamed up a cycling race that, instead of taking place between two fixed points, would cover the whole of France. It was a time, before the age of the automobile, when bicycles, once a luxury of the rich, were becoming more widely available and cycling was emerging as a popular pastime. The Tour from the first was an opportunity for rival bicycle and tire manufacturers to promote their products, and it was among the earliest sporting events to be commercially sponsored.

The first race, in 1903, was between 60 riders and won by Maurice Garin, a chimney sweep. Today, the winners – men like the Frenchman Jacques Anquetil, five times a winner in the 1950s and 1960s, and the Belgian Eddie Merckx, who also won five times a decade later – become national heroes. Over the years the field has become more international and in the 1980s Irishman Stephen Roche and American Greg LeMond became the first riders from outside continental Europe to hold the coveted yellow jersey for the race leader.

A national event Traffic ceases on French roads, and towns and villages take on a festival air when the Tour de France passes through.

LIBERALISM AND ORDERLINESS

A SPIRIT OF INDEPENDENCE · CHANGING ATTITUDES · MINORITY QUESTIONS

"God created the world, but the Dutch made Holland." The adage sums up the mythic importance that their relationship with the land holds for the people of the Low Countries, particularly the Netherlands, some one-fifth of which has been reclaimed from the sea. This relationship has played a central part in shaping their culture: the painters who made the art of the Low Countries preeminent in 16th- and 17th-century Europe took their inspiration from the flat, unadorned landscape. Traditionally, for all the people of the Low Countries, the sense of cultural identity is rooted in the local community – the town or province. The region is divided linguistically; French being the predominant language of the south, and Dutch (or Flemish) the language of the north – a division that does not match national boundaries.

COUNTRIES IN THE REGION

Belgium, Luxembourg, Netherlands

POPULATION

Netherlands	14.8 million
Belgium	9.88 million
Luxembourg	37,100

LANGUAGE

Country with one official language (Dutch) Netherlands

Country with two official languages (French, German) Luxembourg

Country with three official languages (Dutch, French, German) Belgium

Local minority languages are Frisian and German (Netherlands; Luxemburgish (Luxembourg). Other languages spoken in the region include English, Italian, Spanish (Belgium); Italian, Portuguese (Luxembourg); Arabic, Creole, English, French, Indonesian and Turkish (Netherlands)

RELIGION

Belgium Roman Catholic (90%), nonreligious and atheist (8%), Muslim (1.1%), Protestant (0.4%)

Luxembourg Roman Catholic (93%), nonreligious and atheist (5%), Protestant (1%)

Netherlands Roman Catholic (36.2%), nonreligious (34.7%), Protestant (26.4%), Hindu (1%), Muslim (1%)

A SPIRIT OF INDEPENDENCE

The cultural development of the Low Countries has been strongly influenced by the natural landscape. The flat lands of the north and west were once an area of near impenetrable marsh, crisscrossed by countless streams and rivers, that made travel almost impossible except by water. The hills of the southeast were densely forested until comparatively recent times, also making communications difficult. As a result, settlements grew up in isolation on small hills among the marshes, and along the rivers and estuaries of the Rhine, Meuse and Schelde delta. This led to the development of equal and independent cities throughout the region, each serving its own provincial community, and has shaped the very strong sense of local identity that continues right down to the present day.

Ancient settlement

The Low Countries were originally settled by successive waves of people from central Europe, the first of whom moved into the region about 10,000 years ago. When the armies of the Roman empire marched

A link with the past The spring carnival at Binche in Belgium has been celebrated since the 14th century. The fun lasts for several days; on Shrove Tuesday – *Gilles* – boys and men in elaborate costumes with ostrich plumes on their hats parade through the main square, pelting the crowd with oranges.

into the region 2,000 years ago they found two distinct areas of settlement: north of the Rhine were a number of Germanic peoples, the most important of which were the Frisians, and to the south were the Celts. The Frisians threw back the Roman advance, but the Celts were overrun and soon assimilated into the Latin-speaking culture of the Romans; their descendants are the French-speaking Walloons of southern Belgium.

With the breakup of the Roman empire in the 5th century, the Franks – a Germanic people who had settled between the Meuse and Schelde rivers in central Belgium – became the dominant power in the region, establishing their rule over present-day Belgium and Luxembourg, northern France and western Germany. They were the ancestors of the Netherlandic or Dutch-speaking population of the Low Countries who today occupy nearly all the Netherlands and the northern (Flemish) half of Belgium.

The older Frisian population was eventually pushed by occupying forces to the region's northern extreme, and today are settled in the flat coastal lands of Friesland and on the West Frisian Islands, where Frisian, a distinct language, is still in common usage. German and its dialects are spoken in some parts of eastern Belgium, on the border with Germany, and in Luxembourg.

National identities

After the Frankish empire collapsed, about 1,000 years ago, the Low Countries were left as a mosaic of small, more or less independent lordships, many of which survive as provinces. The old duchy of Limburg, for example, now exists as two separate provinces on either side of the Dutch–Belgian border. People on both sides are Dutch-speaking, using the same dialect. The border that divides them is political rather than cultural and they regard themselves as Limburgers.

This border is the result of conflicts arising from the 16th-century Reformation, when deep religious controversy split western Christianity, causing the separation of the reforming Protestant churches from the Roman Catholic church. By this time, the Low Countries had come under the rule of Spain, then the foremost defender of Roman Catholicism. The northern half of the region embraced Protestantism and rose up against their Spanish overlords and established the Netherlands (then known as the United Provinces) as an independent state in 1648. Their struggle for national identity, fought under the guise of religious freedom, was made possible by the commercial prosperity they had gained as a maritime power, and was a time of great cultural activity in the arts, architecture, and science.

The southern half of the Low Countries, including both Dutch-speaking Flemings and French-speaking Walloons, remained predominantly Roman Catholic. It has stayed as a separate political entity ever since, although Belgium was not established as a sovereign state until 1839; Luxembourg not until 1867. The people of the Low Countries are consequently latecomers to the idea of national sovereignty. Their provincial loyalties make them less jealous of national differences, reflected in the role all three have played since World War II toward creating a common European culture.

CHANGING ATTITUDES

Now that the marshes have been drained, the rivers bridged, and dikes built to prevent the sea from flooding the land, the flatness of the landscape, which once made overland travel so difficult in the region, is a positive advantage. There are more than 40,000 km (24,850 mi) of expressway per 1,000 sq km (386 sq mi) of land in the Low Countries – far more than anywhere else in Europe. Improvements in transportation and communications since World War II have led to considerable movement of population. In combination with the popularity and influence of the mass media, particularly television and movies, this has brought about a homogenization of the traditional local cultures that once flourished in different parts of the region.

As people become increasingly mobile, they travel ever greater distances to work

Flourishing arts scene *(above)* The internationally renowned Netherlands Dance Company is noted for its innovative productions. The arts are given a high priority in Dutch society; generous government subsidies support the artistic community at national and local levels.

Sex for sale *(right)* A brightly decorated street front advertises a live sex show in Amsterdam's red light district – a beacon for visitors from less sexually permissive countries. Dutch tolerance is rooted in their belief in individual freedom and morality.

– thousands commute daily across the national borders within the region and to workplaces in northern France and northwestern Germany. There is a growing willingness for people to take up permanent residence in another country, and even to assume a different nationality. The administrative and judicial headquarters of the European Community are situated in Brussels and Luxembourg, which are centers of large multinational communities, and a great number of international industrial companies also have their centers in the region.

All these factors contribute to the emergence of a modern cosmopolitan culture, especially in urbanized areas, which is greatly affecting social attitudes. The greatest change in the way people behave, particularly in the Netherlands, has come about through the decline in religious practice. Since World War II, for example, the Dutch Reformed Church has lost a third of its members, and the number of people ascribing to no faith has risen by half. This has led to the gradual dissolution of the rigid social structuring that formerly grouped people by religious and political conviction – whether they were either Roman Catholic or Protestant, or whether they were "humanist" (which included members of the Liberal or Socialist parties) – determined where they were educated, what newspaper they read, who they married, and what political party they voted for. As society becomes increasingly secularized, these constraints no longer affect people's behavior or ambitions.

Heritage as commerce A street vendor, dressed up in traditional Dutch costume complete with wooden clogs, disconsolately waits for buyers for her remaining souvenirs. Tourism is an important source of income; in some areas traditions are only kept alive for the sake of visitors from other countries.

A civil culture

Nevertheless, certain distinct regional and cultural attitudes still exist. The Dutch sum up their cultural values in the words *deftig* and *gezellig*, meaning "respectable" and "cosy". These characteristics, combining formality with friendliness, are the foundation of the country's liberal attitudes toward personal conduct, setting up civil order rather than morality as the basis of freedom and restraint.

Amsterdam is notorious for its tolerance of types of public behavior that are likely to be condemned in other Western capital cities, such as the smoking of marijuana. The number and variety of its sex shops and strip shows is also well known, as is the fact that prostitutes are permitted to advertise themselves in their brightly lit windows. These aspects of cultural toleration are founded on a genuine political liberalism. Belgian civic culture, on the other hand, might be described as "semibureaucratic" in its orderliness. Encounters with officials, shopkeepers, tram conductors, nightclub bouncers and anyone else with a position, however humble, must be handled with dignified formality. This social habit of actively acknowledging respect for one's fellow citizens is a key element in forming and maintaining the spirit of tolerance that distinguishes the culture of the Low Countries.

Tradition and tourism

Cultural assimilation has been strongest in the cities: in country districts people have been able to preserve more of their traditional cultures. Small farming communities near Staphorst in the province of Drenthe in the northeastern Netherlands, for example, still dress in traditional costumes and observe the sabbath, Sunday, as a day of rest. In southern Belgium, where Roman Catholicism is still strong, many people take part in religious festivals and processions. During the summer months in the Belgian province of Hainaut, a stronghold of Walloon culture, many towns and villages hold marches, when holy relics are protected from would-be bandits: at Ath there is an annual procession of giants and at Mons, St George battles fiercely once more against the dragon.

Elsewhere tradition has degenerated into mere tourism. In Volendam and Marken, just north of Amsterdam, ordinary citizens put on traditional dress to attract visitors and sell them souvenirs. Here the traditions of provincial life have become fossilized as "folk cultures". They are regarded with affection as part of the national heritage, but are treated as museum exhibits.

DUTCH – A DECLINING ASSET?

In global terms, Dutch is a relatively insignificant language, spoken by very few people outside the Netherlands. Since most people acquire great fluency in English, the most widely spoken second language, in order to communicate with outsiders, Dutch books and movies have to compete directly with English-language products. This problem is common to many minor languages and particularly affects the movie industry. Even though two government-funded bodies exist to sponsor Dutch movies, one-third of the money available goes toward the making of art movies, with very limited distribution. Limited budgets do not allow these bodies to invest enough capital to cover the costs of moviemaking on a scale remotely comparable to films from the United States.

Yet Dutch audiences clearly want to watch precisely the sort of lavish Hollywood productions that Dutch moviemakers are unable to emulate. In 1988 only two Dutch language movies figured in the top ten box-office hits, and these two accounted for three-quarters of the total audience for Dutch language movies.

A similar dilemma confronts Dutch literature. Bookstores are full of titles by British and American authors, and many young Dutch novelists choose to write in English to reach a wider audience. Even in its own country, Dutch-language cultural output is uncommercial – and without a wider international audience it is likely to remain isolated and parochial.

Tastes from outside *(above)* Immigrants from the former Dutch colony of Surinam in South America with a tempting display of some of their national dishes at a festival in The Hague. Incomers from former colonies have been more readily accepted than workers from Turkey and Northern Africa who arrived in large numbers to fill gaps in the labor market.

MINORITY QUESTIONS

Although an increase in affluence has brought growing uniformity to the culture of the Low Countries over the last 40 years, there are still some areas where deep-seated cultural differences between groups, or between a minority and the central administration, spark into occasional conflict. The best known of these is the distinct language division in Belgium between the Dutch-speaking Flemings of the north and the French-speaking Walloons of the south.

Most of the time, cultural differences between these two peoples are expressed in quite unexceptionable ways: the Flemings, for example, show a preference for drinking beer, the Walloons for wine. But resentments below the surface can all too easily flare up into hostility and riots when one group feels the other is being favored economically or politically. In 1980 separate executive councils were set up to administer the two communities, a political measure that went some way toward ending the language riots of the previous two decades, but many people consider that the establishment of some kind of federal state is the only way to provide a longterm solution to the problem that will satisfy both sides.

Other language minorities

The persistent rumblings of the Fleming-Walloon issue tend to overshadow the relatively peaceful settlement of the German language question. The "Eastern Cantons" around Eupen and St Vith, on the border with Germany, were only incorporated into Belgium by a dubious referendum after World War I. For many years their German-speaking communities were reluctant to consider themselves as Belgian. However, as a by-product of the 1980 political settlement between the Flemings and Walloons, German was recognized as a national language, and the Eastern Cantons were given the same considerable degree of cultural autonomy that Flanders and Wallonia now enjoy.

In the Netherlands, the Frisian language – more closely related to English than to Dutch – is spoken by some 300,000 people in the province of Friesland. It was the only language spoken and written in Friesland until the 16th century, when the area became a province of the Netherlands, and fell under Dutch administration. Frisian subsequently came to be regarded as a peasant language, but nationalist movements in the 19th century revived its wider use, and the survival of the Frisian language has since become the central issue in the political struggle for greater regional autonomy.

By and large, this struggle has been a peaceful and constructive one, although a tactic of painting over Dutch place names was adopted by Frisian nationalists in the mid-1970s. Frisian is now a compulsory subject in all elementary schools within the province. All members of the provincial administration are expected to be able to speak the Frisian language, while some municipalities also use it in their communications and decrees, and display the names of local towns and streets in Frisian as well as Dutch.

Influx of foreigners

A potentially greater problem of accommodation is posed by the large numbers of foreign workers resident in the region. Apart from the skilled technicians and other professional workers from the northwest of Europe, particularly France and Germany, who share the same cultural interests and aspirations as their Dutch and Belgian neighbors, these fall broadly into two categories. Many migrant workers are from Portugal and Spain, within the EC. They have settled mainly

Time to gaze Contented customers enjoy a leisurely glass of beer or coffee and cakes at a street café in Brussels. Many cafés sell delicacies such as the hand-made chocolates for which Belgium has become famous around the world.

THREE LANGUAGES IN LUXEMBOURG

"Mir woelle bleiwe wat mir sin – we want to remain what we are": so runs the national motto of Luxembourg in the official language of the duchy, Lëtzebuergesch or Luxemburgish. Yet more than a quarter of the population of the tiny state is made up of foreign residents, a fraction that is likely to increase as Luxembourg's role in the multilingual administration of the EC continues to extend.

All native Luxembourgers speak Lëtzebuergesch, enabling them to maintain a strong sense of national identity. Nevertheless, its use is restricted to informal conversational exchanges: it is not spoken in court, in church, in school, or any other area of formal social life – in these situations either French or German is used, depending on social class. French, traditionally the language of the upper class, is used in the administration and by the professional classes. German is the language of the working population. Posters advertising the visit of an operatic company from Paris or Vienna will be written in French; cut-price bargains will be posted up in supermarket windows in German. Although Lëtzebuergesch is gradually making itself felt as a language that may be used in situations where French is more commonly employed at present, there are problems: many Luxembourgers find it a difficult language to read and write. Only one weekly television program is in Lëtzebuergesch; the rest are in French. The four daily newspapers are in German. However, Lëtzebuergesch is well established in the popular performing arts, especially, for example, in the duchy's lively tradition of satirical theater cabaret.

in southern Belgium and Luxembourg, where their common practice of Roman Catholicism provides easier assimilation and an important cultural link.

However, many of the foreigners resident in the region are from cultural backgrounds very different from that of the host population. Some were originally immigrants from former colonies who settled here before or shortly after independence in the 1950s and 1960s – Indonesians and Surinamese in the Netherlands, and Zaireans in Belgium. They are now far outnumbered by "guest workers" from Turkey and North Africa who first began to arrive to provide cheap, unskilled labor in several European countries during the economic boom years of the 1960s and 1970s.

As economic recession led to rising unemployment during the 1980s, resentment began to be expressed against the guest workers, who nonetheless continued to arrive in the Low Countries in their thousands. Many people wished to see an end to the influx, and both the Dutch and Belgian governments offered considerable cash incentives to persuade migrant workers and their families to leave, even promising to pay them social security after returning to their home country. Few, however, took up the offer. Cultural integration of these groups proceeds relatively slowly. Most non-European guest workers are Muslims, and the rise of fundamentalist Islam has led many of them to demand a greater degree of cultural separation, with recognition of religious and linguistic differences. To meet these aspirations, the first Muslim school within the Dutch educational system was established in Rotterdam in 1989.

Art and money in the Netherlands

Beyond value (*above*) The superb collection at the van Gogh Museum, Amsterdam is a priceless part of the Dutch national heritage. Sadly, the staggering prices that van Gogh's works now fetch on the world art market have made it almost impossible for the museum to add to its collection.

The golden age of Dutch painting (*right*) "The Idlers" by Jan Steen is just one example of the spectacular burst of Dutch artistic achievement that took place during the 17th century. The Dutch school typically depicted scenes from everyday life; Steen was a tavern keeper who painted many studies of bourgeois merry-making.

In 1990 a painting entitled "Portrait of Dr Gachet" by the Dutch artist Vincent van Gogh (1853–90) was sold for the then world record price of $82.5 million to a Japanese collector, who considered the price was "very reasonable". Three years previously van Gogh's "Sunflowers" was acquired by Yasuda Fire & Marine for $33.9 million and his "Irises" went to the J. Paul Getty Museum in California for an undisclosed sum, probably even more. The artist himself sold only one painting in his lifetime and at the time of his death was virtually unknown inside or outside his country. Now regarded as the greatest Dutch artist after Rembrandt (1606–69), the astronomical prices that his paintings fetch today mean that they are outside the reach of the Dutch state, let alone individual Dutch collectors. The works of van Gogh – like those of many Dutch Old Masters – are being lost to the nation.

The irony of this hyperinflation in the art world is that it was in the Netherlands that the first commercial art market was established in the early 17th century. The consolidation of the Dutch Republic saw an increase in the prosperity of many of its citizens, which expressed itself in a readiness to acquire paintings to adorn the walls of their neat town-houses: until this date the only patrons of art in Western Europe had been the church and the aristocracy.

Popular art thus had its origins in the Netherlands. Paintings and drawings, often depicting sentimental themes of everyday life, were produced by a small group of gifted commercial artists; the bourgeois emphasis can also be seen in the "Dutch interior" paintings of artists like Jan Vermeer (1632–75) and Pieter de Hooch (1629–81), where the rooms in which the citizens hung their pictures became artistic subjects themselves.

For two centuries there was an active market in Dutch art, and the connection between fine paintings and respectability became well established. Even today, the acquisition of paintings such as those of van Gogh is not necessarily a good financial investment – prices can rise and fall; fashions change – and certainly gives no better return than government securities, for example. What counts still is the immense personal satisfaction of owning a culturally desirable object.

The state as patron

It was the wars of the late 16th century that produced the economic and social conditions that fueled the boom in the Dutch art market. This was far from being the case after the devastations of World War II. In the postwar years, Dutch artists found it so difficult to make ends meet that in 1949 the state set up the *Beeldende Kunstenaars Regeling* (BKR, or Visual Arts Arrangement) to help them.

When it was wound up almost 40 years later, this subsidy system was paying individual Dutch artists up to $16,000 a year. In return, the artists were required to supply a certain number of art works to the government. In this way the government accumulated over 220,000 works of art – most of them lying piled up unseen in vaults – and in 1987 paid out a total subsidy of $34 million.

Although the state no longer subsidizes artists in quite such an arbitrary fashion, it is still the greatest patron of the arts in the country, spending an estimated $33 per person per year on them in 1990. This compared with $12 per artist in Britain and a mere 71 cents in the United States. Nevertheless, artists must now appeal directly to the commercial market to survive, and this means selling not only to collectors but to private individuals and business organizations. A price of $2,000 or so for a painting will still make the ordinary Dutch buyer gasp, but banks and offices, which have large wall space and open areas for people to move in where pictures can be attractively exhibited, may well succeed to the role of patron that was once the preserve of the middle classes.

The Oranje

Soccer, or football, is the world's most widely played team sport, and there are organized leagues in over 140 countries. It is also popular as a spectator sport, and the World Cup – held every four years – attracts the largest television audiences of any single sporting event. It is not always the largest or richest countries that win the competitions. In the past, the South American countries – Uruguay, Brazil and Argentina – held sway, while from 1966 to 1974 it was the Netherlands that dominated the world game. The Dutch teams of Ajax Amsterdam and Feyenoord Rotterdam won the European Cup competition four years running, while the international side reached the World Cup final twice in the 1970s, though never winning.

As in other European countries, however, Dutch soccer entered a difficult period in the 1980s. Several homegrown star players were induced to play for the well-financed clubs such as the Italian League. At home, older clubs such as Ajax and Feyenoord failed to sustain their brilliant records, and only PSV Eindhoven – the club backed by the electronics firm Philips – could find the money to compete with the Italians.

At the same time, the Netherlands developed a reputation for football hooliganism – a growing problem throughout European soccer – as organized groups of fans fought with each other inside and outside the grounds in a violent "shadow" competition. The 1988 European Championships threatened to become a showdown between Europe's rival hooligan armies, but tight policing and surveillance prevented any pitched battles, and the Dutch fans, resplendent in their orange shirts and scarves, and sometimes with painted faces, were well-behaved. After all, their side won.

Soccer tribalism A Dutch fan, his face emblazoned with his team's colors, yells support.

LAND

THE PAST IN THE PRESENT

THE PEOPLE OF IBERIA · A LAND OF CONTRASTS · ADJUSTING TO CHANGE

Spain and Portugal's diverse physical environments have helped to shape their cultural variety: the way of life of people in the rugged, rainy mountains of the north is very different from those of the high, harsh plateau of the Meseta or the gentler landscapes of the coastal plains. A history of invasion, and of reconquest, has also played its part – Roman aqueducts, Arab tiled courts, and fortress-like cathedrals are all striking reminders of the past, which has left a less visible legacy in the region's linguistic and cultural diversity. Spain and Portugal's former overseas empires helped to spread their influence worldwide. Rapid modernization has had inevitable impact on traditional cultures but the separatist movements of Catalonia and the Basque Country have helped to preserve local cultural differences.

COUNTRIES IN THE REGION

Portugal, Spain

POPULATION

Spain	39 million
Portugal	10.4 million

LANGUAGE

Countries with one official language (Portuguese) Portugal; (Spanish) Spain

Local minority languages spoken in Spain are Basque, Catalan and Galician

RELIGION

Portugal Roman Catholic (94%), nonreligious (3.8%)

Spain Roman Catholic (97%), nonreligious and atheist (2.6%), Protestant (0.4 %)

THE PEOPLE OF IBERIA

The Spanish and Portuguese peoples of today display considerable ethnic and cultural uniformity. They share the characteristic physical appearance of Mediterranean peoples – short stature, olive complexions, dark hair and brown eyes. However, this homogeneity belies the Iberian Peninsula's history of invasion, settlement and reconquest. In prehistoric times the peninsula was populated by the Iberians; little is known of them, but they were possibly of North African origin. Celts, originally from central Europe, moved into the western part of the peninsula through the Pyrenees in about the 9th and 8th centuries BC. The Phoenicians (from the eastern Mediterranean), the Greeks and the Carthaginians (a Phoenician settlement in North Africa) in turn established colonies along the southern coasts, but the Iberian and Celtic populations were not subdued until the Roman conquest (from 211 BC). Celtic influences remained strongest in the outlying northwest region of Galicia, which has retained a distinctive seafaring culture to the present day.

At some date in prehistory the Basque people became settled in the western foothills of the Pyrenees. Their origins are obscure – they are not known anywhere outside the western Pyrenees. They preserved their identity intact from the Celts, and similarly resisted assimilation by the Romans.

Cultural invasions

In contrast, the central and southern parts of the peninsula became thoroughly Romanized and shared in the prosperity of the empire; Christianity became the dominant religion. As the power of the Roman empire began to disintegrate in the 5th century AD the peninsula was left defenseless and, by 467, had been conquered by the Visigoths, a Romanized Germanic tribe. Their domination of the peninsula ended with the arrival of the Muslim Arabs (or Moors) in 711.

This Moorish invasion dominates the history of the southern half of the peninsula. By the mid 10th century Muslims were in the majority, and the Arab kingdoms in the south (known as al-Andalus) were renowned centers of learning and culture. Relics survive in the university, palaces and gardens of Granada, the last part of al-Andalus to fall to Christian Spain, in 1492.

The influence of Moorish architecture is commonly seen in the south, and Arabic traits are still alive in the popular arts, particularly ceramics. In remote parts of Andalusia women wore the veil for many years after the expulsion of the Moors, and the custom of separating courting couples to converse through a metal window grille survived into the 20th century.

Three elderly men (*right*) sit in the town square of Vejer de la Frontera in southern Spain. Evidence of the Moorish occupation of the peninsula is seen in the architectural detail of this ornate fountain, a reminder of the Arabs' love of playing water. The sound of water can also give the illusion of coolness.

Spain's Catholic past (*below*) A solemn procession at Santiago de Compostella in northwestern Spain re-enacts the events of Christ's crucifixion. During the Middle Ages Santiago, which houses the relics of St James the Apostle, credited with bringing Christianity to Iberia, was a center for pilgrims rivalled in importance only by Rome or Jerusalem.

Expulsion and consolidation

Cultural and religious tolerance had been among the most remarkable features of medieval Spain. Indeed, St Ferdinand, the 12th-century king of Castile, in central Spain, ruled over Christian, Arab and Jewish subjects, and styled himself "King of Three Faiths". The success of the crusade to reconquer Spain from the Moors ended this tolerance. Jews and Moors who refused to become Christian were expelled from Spain in 1492, and from Portugal in 1497.

Nominal conversion was not enough; inquisitions were set up by the clergy to purge these "New Christians", and as a result Jewish and Moorish minorities were removed from the peninsula, though isolated communities survived for some time in remote mountain districts. However, the Gypsies, who arrived in the peninsula at about this time, though of Asiatic origin, seem to have escaped the attention of the inquisition.

The reconquest of Spain was carried out by the united kingdoms of Castile and Aragon, and its subsequent history was one of political consolidation and cultural assimilation. The greatest challenge to Castilian domination came from the Catalans in the east of the region. During the Middle Ages the kingdom of Catalonia had risen to prominence as the hub of a Mediterranean trading empire that included the Balearic Islands, where Catalan is still spoken today. Catalonia's power ended just as Castile began its period of greatest glory, but cultural and linguistic differences were maintained. Despite Spain's ethnic homogeneity, these exist to this day.

Following their establishment as independent kingdoms, both Spain and Portugal took a leading role in the European age of exploration, and rapidly founded overseas empires stretching from the Americas to the Philippines. As a result, Iberian cultural influences are now disseminated worldwide: Spanish remains the dominant language of Central and South America, for example, and there are other close links. Both countries' Atlantic island territories – the Azores and Madeira in the case of Portugal, and the Canaries in the case of Spain – colonized in the 15th century, are culturally indistinguishable from the mainland. The Spanish towns of Ceuta and Melilla on the coast of North Africa also have great affinities with Spain.

A LAND OF CONTRASTS

Sausage making in a Spanish kitchen *(above)* The food of the region has its origins in simple peasant cooking – garlic is a favorite ingredient. Highly spiced sausages are often served in the popular *tapas* (snack) bars.

A crowd of men and boys *(right)*, many of them wearing white with red scarves and berets and clutching rolled newspapers as their only weapon, gather in the streets of Pamplona before the bulls are released to run to the bullring.

As a consequence of the reconquest, Spain and Portugal today have near-complete religious unanimity. Yet, although the overwhelming majority of the peninsula's inhabitants are baptized members of the Roman Catholic church, there are considerable regional discrepancies: religious practice is traditionally much higher in the north than in the south. In part, this derives from the south's centuries-long subjection to Islamic rule, as opposed to the north's almost unbroken Christian tradition.

This divide also reflects regional contrasts in cultural attitudes and ways of life. A conservative outlook and distrust of change are characteristic of the small, isolated peasant communities that predominate in the mountains of northern Spain and Portugal, whereas the great aristocratic estates *(latifundia)* of the southern plateaus were traditionally worked by landless day laborers, who were always ready to question religious and political authority.

Among the most isolated of the northern populations are the pastoral farmers of the Basque Country who graze their livestock on summer pastures in the high mountains of the Pyrenees. Traditionally, families and kin groups are associated with a particular house, often an isolated farmstead, where the male head of the household is preeminent. As elsewhere in Spain, land is handed down from parents to children, sons and daughters inheriting equally, though a son will normally take over the family home. Among the fishing and seafaring communities of Galicia, the women traditionally tend the fields while the men are away at sea, leaving domestic tasks and child-rearing to the elderly and creating an association of women and household that is also found in northern Portugal. Throughout the peninsula it is common practice for children to take both their parents' surnames but, whereas in Spain the father's name comes first, in Portugal the maternal name takes precedence.

Patterns of daily life

The diversity of the peninsula finds expression in virtually every area of daily life, even what people eat. The centuries of Arab occupation have left an indelible mark on southern cuisine. *Paella* – the characteristic dish of Valencia, in the southeast, consisting of chicken and seafood cooked with saffron-flavored rice, strongly resembles the saffron rice dishes of North Africa and the Middle East.

Other culinary borrowings have influenced local cooking: the Mediterranean diet of the south makes heavy use of the capsicums and tomatoes of the New World and, inland, the Castilian staples of game, roast meat and pulses are invariably accompanied by potatoes, also brought here from the Americas. Food is taken especially seriously in the French-influenced Basque Country. Here, all-male dining clubs meet regularly in order to try new dishes and, in some cases, sing their way through a meal.

Despite the growing popularity of hamburgers and other fast foods, Spain's eating habits remain distinctive. The main meal of the day is taken between 2 and 4 in the afternoon. A lunchbreak of two, even three, hours is customary, leaving time for the traditional *siesta*. Although this afternoon nap is now often taken only during the hottest season, the long lunchbreak persists; shops and offices do not reopen until 4 p.m. and close again about 8 or 9 at night – one reason why the evening meal is rarely eaten before 10 p.m.

Most popular in the south of Spain, bullfighting attracts *aficionados* all over the peninsula, even though many now advocate the Portuguese form of fighting, which spares the bull's life. However, the national sporting passion today is soccer. Matches have long provided an outlet for nationalist feeling, and rivalry between Barcelona – the leading Catalan team – and Real Madrid, the leading Castilian team, is particularly fierce.

Folk song and dance is far from being just a cultural curiosity in Spain. People still dance in the streets and squares on highdays and holidays, and the elegant

FESTIVAL OF BULLS

Every year, from 6 to 14 July, bulls are released early each morning in Pamplona, the capital of Navarre, in the north of Spain. They chase the men and boys of the town through the streets of the city in celebration of the feast of St Fermín, or Ferminus, a mythical Roman martyr who is credited with the conversion of Amiens in northern France. His cult was probably imported to Navarre from France during the 12th century, when Navarre passed by marriage to a succession of French rulers. Today he is the patron saint of Navarre and St Fermin's *fiesta* is of great national and cultural importance for the Navarrese.

This fiercely independent Pyrenean region, which was annexed to Spain only in 1516 and not fully incorporated until 1833, is the home of a reactionary political tradition, Carlism, that provoked two civil wars in the 19th century. Although Carlism has not been a significant political force since the 1930s, each July many Navarrese don the distinctive Carlist red beret and travel to Pamplona for the celebrations. San Fermín is the best known of the bull-runnings, but it is not unique: others are held in several towns in the neighboring Basque Country, in some of which women take part.

In all cases, bull-running is only a part of the festivities. Dancing, music-making, fireworks, as well as bullfights and religious services all combine in the *fiesta*'s mixture of sacred and secular celebration. Bullfighting, popular today in southern France as well as Spain and Portugal, probably has its origins in the animal combats and circuses of the Roman empire.

steps of the southern *sevillianas* are even to be seen in the discotheques of Madrid and Barcelona. The sound of the guitar characterizes both popular and classical music throughout the region. Along with the tamborine and castanets, this is a legacy of Arab music. Flute and drum bands in the northern Basque towns and bagpipe playing in Galicia and Portugal reflect other local musical traditions.

Differences of language

Although there are only two main dominant languages – Portuguese and Spanish – which correspond to national boundaries, a range of minority languages and dialects are spoken within the peninsula, all of which, with the exception of Basque, are derived from Latin, a legacy of the Roman occupation. In Spain, the dominant form of Spanish is Castilian *(castellano)*, which encompasses diverse regional accents and the southern dialect of Andaluz, occasionally classed as a discrete language. Virtually everyone in Portugal speaks Portuguese, which developed from Galician (*gallego*). Today, however, the older language survives in Portugal only as a dialect. A tradition of vernacular literature has been maintained, but *gallego* is now spoken almost exclusively in Galicia.

In contrast to the linguistic uniformity of Portugal, more than a quarter of Spanish people speak a minority tongue as well as, or instead of, Castilian. The most widespread of these is Catalan and its related dialects – *mallorquin,* the language of the Balearic Islands and *valenciano,* spoken in Valencia, though the latter is strongly influenced by Castilian. It is very similar to Occitan, the language of southern France, both having their roots in a common medieval past. The language of the Basques (known as Euskera) is the oldest living language in Europe, and was spoken in the peninsula long before the Roman occupation. It does not belong to the Indo-European family and appears to be unrelated to any other known language.

ADJUSTING TO CHANGE

The political campaign for separate identity fought this century by both Catalonia and the Basque Country has centered on the issue of language. On revoking the autonomy granted to both regions by the Second Republic (1931–39), General Franco (1892–1975), Spain's right-wing leader, banned the public use of Catalan and other minority languages, even in the few remaining areas where they were the only spoken language. Basque and Catalan speakers faced severe fines for using their native language, and schoolchildren in the regions were instructed to "speak in Christian". Although Franco was a native Galician, the use of *gallego* was similarly discouraged, though never prohibited.

Cultural repression took many forms. Displays of the Catalan and Basque flags were forbidden; Catalans were forbidden to dance the *sardana*, their characteristic communal dance, performed in a ring by both men and women to music played on traditional instruments; Basque names were Castilianized, and there were even cases of Basque inscriptions being erased from tombstones.

Since Franco's death in 1975, and the ensuing transition to democracy (1975–82) Spain's minority languages have been rehabilitated. A survey conducted in the mid 1980s found that, among members of the national minorities living in their native regions, the vernacular languages were spoken or understood by 46 percent of Basques, 86 percent of Galicians and a remarkable 97 percent of Catalans. All schools in Catalonia are now bilingual,

The water pump (*top*) is a popular place for women to exchange news and gossip as the clothes are washed. Such focal points play an important role in binding communities together, but do not always survive amid the pace of city life.

City life (*above*) gives unmarried women greater freedom than before. These girls have gathered in front of the cathedral in Barcelona to dance the *sardana*. This traditional dance, performed in a ring, is closely bound up with Catalan national identity and was banned under General Franco (1892–1975).

THE ANGLICIZATION OF THE COSTA DEL SOL

In the 1960s, tourists from northern Europe, particularly Britain, discovered Spain's Mediterranean coasts. By 1973 the appeal of sun, sea and sangria was such that more than 30 million tourists a year were visiting Spain. The Costa del Sol, on the south coast, was one of the first areas to be developed – new highrise resorts such as Torremolinos sprang up at regular intervals along the coast, completely swamping the small fishing villages and transforming the lives of their inhabitants. Significant numbers of British visitors, attracted by the good climate and cheap living, decided to settle, either permanently or semipermanently, and bought holiday villas or retirement homes.

Although the first northern visitors had caused considerable culture-shock – the sight of scantily clad girls on the beach brought much condemnation – television has since played a much more important role in accustoming the people of rural Spain to international (particularly American) influences. Longterm British residents are now primarily seen as a source of employment for the local population, working as building laborers and bartenders, domestic servants and hotel maids.

There is also a burgeoning professional sector, fueled by the housing market and the demand for banking facilities. As a result, a parallel English-speaking culture has developed: estate agents displaying house particulars in English are a common sight along the coast, and British residents have their own newsletter, *Lookout*.

A British-based culture also exists in Gibraltar, the tiny British colony on Spain's southern coast. Most Gibraltarians (two-thirds of the population, which also includes British military personnel as well as many immigrant workers) are of mixed Genoese, British, Spanish, Maltese and Portuguese descent. English is the official language, but Spanish is the language spoken in most homes.

the University of Barcelona teaches almost exclusively in Catalan, official administrative business is carried out in Catalan, and there is a flourishing newspaper and book publishing industry.

Unlike Catalan, which has a long literary history stretching back to the medieval period, Basque has until comparatively recent times been primarily a spoken language. Its future seems more uncertain, due largely to the problems non-native speakers have in learning this difficult and awkward tongue. Until recently its use was confined to the rural areas and small towns and villages of the Basque Country. However, it was revived at popular level and among intellectuals as part of the assertion of Basque identity during the separatist struggles of the 1960s and 1970s, and the teaching of Euskera is today becoming more widespread in Basque schools.

Rapid modernization

During the 1960s and 1970s both Spain and Portugal underwent rapid economic and social change, following the dramatic expansion of the service sector of the economy, largely as a consequence of the massive increase in tourist development along the coasts of Spain and on the Portuguese Algarve. At the same time, great increases in the urban population, particularly in the great conurbations of Madrid and Barcelona, accentuated tensions between the modern and traditional worlds. While the growth of the urban and tourist markets has led to a great increase in demand for traditional craft goods such as ceramic pots and tiles, this new mass market is resulting in a homogenized craft industry that is beginning to erode local traditions, whether of pottery design or architectural style.

Although much new wealth has been created in the cities, many areas of the peninsula, particularly in Portugal, are among the poorest in Europe. The shortage of employment opportunities on the land led many people, especially men, to emigrate in search of work: in the late 1960s more than 100,000 Portuguese emigrated every year; by the mid 1980s the figure was less than 15,000. However, although numbers of permanent emigrants have fallen significantly, it is still common for people to seek temporary employment abroad.

The absence of men for considerable periods, which is a particular feature of life in northwestern Spain and northern Portugal, inevitably affects family structures. However, although the wife acts as head of the household in her husband's absence, making decisions and controlling family finances, she is disadvantaged by his much-increased economic power when he returns. In the long run, therefore, work abroad probably strengthens, rather than subverts, the man's traditional patriarchal role at the head of the household.

In the countryside, ways of life continue much as they have done for many centuries, with the addition of modern conveniences such as tractors and televisions. The bustling, modern cities, however, are very different places, and many young people, both men and unmarried women, move to the cities for work (either at home or abroad), at least for a few years.

A more liberal society

The experience of urban life has, over the last thirty years, led to greater social freedom, particularly for women. Although the Franco regime expected women to remain at home and encouraged large families by offering prizes to those producing several children, the modernization of the Spanish economy and society made such a policy untenable into the 1970s. The legalization of birth control in 1978 merely reflected common practice; the average size of Spanish families had already fallen from 4 in 1960 to 3.02 in 1975. However, although divorce is now legal, the divorce rate remains low. While some couples choose to live together before or instead of marrying, this option is far less common than in northern European countries.

Few women worked outside the home before 1970 – an inevitable consequence of the Franco regime that made women legally subordinate to their husbands. Growing economic affluence and the return to parliamentary democracy freed ever-increasing numbers of women to enter the work force. As they did so, feminism became a significant political force and, compared to other European countries, Spanish women are now well represented in both parliament and the government.

Another related effect of increasing urbanization is a decline in formal religious practice. Since Franco's death the church has distanced itself from playing an active role in politics, and so anticlericalism, both urban and rural, has declined. The Roman Catholic church is now formally disestablished in both countries, and conflicts with the state are therefore few. However, divorce and abortion – traditionally forbidden by the church – and religious education remain sensitive issues among many sections of the population.

The myth and reality of Spain's Gypsies

Spanish Gypsies (*gitanos*) are romantic figures in European art and literature, epitomized in the exotic and colorful figure of Carmen made famous in the popular opera of Georges Bizet (1838–75), who dies outside the bullring in Seville, killed by a spurned lover. Passion and violence, those two most noted Gypsy traits, flourish under the burning Andalusian sun, but these enduring folkloric stereotypes bear little relation to the historical experience of Spanish Gypsies.

A nomadic, tribal people, originally from the Indian subcontinent, the Gypsies appeared in Western Europe in the early 15th century and soon reached Spain. Their lack of a rooted religious faith meant that their nominal conversion to Christianity was quickly achieved. Although they escaped the early attention of the inquisition, they later became subject to official persecution, and severe penal laws were enacted against them, together with repeated orders of expulsion. However, the Gypsies were never driven from Spain, although many took refuge in the mountains, forming bandit and smuggler bands.

The traditional Gypsy occupations of horse-trading, fortune-telling, peddling and begging are all common among wandering peoples. This nomadic lifestyle was the particular target of legislation and – in contrast to other European Gypsy populations – most Spanish *gitanos* live in settled communities. Nearly all of Spain's southern cities have sizable Gypsy quarters, the most famous of which is the Triana quarter of Seville, much glamorized in song and legend. However, Gypsy neighborhoods are invariably found in the cities' poorest, most rundown areas. Many Gypsies still make their living by begging, and modern ghetto occupations such as drug-dealing are increasing.

A rich cultural tradition

Gypsy culture has been a major influence in southern Spain. Traditions and festivals are still observed – each year the Gypsies of southwestern Spain make a pilgrimage to the shrine of Our Lady of the Dew at El Rocio, near Huelva. This is one of the most spectacular *fiestas* in the Spanish calendar. Traditional dress is worn, and many pilgrims travel on horseback or in decorated caravans.

Much typical "Andalusian" culture is of Gypsy origin, above all the arts of flamenco – vigorous rhythmic dancing, accompanied by guitar-playing and the distinctive *cante hondo* ("deep song") style of singing performed by the men. Women dancers often wear the traditional flounced "flamenco" dresses.

Gypsy performers dominate professional flamenco, and talent is widely believed to be a matter of blood: only "pure" Gypsies have the true ability to dance, sing or play flamenco. This emphasis on purity of blood is also found among non-Gypsy Spanish communities, though here Gypsy blood is defined as "bad blood". The expression *limpieza de sangre* (literally, "cleanness of blood") signifies the absence of Gypsy, Jewish or Moorish ancestry.

Among the *gitanos*, racial considerations are a factor in family relationships. Gypsies usually marry within the community: an old proverb enjoins Gypsy men to "have nothing to do with a faulty mare, nor a woman that is not pure-blooded". Female virginity is much prized, and Gypsy girls often marry very young by European standards, as early as 14. Despite the common cultural stereotype, Gypsy women are rarely promiscuous. Divorce is much frowned upon, and few Gypsies are involved in prostitution: a marked contrast to the easy virtue of Bizet's Carmen.

Far from the romantic image (*above*) of Gypsy life, these boys are playing in the squalor of a camp on the outskirts of Seville.

Annual pilgrimage (*right*) A line of decorated caravans makes its way to the shrine of Our Lady of the Dew, near Huelva, southwest Spain.

Gypsy girls and women (*below*), pilgrims to the shrine, wear traditional dress. Flounces and bright colors are typical of the Gypsy "flamenco" style associated with the culture of southern Spain.

NEIGHBORS AND RIVALS

ANCIENT CIVILIZATIONS · SAINTS AND SOCCER HEROES · PRESSURE TO CHANGE

The peoples of Italy and Greece, divergent in many ways despite being close neighbors, are united by a common Mediterranean climate and by a mountainous physical environment that is always difficult, often harsh and sometimes even dangerous and has strongly influenced the development of their peasant societies. For centuries both were at the center of European civilization; ancient Greek and Roman philosophy, politics, architecture and drama continue to influence other societies around the world, and Latin – the language of the Romans – has been a major influence on many European languages. In recent years both countries have had to contend with the stresses of rapid modernization, which has produced strong contrasts in lifestyles between endemic rural poverty and urban affluence.

COUNTRIES IN THE REGION

Cyprus, Greece, Italy, Malta, San Marino, Vatican City

POPULATION

Italy	57.5 million
Greece	9.99 million
Cyprus	687,000
Malta	349,014
San Marino	22,746
Vatican City	1,000

LANGUAGE

Countries with one official language (Greek) Greece; (Italian) Italy, San Marino, Vatican City

Countries with two official languages (English, Maltese) Malta; (Greek, Turkish) Cyprus; (Italian, Latin) Vatican City

Other languages spoken in the region include Albanian, Macedonian and Turkish (Greece); Albanian, Catalan, French, German, Greek, Ladin, Sard and Slovene (Italy)

RELIGION

Cyprus Greek Orthodox (80%), Muslim (19%)

Greece Greek Orthodox (97.6%), Muslim (1.5%), other Christian (0.5%)

Italy Roman Catholic (83.2%), nonreligious and atheist (16.2%)

Malta Roman Catholic (97.3%), Anglican (1.2%), other (1.5%)

San Marino Roman Catholic (95%), nonreligious (3%)

Vatican City Roman Catholic (100%)

ANCIENT CIVILIZATIONS

The ancestors of today's Greeks probably arrived in the region about 4,000 years ago. A distinctive civilization, known as the Minoan, developed on the island of Crete between 2000 and 1400 BC. It was succeeded by the Mycenean civilization on mainland Greece. One of the greatest pieces of classical literature, Homer's epic poem the *Iliad,* vividly describes the Myceneans' war against the Trojans.

The peak of early Greek civilization was reached in Athens between c. 500 and 400 BC. Much of modern Western philosophy, architecture, poetry, sculpture, and mathematics have their roots in this period. Political ideology, too, has been profoundly influenced by the ancient Greek concept of democracy, which, however, excluded women and slaves. The unit of social and political organization was the city-state, or *polis*. The Greek city-states founded trading colonies all around the eastern Mediterranean and as far as Sicily, Italy and southern France. As the city-states fell into decline, Alexander the Great (356–323 BC), the warrior-ruler of Macedonia bordering the northwest Aegean, carried Greek civilization deep into Asia as far as India.

The Romans, who took over the domination of the Mediterranean from the Greeks, began as a minor people settled in the hills above the river Tiber in central Italy. From there they progressively conquered the entire peninsula and then expanded into Greece. During the first three centuries AD, under the emperors, Roman power extended all around the Mediterranean and Middle East, and into northwestern Europe as far as the British Isles. Originally a people of peasants and soldiers, the Romans were the heirs of the Greeks in more ways than one, taking over their art, literature, architecture and pantheon of gods, and adopting and developing their form of urban civilization.

Christianity became the official religion of the empire in the 4th century. Too large to control effectively, and faced with insuperable pressure from the peoples on its borders, the Roman empire eventually collapsed in the 5th century. Among its lasting legacies are the Romance languages, derived from Latin, the language spoken and written by the Romans, still widely spoken in Europe today.

St Peter's basilica, at the heart of the Vatican City in Rome, is the world center of the Roman Catholic religion. The Pope celebrates mass, the most important rite of the church, at the high altar beneath the massive central dome.

Fragmentation and renaissance

The Roman, western part of the Roman empire collapsed first and fragmented in Italy into a collection of warring city-states linked only by religion (the Pope, the head of the western Catholic church was based in Rome) and the Italian language. During the 15th and 16th centuries, Italy experienced a period of great cultural activity (the Renaissance) when the cities of Florence, Venice and Rome became the centers for a spectacular flowering in scholarship and literature, science, and, particularly, the arts. The inspiration for this cultural revival lay in the rediscovery of Greco-Roman art.

Conquest and recovery

The Byzantine, eastern part of the Roman empire, which had been ruled from Constantinople from 323 AD, struggled on for another 1,000 years as the focus of Greek culture, language and Orthodox Christianity. But by 1453, when the Muslim Turks, originally from Central Asia, captured Constantinople (which they called Istanbul), they had already conquered most of Greece. For the next 400 years, the Greek people were ruled by the Ottoman Turks. The struggle for independence took from 1831 until 1913, and the modern boundary with Turkey was only established in 1923 when mass exoduses of people took place in both directions. For many Greeks, Istanbul is still in alien hands, while a small Turkish minority is trapped inside Thrace, in northern Greece. Turkish occupation has very few discernible traces, though the liking for spiced foods and sweetmeats, for strong, sweet coffee, and the daily male ritual of a visit to the village café for cards and argument are found throughout the former Ottoman empire.

Greek–Turkish rivalry lives on in the island of Cyprus. Some 80 percent of the population is Greek-speaking, and many of them seek union with Greece. This is fiercely resisted by the Turkish minority that, since 1974, has lived within an enclave in the north of the island. The wall that passes through the center of the capital, Nicosia, symbolizes the failure of the two communities to find common agreement. The population of Malta, the other independent island state of the eastern Mediterranean, is a mix of Arab, Italian and British influences, reflecting its history of colonization.

SAINTS AND SOCCER HEROES

Italy is twice as big as Greece and has a much larger population. A shortage of productive farming land and endemic rural poverty in both means that each has provided a steady stream of emigrants throughout the 20th century, particularly to the United States, Australia and South America. More than a third of the total Italian population emigrated, sometimes for a few years, sometimes for life, in the 100 years since the late 19th century, and typical Italian foods such as pasta, pizza and espresso coffee have now become almost universal.

Two languages and two churches

The close relation of the Italian language to Latin means that speakers of other

Latin-derived languages find it relatively easy to learn. It has an expressive, musical quality that lends itself to both song and literature – not for nothing is it the language of international opera.

Italian is the first language nearly everywhere in the country. However, there is a significant community of German speakers in the Alto Adige in the north of Italy; a smaller number speak Ladin, related to Swiss Romansh. Slovene is spoken by a minority around Trieste on the north Adriatic coast. In addition to Italian, many Sardinians speak Sard, a language whose origins are unknown. Catalan, Spanish and Arabic are also found here – a relic of the island's long history of invasion.

Greece is still more linguistically homogeneous, albeit with variations in regional dialects. It has its own alphabet. Modern Greek has been hampered by having two forms: *katharevouse*, a difficult mixture of ancient and modern Greek invented in the early 19th century, was the official language until 1976; the simpler spoken tongue, *demotiki*, has now replaced it in all spheres.

The churches of the two countries are also quite distinct and for long periods of their history were often implacable enemies. The original separation of the two was a consequence of the division of the Roman empire in 323. The Roman Catholic church followed the authority of the popes in Rome, while the Byzantine or eastern part of the empire acknowledged the authority of the patriarchs in Constantinople. There were also important differences of belief. Relations between the two were completely ruptured in 1054 and not restored until 1963.

The Greek Orthodox church was the repository of Greek national identity during the years of Turkish domination and the struggle for independence. Still deeply embedded in national culture, it is a strongly conservative force in Greek society. By contrast, the wider international outlook of the Roman Catholic church was not in sympathy with the nationalist movement in Italy. The papal states had to be forced into joining a united Italy in 1871, and the Vatican City remains an independent enclave. Although a high proportion of both Greeks and Italians claim membership of their national churches, active participation is much greater in Greece. The declining influence of Roman Catholicism in Italy may be measured by the contemporary birth rate of 1.3 children per woman, which is one of the lowest in the world. Family planning, in the past, was extemely rare in Italy (possible due as much to its implied attack on male virility as to the teachings of the church) but now it is very widely practiced.

Traditional communities

Until recently, the ordered stability of the family, often supported by the church, was the foundation of the traditional peasant cultures of Greece and Italy. Faced for many centuries with governments that seemed arbitrary, corrupt and distant, the secure and unchanging nature

Classical decay *(above)* Greece and Italy live in the shadow of their illustrious pasts – both a burden and a source of pride. Too many visitors and air pollution are damaging the Parthenon in Athens.

A national obsession *(right)* Opera was invented in Italy, and embodies the Italian passion for drama and musical expression. The Roman amphitheater at Verona makes a spectacular setting for a staging of Verdi's *Aida*.

Still a part of daily life *(below)* A Greek Orthodox priest stops for a chat in a village café. Unlike the Roman Catholic clergy, Greek parish priests (but not monks) are permitted to marry.

RESTORING PAST PRIDE

The Parthenon, the temple to the goddess Athena that was built between 447 and 432 BC and is the crowning glory of the Acropolis at Athens, is the most famous example of classical Greek architecture. It is both a symbol of intense national pride and a major tourist attraction. During its checkered history it has also been a Christian church and a Turkish mosque. In 1687 its very existence was threatened when gunpowder, being stored in the temple by the Turkish authorities, exploded.

Today its future is once again in question. Athens' heavily polluted atmosphere and rust from iron joints unwisely implanted earlier in the 20th century have caused structural damage. The enormous tasks of cleaning the marble, halting the erosion and replacing the iron joints with non-rusting titanium are already in progress. However, debate continues about whether the Parthenon should be preserved in a ruined state or restored to its original form.

Statues and sculpture taken from the Parthenon over the years are currently in museums around the world. Controversy surrounds the largest of these collections, the Elgin Marbles, now kept in the British Museum in London. The Greek government hopes that these will be returned and housed, along with copies of the remaining statuary, in a new Acropolis museum. These aims, however, require great intellectual, technical and financial resources as well as political goodwill if they are ever to be carried out.

of traditional society, offered a refuge from the uncertainties of public life. In some areas of southern Italy and in the mountain regions and islands of Greece, the age-old rhythms and customs of the Mediterranean village are still played out, apparently undisturbed by the outside world. The main meal is eaten in the middle of the day when all the family will gather to share it. After the afternoon siesta, the men come together to debate the news of the day, while the women bring chairs out into the street.

In Greece children are still named after their grandparents and names often carry echoes of past classical glory, such as Orestes and Aristotle. In Italy, children are still almost invariably named after saints. Relics and statues of the saints are readily invoked in the face of disaster, especially in southern Italy. In rural Greece, the name-day of every saint is celebrated in his or her particular chapel with a service and a party. Easter is the major event in the Orthodox calendar, celebrated with the roasting of lambs and traditional dancing. Other festivities include the New Year blessing of the sea at Piraeus and a number of other places.

In northern and central Italy, and to a much lesser extent in Greece, there is a parallel, equally venerable urban culture. The long histories of Italy's great cities – among the most beautiful in the world – have fostered a determined spirit of independent vitality among their citizens, and many of their architectural gems, such as those of Siena, Urbino, Venice and Florence, have escaped the depredations of an industrial revolution. Civic and local pride is strong in these towns, reflected in the retention of particular customs, such as the *corsa del palio*, the twice-yearly horse race run around the central square of Siena. Today it finds new expression in the strength of support, stopping little short of adoration, given to Italian soccer teams such as Inter Milan, AC Milan and Juventus (Turin).

Soccer mania may have its roots in the gladiatorial contests that were the focus of Roman urban life. It also derives from the value given to theatrical display. Physical strength, smartness and guile all win admiration on the soccer field. The most skillful players earn vast salaries and are idolized as true heroes as long as their sporting prowess lasts. Soccer may thus provide a rapid rise to success for boys from the poorest urban or rural families.

PRESSURE TO CHANGE

Since World War II Italy has made steady, and sometimes spectacular, progress in science, industry and technology. The city that best epitomizes the new face of Italian urban materialism is Milan, since 1970 one of Europe's boom cities. Its glassy skyscrapers show the mark of American influences, but a wealth of talented designers have created an international reputation for Italian design, working with flair and innovation in a number of fields from footwear and clothing, jewelry and cosmetics, to furniture and lighting technology. Greece has been much slower to react to modern developments. In both countries, however, inequalities between the regions and conflict between traditional and modern, urban values are the cause of considerable tension within society.

Three elderly women (*above*) from a village in the Greek Peloponnese pause to rest in the sun. In both Italy and Greece widows – even comparatively young ones – wear black throughout their remaining lifetimes as a sign of bereavement. Death is both respected and treated with familiarity; the graves of relatives, often carrying a photograph of the deceased, are well cared for, and visits to the cemetery are treated as important family occasions.

Engineering excellence (*right*) innovative design, imaginative styling and the quest for speed are all united in the Ferrari sports car, painted in the gleaming red colors of the Italian racing team. The Ferrari car is a potent symbol of Italy's remarkable economic success following World War II, which has done so much to change the face of Italian society.

Societies in transition

As postwar industrial developments got underway cities such as Milan, Turin, and Athens expanded dramatically, drawing young, capable workers from the countryside. Here they lived in uprooted, isolated communities, not always readily accepted by local inhabitants.

In the last decades of the 20th century, Italy ceased to be an exporter of people and became instead an importer. By 1990 an estimated 1 million people from Africa, particularly Senegel, Tunisia and Ghana, had come there to work, either as agricultural laborers in the south, or in small northern industries, while many found work as traveling vendors. Their growing numbers sparked off a hitherto invisible racism, and in the north regional political parties such as Lega Lombard and Lega Venta mobilized a popular movement of feeling that was directed against both African immigrants and southern Italians.

Clash of values

The *Messogiorno*, the hot, poor region in the south of Italy, has struggled with the problems caused by rural depopulation and an aging population for many years. With the decline of rural services, communities became so isolated that whole villages in mountainous parts of southern Italy and the Greek islands were completely abandoned. Similar problems, less well-known but equally significant, also exist in parts of northern Greece and most of the Aegean islands.

A MIRROR TO SOCIETY

Since World War II, the Italian cinema has been the strongest and most varied in Europe. Scarcely a year elapsed without an Italian movie winning a major award at the International Festivals at Cannes (France) and Venice, or an Oscar for the Best Foreign Film at the Academy Awards. Italian directors and several actors and actresses have become world famous in spite of the handicap of making movies in a language with limited international currency. The Italian cinema has faced up to the problems confronting Italian society in a way perhaps unequalled by any other institution in the country, producing a sustained stream of intense social and political commentary.

The poverty of urban and rural deprivation and the despair of ordinary people faced by a corrupt or autocratic system in both the Fascist era of the 1930s and the postwar period are represented by movies such as *The Bicycle Thieves* (Vittorio De Sica), and *La Strada* (Federico Fellini). In the early 1960s the emptiness and corruption of the modern, moneyed middle classes were dissected in *La Notte* and *L'Avventura* (Michelangelo Antonioni), and most famously of all in Fellini's *La Dolce Vita*. Such themes merged with the problems of the growth of Italy itself in, for example *The Leopard* (Luchino Visconti) and *Christ Stopped at Eboli* (Francesco Rosi). A later development was the trend for Italian cinema to analyse its own role as a creator of myth and reality. This inward-looking preoccupation had noticeable success in Tornatori's *Cinema Paradiso.*

Such profound reflections on the contemporary scene by Italian directors are leavened by self-mockery and often black comedy. This is why, however serious the message, Italian movies are invariably so enjoyable.

Cinematic neorealism Giuletta Masina in a scene from Frederico Fellini's *La Strada*, made in 1954.

Migration has had wide-reaching effects on the structure of family life, altering the balance between the generations. Left in the home village, members of the older generation are dependent on the money-check sent back by sons or daughters in Milan or Athens, Australia or the United States, and no longer exercise the direct authority they used to. Extended family ties have severed as young families in the cities set up on their own – though many still return to the family home for the weekly Sunday meal, if the distance permits, or at Easter and other annual holidays, bringing with them new ideas and forms of behavior.

In the traditional village community, relations between the sexes are strictly regulated, and unmarried men and women are unable to mix freely. In the warm Mediterranean climate, this has led to the custom, in small towns and villages, of the evening parade when separate groups of young women and men walk up and down the main street or square. Women remain under the protection of their fathers and brothers until marriage, then of their husbands. Their role in adult life is restricted to that of wife and mother.

As the more liberated behavior of the cities gradually percolated back to the countryside the traditionalists feared the threat to conventional morality. Greater sexual freedom and an increase in contraception, the growing strength of communism and the materialist values of urban society, all seemed part of a package of modern ideas from which it was impossible to select only the desirable items, and which challenged the very heart of the social order.

Perhaps the most obvious challenge to traditional values came from the growth of mass tourism. Each year, Italy receives 40 million tourists, and Greece 8 million – almost as many as their total populations. While this has brought financial benefits and increased employment opportunities in many places, particularly some of Greece's smaller islands, initial exposure to the more unrestrained behavior and conspicuous wealth of their visitors had dramatic impact. The older people often found these new ways distasteful, but the young frequently aspired to the same freedom. To this end they were increasingly prepared to act against the wishes of their parents, while younger women started to question their family orientated role. Both Italy and Greece now have active feminist movements.

Because mass tourism in Greece took off very suddenly in the 1970s, the problems it created have yet to be resolved. Uncontrolled tourist development on Greek islands such as Rhodes has resulted in huge hotel complexes being built near ancient sites or small fishing villages. As a result buildings and people who seem to belong to different worlds are found in close, not always happy, proximity.

Yet, in contrast to Italy, many aspects of Greek culture have proved to be resilient to international influences. Traditional Italian melodies seem now to be sung only for the tourists. Western pop has otherwise taken over completely as the popular form of music in Italy. In Greece the invasion is altogether less marked. Traditional music, with its haunting rhythms and musical structures that seem almost oriental and may be a legacy of the long period of Turkish occupation, can be heard almost everywhere, performed on traditional instruments.

Sicily – living with the past

At the hub of the Mediterranean world, Sicily links Italy with Africa and separates the eastern and western basins of the inland sea. However, instead of dominating the Mediterranean, the beautiful, sundrenched island of Sicily has been invaded and colonized by a succession of peoples for 2,500 years. Today its architecture and landscape show evidence of an extraordinary range of cultures – Greek, Carthaginian, Roman, Arab, Norman, Spanish and eventually Italian.

The Greeks founded most of the cities, leaving fine ruins in Syracuse, Taormina, Agrigento, Selinunte and, particularly, the beautiful temple of Segesta. The Romans, who concentrated their efforts in Sicily on turning it into a grain-producing area to feed the people of Rome, left surprisingly little tangible evidence of their long rule. Following the division of the empire, Sicily was fought over by east and west before falling victim to the Arabs who invaded from North Africa in the 8th century. They made Balerm (Palermo) their sumptuous capital, but of its 300 mosques, only a couple survive today. The legend of Orlando Furioso, which tells the story of the expulsion of the Arabs by the Paladins, or Christian Knights, has long been the mainstay of Palermo's traditional puppet theater; until recent times, scenes from the drama decorated every Sicilian cart.

The Normans, who migrated here from northern France in search of land for colonization, expelled the Arabs in 1061, thereby giving Sicilian culture a new burst of vitality. Greek, Latin and Arabic influences were all encouraged, and blended with Norman styles in the cathedrals they built. For more than 200 years, from 1189, Sicily was plunged in dynastic quarrels, being held in turn by the Hohenstaufen emperors of Germany, the count of Anjou and the kings of Aragon. From 1416 it became, with Naples, part of the Spanish crown, and this lasted until Italian unification. It was a period of neglect, when absentee landlords ruled over large estates (*latifundia*) worked by peasants.

The unification of Italy must have seemed like a new dawn to the people of Sicily, but it was not to be. The continuing poverty of the majority of the islanders meant that for the ablest – or most desperate – the only route to a better way of life was emigration to Milan, Turin or the United States. Remittances from the emigrants to their families in Sicily were an important source of income and a rare occasion for celebration.

Poverty and honor

Life is hard on Sicily, even today. For those Sicilians who are employed, work frequently involves exhausting hours of

Finely decorated carts (*above*), showing scenes from the classic battles of chivalry, were once common on Sicily, but now serve as tourist attractions.

The heat of the day (*right*) Men and animals in a timeless pose of exhaustion after labor. Life in Sicily remains harsh.

Freedom on wheels (*below*) In Sicily, as in small towns and villages throughout Italy, the noise of motorbikes and scooters is a hazard of modern life.

hot toil and a long trudge back to the village each evening. The 2,500 hours of sunshine each year brings frequent drought, and nature has also made Sicily dangerous: earthquakes are common, and in 1908 killed 100,000 people in and around Messina. Lava flows from Mount Etna have twice destroyed the town of Catania, despite invocations to St Agatha whose veil, when displayed to the lava, is supposed to halt it and so protect the vulnerable villages.

The neglect by centralized government that many Sicilians experience, has led them to create their own form of self-reliance, though this has had the effect of reinforcing their isolation and backwardness. The origin of the *onorate societa* (the honored society), or Mafia, goes back to the 18th century. After unification it became the backbone of Sicilian life, fulfilling the Sicilian desire for respect, honor and power, as opposed to wealth.

Despite periodic attempts by central government to dismantle it, the Mafia today allegedly controls the whole of Sicilian life, particularly in the west of the island, involving itself in theft, drugs, and even the running of cemeteries. Those who break the *omerta* – the code of silence and honor – are simply killed; it is, perhaps, no wonder that black clothes are still so much in evidence on the island. Tourists, it is said, have little to fear from the Mafia, however, and may even be protected by its existence from petty crime. They, at least, are free to enjoy the fine beaches, beautiful scenery and the splendid remains of Sicily's ever-present past.

COMING TOGETHER

MIXING OF PEOPLES · UNIFORMITY AND DIFFERENCE · LIVING WITH PROSPERITY

Germany, Austria and Switzerland lie at the historic crossroads of European culture. The southern parts of the region fell within the Roman empire, and it was in the Black Forest, at the center, that the Germanic peoples, who originated in the north, halted its further advance. The north–south split between Protestantism and Roman Catholicism also runs through the region. In the east, which lacks natural frontiers, the German-speaking peoples have at times been subject to invasion, or have sought to exert influence over the neighboring Slavs. Austria once dominated the peoples to the southeast, and German influence extended into Poland. In this century Germany's eastward expansion led to defeat in World War II and to the political disunity of its people until reunification brought new challenges in 1990.

COUNTRIES IN THE REGION

Austria, Germany, Liechtenstein, Switzerland

POPULATION

Germany	77.7 million
Austria	7.6 million
Switzerland	6.6 million
Liechtenstein	28,181

LANGUAGE

Countries with one official language (German) Austria, Germany, Liechtenstein

Countries with three official languages (French, German, Italian) Switzerland

Other languages spoken in the region include Hungarian, Italian, Serbo-Croatian, Slovene (Austria); Dutch, Greek, Italian, Polish, Portuguese, Russian, Spanish and Turkish (Germany); Romansh (Switzerland)

RELIGION

Austria Roman Catholic (85%), Protestant (6%), nonreligious (9%)

Germany Protestant (47%), Roman Catholic (36%), other and nonaffiliated (11.3%), nonreligious (3.6%), Muslim (0.02%)

Switzerland Roman Catholic (49%), Protestant (48%)

MIXING OF PEOPLES

Between 6000 and 4000 BC a knowledge of agriculture, which had originated in the Middle East, gradually began to spread into Central Europe, carried northward along the river corridors. In time it came to displace the hunter–gathering way of life of earlier peoples. There is considerable archaeological evidence to show that groups of settled farmers were linked by extensive trading networks.

A great expansion of population took place in the 1st millennium BC, with heavier land being opened up for farming. Iron-working developed in the area of the Rhine–Danube axis in the center of the region, sparking off a series of migrations among the Celts on the western edge of this area. During the 6th and 5th centuries BC their iron-based culture spread throughout the southern half of the region and beyond it as far as the Atlantic coast in the west and present-day Czechoslovakia in the east.

By the 1st century BC the Celts were succumbing to pressure from two directions. The Roman empire began to extend its power north of the Alps and established provinces in present-day Switzerland and Austria. At the same time Germanic peoples from Scandinavia and the Baltic gradually expanded into the territories settled by the Celts east of the Rhine and threatened the northern frontier of the Roman empire, preventing it from expanding beyond the Black Forest.The irruption into Europe of mounted nomads from the plains of Central Asia threw the Germanic peoples into confusion, pushing them southward into the territories of the Roman empire. Under this weight of people the empire began to crumble in the 4th and 5th centuries AD. In the vacuum left by its demise, the Franks, a Germanic people settled west of the Rhine, rose to prominence. Eventually, under Charlemagne (742–814), they united Western Europe in an empire that stretched from the Atlantic coast of France to present-day Bavaria in southern Germany, and included a large area of northern and central Italy. The Saxons in the north of Germany were conquered and converted to Christianity.

Growth and prosperity

This empire quickly broke up after Charlemagne's death, and the territories east of the Rhine fragmented into a loose federation of feudal states and duchies. Although they were under the authority of the German emperors, power was locally based, especially in the eastern borderlands. Renewed pressure in the east from the Magyars, whose raids extended deep into the region, was halted by their defeat by the Bavarians at Lechfeld in 955: by the 11th century the Bavarians had completed the settlement of the Ostmark, an area roughly equivalent to modern Austria.

Rapid population growth throughout the region put increasing pressure on land. As a result the lands to the east beyond the Elbe and the Saale rivers, occupied by the Slavs, were invaded and colonized between the 12th and 14th centuries. Their peoples were gradually assimilated into the Germanic culture with a system of planned towns.

By the end of the medieval period a number of relatively large and organized states had emerged in the region, such as Brandenburg (in the northeast) and Bavaria. Austria, under the Habsburg dynasty, had become the center of an empire in the east that included present-day Czechoslovakia and Hungary, and parts of Yugoslavia, and lasted until 1918. For much of this time it stood as a line of defense against the further extension of Turkish power in southeastern Europe. Resistance to Habsburg rule by the people of the Swiss cantons, or mountain communities, led to the emergence of Switzerland as an independent power by the 14th century. In the western parts of Germany a large number of smaller states remained, of which the Alpine principality of Liechtenstein is today the single surviving example.

Unity and division

Germany's 17th-century war of religion (the Thirty Years' War, 1618–48) between Roman Catholics and Protestants reinforced differences between these minor states, with some rulers taking one side and others allying with the other. Although the postwar settlement confirmed the independent status of 300 of these states, it inflicted enormous damage – the German population was reduced from approximately 21 million to 13.5 million. During the 18th century Prussia (the former state of Brandenburg) emerged as the dominant political power in Germany. Rapid industrialization in the 19th

Money and religion *(above)* A Swiss investor looks anxiously for news of the latest trading prices in a bank window. Reflected in the plate glass above him are the twin towers of the Roman Catholic abbey church of Einsiedeln, a famous pilgrimage center.

Memory of times past *(left)* An elegant couple take coffee in a Viennese cafe, the walls of which are hung with portraits of Austria's past soldiers and statesmen. Coffee, often served with lashings of cream, is virtually a national institution.

century led to its economic supremacy, and after victories against the Austrians in 1866 and the French in 1871, the unification of Germany was achieved under the leadership of Prussia.

The political upheavals of the 20th century meant that this unified German state lasted only until 1945. The post-World War II division of Germany into two separate states and two different political systems endured until 1990, when the failure of communist rule in the German Democratic Republic (East Germany) brought about the reunification of the peoples of East and West Germany.

Like the Germans, Austrians have also had to come to terms with the loss of imperial power in the 20th century; again like them, they have turned military defeat into postwar economic prosperity. The collapse of the communist regimes of Eastern Europe in 1989–90 held out for Austria the possibility of renewed political and economic influence in its former territories in Czechoslovakia, Hungary and Slovenia, with which it still had close cultural ties.

UNIFORMITY AND DIFFERENCE

One factor that unites the people of Central Europe is the use of the German language, almost totally predominant in Austria, Germany and Liechtenstein, and the most widely spoken language in Switzerland. Although standard or "high" German (*Hochdeutsch*) emerged in the 19th century as the language of administration, business and education, dialects are still widely spoken in Germany, and are a means of instantly placing individuals – Saxons have a flat, rather monotonous accent, Swabians speak with a drawl, and Berliners communicate with each other in a rapid slang.

Linguistic minorities in Germany include a group of Slav-speakers, estimated at about 30,000, in Lausitz, an area in the east on the border with Poland. There are some Danish-speakers in the north of Schleswig-Holstein. Linguistic diversity was increased by the influx of foreign, or "guest", workers from southern Europe and Turkey between the mid 1950s and the mid 1970s, and by the numbers of people, particularly from Iran and Sri Lanka, who have sought political asylum.

The Alpine countries of the region are more varied linguistically. Austria has small groups of Croat and Hungarian-speakers on its eastern boundary, and Slovenes in the southeast. Most people in Liechtenstein speak a local dialect (Alemanni) in addition to German, and a small community of Walsers, immigrants from the Swiss canton of Valais in the 13th century, still use a distinctive form of the language.

Linguistic diversity is especially high in Switzerland. Although some 65 percent are German-speaking, there are substantial blocs of French-speakers (18 percent) in the west of the country and Italian-speakers (10 percent) in the south. Romansh, a Latin-derived dialect, is spoken by less than 65,000, and Slovene by a smaller number yet in the east. These linguistic divisions do not match the way the Swiss people separate on religious lines, nor the rural–urban divide, and consequently do not weaken national unity and cohesion.

Aspects of religion

It was in Germany that the former monk Martin Luther (1483–1546), initiated the Reformation, which left the church throughout Western Europe bitterly divided between Roman Catholics and Protestants. Luther's actions, however, had more than purely religious significance. He was a prolific and accomplished writer: his translation of the New Testament books of the Bible into a particularly clear German vernacular style had considerable influence on the later development of German language and literature.

The religious wars in Germany did not end in victory for either side but with an agreement that the religion of every state, however small, should be determined by the form of religion practiced by its ruler. The result was to create a mosaic of faith that still endures today, with the Protestant heartlands lying in the north and east. The west is more mixed, with Roman Catholicism increasing in strength toward the south, especially in Bavaria. The recent immigration of large numbers of Turks has produced a significant Muslim minority.

In Switzerland, where new currents of Protestant belief developed in the 16th century, religious controversy also left its mark. Today the country is fairly evenly divided between Protestants and Roman Catholics. In Austria, however, the Habsburg emperors led the fight both against Islam in Europe and against the Reformers in the religious wars, and the country today remains overwhelmingly Roman Catholic.

In common with other Western industrialized countries, church-going is declining throughout the region, though it remains highest among Roman Catholics. For example, in the 1980s, less than 10 percent of the population of West Germany were regular church-goers, but the figure rose to nearly one-third among Roman Catholics. The decline was even more dramatic in East Germany, where religious practice was formally discouraged by the Communist Party. Nevertheless, the Protestant church there remained a focal point for political and intellectual dissent. Church pastors played a key role in the events that brought about the downfall of the communist government in 1989–90.

In the past, Protestantism was associated with the rise of industrial capitalism and was consequently urban-based. Roman Catholic areas were traditionally more conservative, rural, and family-centered. Religion dominated many aspects of life, including where people were educated and how they voted. These differences are becoming harder to detect today. However, Roman Catholics in southern Germany and Austria still exert considerable influence on social policies, especially over education and moral issues such as abortion.

A taste for living

All the countries of the region enjoy a rich variety of food and take pleasure in the social activities of eating and drinking – a legacy of their traditional farming cultures and continued agricultural prosperity. Austria is famous for its cafés where rich cream cakes and strong coffee are served; Switzerland for its dairy-based products, especially its cheese and its chocolate. Germany has tremendous regional variety in food, including countless kinds of sausage.

Much food and drink is consumed at traditional religious festivals. Many of these, associated with village feast-days, are very small, but Carnival (or *Fasching*,

A neon-lit heart *(above)* advertises the entrance to a brothel. Prostitution is legal in Germany; apartments are rent-controlled, and customer charges are regulated.

In praise of beer *(left)* As many as 5 million people attend the 16-day Munich Oktoberfest every year. Germans drink more beer than any other nation - Bavaria alone has 1,000 breweries, and brewing standards are fiercely guarded.

Martin Luther *(below)* attacked what he perceived as faults in the teaching of the Catholic church. He is painted by Lucas Cranach (1472–1553), who was a member of Luther's circle and initiated a new style of religious art in Protestant Germany.

as it is known in Bavaria) is often an elaborate civic display of organized chaos, the cost of which can run into thousands of marks. There are also wine and beer festivals – Germany is renowned both for its Rhine wines and for the quality of its beer.

Bavarians pride themselves on their regional individuality – their checkered flag is often more conspicuous on public buildings than the national flag. This regionalsim is partly a legacy of Germany's former multiplicity of states. Berlin, the new political capital, does not have the cultural predominance of London or Paris. Instead, a large number of moderately large cities act as regional centers – many German cities have internationally renowned opera houses, ballet companies and museums, for example. There are few national newspapers, and hundreds of regional ones. By contrast, Vienna continues to dominate Austrian life as it did in the days when it was the center of a large European empire.

MORE THAN CUCKOO CLOCKS

Harry Lime's remark, made in the 1949 movie *The Third Man*, that in 500 years of democracy and peace the Swiss have produced nothing more memorable than the cuckoo clock, is untrue in more ways than one. The cuckoo clock's origins are to be found in the Black Forest region of Germany. And Switzerland can boast a long list of writers, scientists and artists who have enriched the cultural life of Europe over the centuries. Swiss architects were responsible for some of the finest buildings of Renaissance Italy; the philosophical ideas of Jean-Jacques Rousseau (1712–78), born in Geneva, revolutionized European thought.

Its position at the crossroads between southern and northern Europe, and its multilingual character, have contributed to Switzerland's unique cultural role. At many points in its history it has provided a haven for writers and thinkers suffering persecution for their views. The novelist and political propagandist Madame de Stael (1766–1817), banished by Napoleon from Paris, made her home on Lake Geneva a center of European literary life. During the 1930s and 1940s the rise of Fascism led a number of German, Austrian and Italian writers such as Thomas Mann (1875–1955) and Ignazio Silone (1900–78) to seek asylum there.

In the 20th century, the Swiss-born Le Corbusier (Charles-Edouard Jeanneret, 1887–1965) was a major creative force in modern architecture. The works and ideas of the painter Paul Klee (1879–1940), the sculptor Alberto Giacometti (1901–66), and the dramatists Max Frisch and Friedrich Dürrenmatt (1921–90) have received wide international acclaim.

LIVING WITH PROSPERITY

Central Europe's postwar economic prosperity is thought by some to have created societies that are conservative rather than progressive. Today's cultural climate has been compared unfavorably with the intellectual energy of the 1920s and with the days when Vienna pioneered movements in 20th-century music, expressionist painting, architecture and philosophy. At various times the younger generation has been vociferous in denouncing what they regard as the unremittingly "middle-class" nature of society. This dissatisfaction perhaps lay at the root of the urban terrorism (led by the Baader-Meinhof gang) that rocked Germany in the 1970s; it also inspired the work of some of Germany's postwar artists, notably the novelist Günter Grass, the artist Joseph Beuys and the film director Rainer Werner Fassbinder.

The politics of the Cold War that placed Germany in the frontline of hostilities between East and West for 40 years probably made its people more aware than any other in Europe of the threat of nuclear war. In addition, atmospheric pollution has caused visible damage to its forests, which – celebrated in music and literature – have an almost symbolic link with national identity. Both these factors undoubtedly had great importance in the upsurge of feeling that led to the founding of the Green Party (*Die Grünen*) in the 1980s, and has bolstered the strength of the environmentalist movement.

Many young people have gone further in rejecting the materialistic values of the consumer society, choosing instead an alternative lifestyle. Since the 1980s West Berlin has been the center of a flourishing counter-culture, with its own alternative newspapers, foodshops, theaters and social centers. Drug addiction has become a particular social problem even in Switzerland where the pressures to conform are perhaps even greater than in other parts of the region. Householders, for example, may be liable for fines for failure to clear snow and ice from the sidewalk in front of their houses, and a plethora of local laws may control such things as the placing of washing lines or the color of frontdoors.

In 1991 the leading Swiss writer Max Frisch boycotted his country's 700th anniversary celebrations on the grounds that a hundred years of "bourgeois class dominance" had left the country "cheap and nasty". The celebrations took place in a climate of considerable self-doubt. Some even questioned whether some of the country's most hallowed national institutions, such as its citizen militia army and its role of international neutrality – the very essence for many of what it means to be Swiss – should continue.

The rural conservatism of some of Switzerland's inhabitants means that in two half-cantons women are still not allowed a say in local affairs. Within the region as a whole, however, women have long moved out the realms of *Kinder, Küche und Kirche* ("children, kitchen and church") to which they were traditionally confined by a male-dominated society. Nevertheless, in 1991 German women felt compelled to go on strike for a day to protest against prevailing male attitudes that, despite the fact that more and more women are working outside the home, means that they are still expected to carry out the bulk of household tasks.

New citizens

As their economies expanded in the 1960s, West Germany, Austria and Switzerland all sought to overcome labor shortages by encouraging the immigration of "guest workers", mainly from the Mediterranean countries: Spain, Italy, Yugoslavia and above all Turkey. These workers took on the most menial and poorly paid jobs, such as street-cleaning and catering. It was first assumed that most migrants would return to their countries of origin, but many settled permanently with their families. Switzerland responded to this by changing its

East German army caps *(above)*, now collectors' pieces, are heaped for sale by a Turkish vendor. But for many more, the profits of reunification were slow in coming. In the aftermath of change, forgotten tensions rose again to the surface.

History as performance *(left)* A replica of the Berlin Wall is demolished at a huge openair concert held in Berlin in 1990. Celebrities and others from all over Europe joined thousands of Berliners at this symbolic musical celebration of the city's reuniting.

laws. After 1970, when foreigners comprised 16 percent of the total population and 22 percent of the workforce, it was made impossible for foreign workers to acquire rights of residence.

A number of German cities now have large Turkish communities. Germany's relatively liberal immigration laws have also allowed quite large numbers of political refugees to seek asylum. The presence of these "new Germans" is welcomed by many as bringing an added cultural dimension to society, but increasing competition for jobs and services has fueled racial resentment. Second-generation members of ethnic minorities face continuing discrimination in the search for better jobs.

A PERSECUTED MINORITY

Before World War II the intellectual, scientific and artistic culture of Central Europe owed a great deal to the influence of the Jewish community, many of whom had migrated to the cities of Germany, especially Berlin and Frankfurt, from Russia during the 19th century to avoid religious persecution. The role they played, particularly in the universities and in fields of activity such as art, architecture and music, was out of all proportion to their numbers. Before Adolf Hitler (1889–1945) came to power in 1933, Jews constituted about 1 percent of Germany's population, but they held 12 percent of university posts. In Austria, too, Jews were drawn to the capital from all over the Habsburg empire. Jewish scientists and artists such as the psychiatrist Sigmund Freud (1865–1939) and the composer Arnold Schoenberg (1874–1951) helped to shape Vienna's role as a leader of European intellectual and artistic achievement in the early 20th century.

The implementation of Nazi laws forbidding Jews from holding public office had an immediate effect on intellectual life in both countries. Many Jewish academics and artists fled to Britain or the United States, along with large numbers of non-Jewish sympathizers, who were also persecuted for their liberal views. Those who who did not leave were deprived of their livelihood; thousands would later die in the Nazi concentration camps. Very few Jews returned after the war. Their loss to Central Europe's cultural life has been permanent.

Bridging the divide

Between 1949 and 1989 one part of the region stood out from the rest in its social and political organization. Within the German Democratic Republic of East Germany people became accustomed to all decisions being taken by the ruling Communist Party. Public opinion was monitored by the omnipresent State Security Police (*Stasis*) with its network of informers – estimated to be a third of the population. There was strict censorship and restriction of movement – the Berlin Wall that divided the city stood as a symbol of a regime that sought to coerce its citizens into acquiescence.

The state did, however, provide a wide range of social services. Housing was provided for all, albeit often of low standard, and rents were fixed at 1939 prices. Medical care and education were free, though access to higher education was strictly controlled to serve the objectives of the Party. Childcare facilities and nursery schools were available to all. Reunification in 1990 brought many changes and readjustments to both sides, not all of them easy ones. While East Germans were presented with a wider choice of goods in the shops, they were often unable to pay for them: the need to compete in the market for income, and to pay market rates for services they had previously received free, left many worse off. Greater individual freedom of choice was accompanied by the emergence of social problems, such as organized crime and drugs, endemic in Western societies, that had been suppressed by the rigid control of the *Stasis*.

The opening of the frontiers in 1989 brought an influx of immigrants to the west: arrivals from East Germany amounted to 3,000 a day in early 1990. West Germany was also a magnet for migrants from elsewhere in Eastern Europe, especially for Poles of German descent, who often spoke little German, and for Gypsies from Romania. This sudden movement of population contributed to an upsurge of racist feeling in both halves of the country. Poles on shopping trips were attacked by right-wing extremists at border crossings in the east, and gangs of German and Turkish youths battled on the streets of Berlin.

Christianity – one faith, many churches

Central Europe has played a major role in the evolution of Christianity, for it was here that the fundamental split between Roman Catholicism and Protestantism, which still divides it into two separate bodies of believers, occurred.

Christianity is today the most widely followed of the world's great religions. It is based on the teachings of Jesus Christ, who was born in Palestine, then under Roman occupation, in about 6 BC. Accounts of his life and his death on the cross, by which he is believed to have obtained eternal life for humankind, are derived almost entirely from the gospels of four of his followers, Matthew, Mark, Luke and John, contained in the New Testament of the Bible.

Christianity emerged out of the Jewish religion. It accepted the divine origin of the Jewish scriptures contained in the Old Testament of the Bible, but held that Jesus was the Messiah or messenger whom the Jews believed would come to announce the arrival of the Kingdom of God on Earth. A fundamental aspect of Christian doctrine is the concept of the Trinity: that God exists in three persons – the Father, the Son (who was born on Earth as Jesus) and the Holy Spirit, often represented in Christian art as a dove or flame. Christianity's essential rites (or sacraments) are baptism, the admittance to membership of the faith by a ritual washing with water, and communion (also called the Mass, or the Eucharist), the communal sharing of bread and wine.

From the beginning Christianity was a missionary religion that actively seeks to convert nonbelievers. The early Christians were persecuted by the Romans, who saw the religion as a threat to the authority of the empire. After the conversion of the emperor Constantine in 312, however, it became the official religion of the empire and spread rapidly within its territories around the Mediterranean and in Western Europe.

Christianity survived the collapse of the Roman empire, but it was left permanently divided between the western, Catholic church, based in Rome, and the eastern, Orthodox church, based in Constantinople. The Catholic church is ruled by a succession of popes, or bishops of Rome, who claim to be the successors of St Peter, one of Jesus's original 12 disciples; the Orthodox church by the patriarchs of Constantinople. Differences between the two branches were based on matters of doctrine and belief; relations between them steadily deteriorated until, in 1054, the heads of the two churches refused to recognize each other. The breach lasted until 1963.

As a result, the Catholic church during the Middle Ages consolidated its authority in western and northern Europe, and the Orthodox church in the eastern Mediterranean, including Greece, and in southeastern Europe and Russia. The growth of Islam from the 7th century curtailed the expansion of Christianity in the Middle East, Central Asia and Northern Africa, but the spread of European colonialism around the world subsequently took the religion in its various forms to the Americas, Australasia and many parts of Africa and Asia. Today it has more than 1 billion adherents around the world.

The Protestant churches

During the 16th century the Reformation led to the split of western Christianity between Roman Catholic and Protestant.

The Reformation began in 1517 when Martin Luther, a German priest who had become disenchanted with what he saw as a lack of real spirituality within the church, began to argue against papal authority. His doctrine that the individual man or woman is responsible for his or her own salvation became one of the principal tenets of the Reformed, or Protestant, churches.

The ideas of Luther and others, notably the Swiss reformers John Calvin (1509–64) and Ulrich Zwingli (1484–1531), quickly won supporters in many parts of Europe, particularly Germany, Switzerland, the Netherlands, England and Scotland, and the Nordic Countries. They were fiercely opposed by others. The Roman Catholic emperors of Austria and Spain spearheaded the struggle against the growth of the Protestant faith. The divisions led to wars of religion in France and the Low Countries as well as Germany.

The bitter centuries of confrontation have left deep scars that the more recent spirit of ecumenicism (drawing together) and reconciliation between the Christian churches has not entirely eradicated. Today's sectarian conflict in Northern Ireland, for example, still draws on the memory of past quarrels between Roman Catholics and Protestants. A number of strict Protestant sects, who advocated communal living, austerity and non-violence, sought refuge from persecution in Germany by migrating to North America. Today the United States contains a multitude of different Christian sects and churches.

Passion play *(left)* The village of Oberammergau, in southern Bavaria, is famous for its staging of a play telling the story of Jesus' passion - the last days of his life before his death on the cross. Villagers, chosen for their devoutness of character and upright way of life, take all the parts, and the play has been performed every 10 years since 1634 as a thanksgiving for delivery from the plague. Here the Last Supper, when Jesus shared bread and wine with his disciples, is being enacted.

The Lutheran cathedral *(above)* in Lübeck, north Germany. In common with all Protestant churches, the altar, or communion table, is bare of ornament, the walls are painted white, and the interior is light and uncluttered. Roman Catholic churches, by contrast, have much greater decoration and usually contain many statues of saints, in front of which candles are lit. The pulpit, on the right, is very imposing in this church - doubtless, many rousing sermons have been preached from it.

In celebration of music

Over 2,000 regular annual music and arts festivals are held in Europe, of which the Salzburg Festival is the oldest and one of the largest. Each July and August, this small Austrian city is taken over by up to 200,000 opera fans, two-thirds of them from abroad. They come to see lavish productions of the operas of Wolfgang Amadeus Mozart (1756–91) (who was born in Salzburg), Richard Strauss (1864–1949), and others, performed by the world's top orchestras and finest singers. During the festival there may be as many as 30 productions, some at the Felsenreitschule, an open-air auditorium hewn out of rock, and others at the Grosses Festspielhaus, which has one of the world's largest operatic stages.

Salzburg has developed a reputation for spectacular, expensive productions. However, under the artistic direction of the conductor Herbert von Karajan (1908–89), another native of the city, they gained a reputation for conservative interpretation. His death signaled a period of uncertainty for the festival, which – along with many others throughout Europe – has faced difficulties in reconciling high art with commercial success. Opera is among the world's more costly performance arts, and its top artists command large fees.

The first Mozart festival at Salzburg took place in 1877, 86 years after his death, but it was not until the end of World War I that his works began to be regularly performed. Mozart reached a peak of popularity in the second half of the 20th century. "The Magic Flute" (*Die Zauberflöte*) is perhaps the most popular of his operas. Its tale of the triumph of knowledge and reason, told with a mixture of seriousness and comedy, has lent itself to considerable reinterpretation over the years.

The triumph of reason Sorastro and the priests before the temple, in a performance of *The Magic Flute* at the Salzburg festival.

PEOPLE OF THE BORDER LANDS

A MEETING PLACE OF PEOPLES · RESILIENT CULTURES · COMMUNISM AND AFTER

Lacking natural frontiers, Eastern Europe has long been a meeting place for cultural influences. The region was settled by Slavic peoples from the east and Germanic peoples from the west, and before 1000 AD Christians from western Europe were competing for converts with those from Constantinople, leaving the region divided between Roman Catholicism and Orthodox Christianity. Invaders from Central Asia – Tatars (Mongols) and Ottoman Turks – later introduced Islam in the southeast. East and West have continued to contest influence in the region, most recently during World War II and Eastern Europe's subsequent subjugation to Soviet-dominated communist culture. The collapse of communism in 1989 allowed deep-seated rivalries between ethnic groups and nationalities to resurface.

COUNTRIES IN THE REGION

Albania, Bulgaria, Czechoslovakia, Hungary, Poland, Romania, Yugoslavia

POPULATION

Over 30 million Poland

10 million–25 million Czechoslovakia, Hungary, Romania, Yugoslavia

Under 10 million Albania, Bulgaria

LANGUAGE

Countries with one official language (Albanian) Albania; (Bulgarian) Bulgaria; (Hungarian) Hungary; (Polish) Poland; (Romanian) Romania

Country with two official languages (Czech, Slovak) Czechoslovakia

Country with three official languages (Serbo-Croatian, Slovene, Macedonian) Yugoslavia

Other languages spoken in the region include German (Czechoslovakia, Hungary), Greek (Albania), Romany (Bulgaria, Romania, Yugoslavia), Tatar (Romania), Turkish (Bulgaria) and Ukrainian (Czechoslovakia, Poland, Romania)

RELIGION

Countries with one major religion (BO) Bulgaria; (RC) Czechoslovakia, Hungary, Poland; (RO) Romania

Countries with more than one major religion (A,AO,M) Albania; (EO,M,O,RC) Yugoslavia

Key: A–Atheist, AO–Albanian Orthodox, BO–Bulgarian Orthodox, EO–Eastern Orthodox, M–Muslim, O–Other, RC–Roman Catholic, RO–Romanian Orthodox

A MEETING PLACE OF PEOPLES

The key to the mix of peoples, nations and languages in Eastern Europe lies in its history of migration and invasion. In the 5th and 6th centuries AD, Celtic and Germanic peoples living in the west of the region migrated farther to the west, and the Slavs, originating in the Carpathian Mountains, replaced them. During the 9th century the invasion of the Magyars, a Finno-Ugric people, into Hungary separated the Slavs into two distinct groups who grew further apart with time. In the centuries that followed the western Slavs were drawn both culturally and politically into the orbit of western Europe; the southern Slavs, who were subsequently invaded and dominated by Mongols and Turks, looked to the east.

As a result, the region's Slavs are divided into two major linguistic groups. In the west and north are Poles, Czechs and Slovaks; in the south Bulgarians, Serbs, Croats, Macedonians and Slovenes (the last four found today within Yugoslavia). All these Slavic peoples speak distinct but related languages, which to a greater or lesser degree are mutually intelligible. Hungarian, or Magyar, is related only to Finnish and to two Siberian languages. In the east of the region, Romanians, isolated by the Slav and Mongol invasions, speak a Latin-derived language that is a relic of the Roman empire. Albanians are descended from the Illyrians, an ancient people settled in the mountains of the southwest from at least 1000 BC. Their difficult and remote terrain has enabled their language and culture to escape assimilation.

A common history of invasion and linguistic and cultural closeness did not necessarily result in friendship between the emergent nationalities of the region. Serbia (today part of Yugoslavia) and Bulgaria share the Eastern Orthodox religion, the Cyrillic alphabet and many food customs. Yet Serbians and Bulgarians have a 1,000-year history of rivalry, and both looked back to a Golden Age before they were eclipsed by Turkey. After regaining their independence in 1878, they each tried to incorporate, at the other's expense, lands and territories that they had ruled at their peak. Finally, in 1918, these conflicting claims were decided in favor of Serbia on its becoming part of Yugoslavia.

Cultural rebirth

For 500 years or so before the beginning of the 20th century political power in Eastern Europe was contested between the four competing empires of Austro-Hungary, Russia, Turkey and Prussia (later Germany), who imposed their own languages and alphabets. As nationalist movements opposing these occupying powers took shape in the 19th century, the peoples and groups of the region began to rediscover their languages and cultural traditions. Music and literature were revitalized. Folk tunes were the inspiration for internationally renowned composers such as Franz Liszt (1811–86), Béla Bartók (1881–1945) and Zoltán

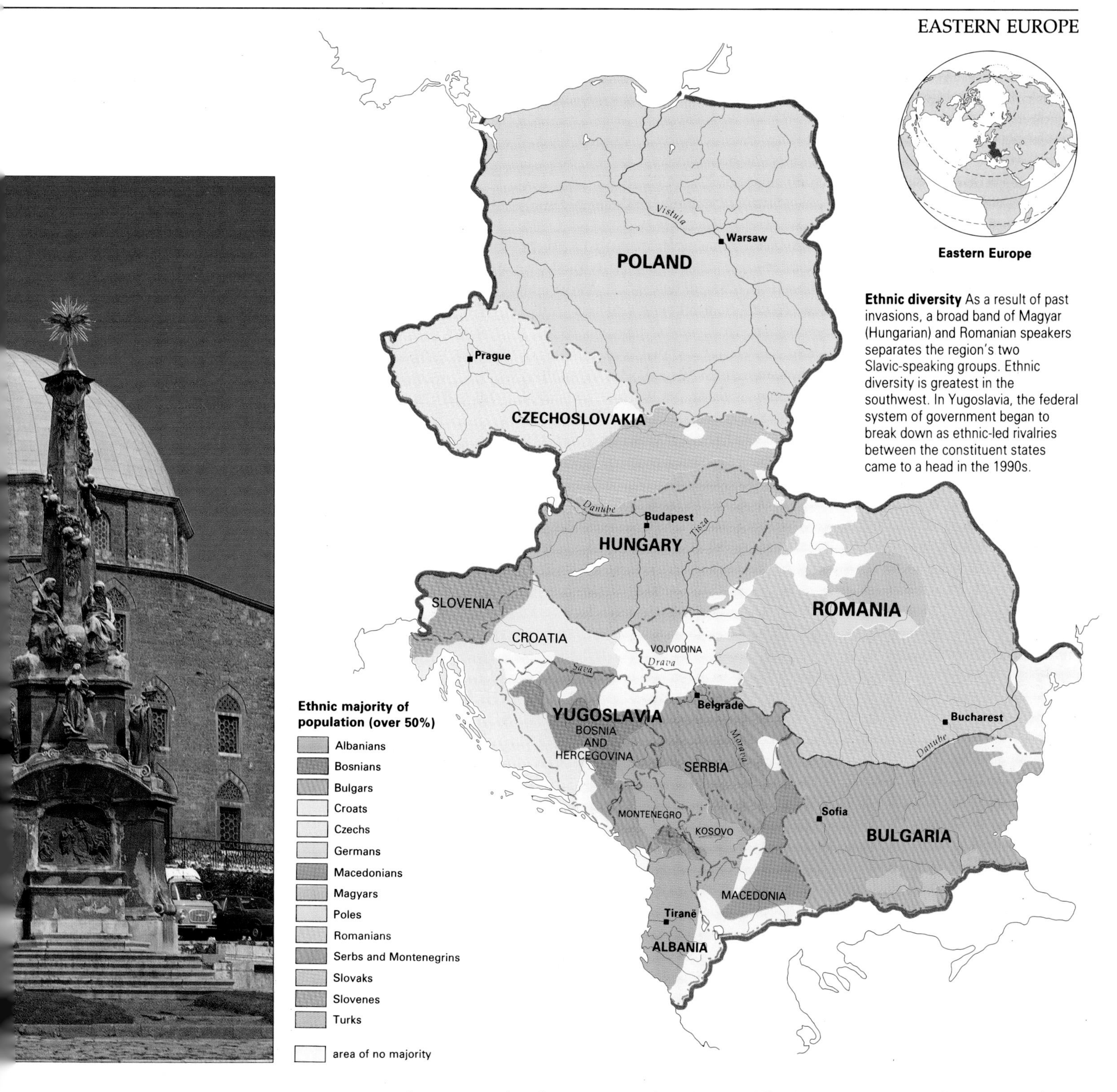

Ethnic diversity As a result of past invasions, a broad band of Magyar (Hungarian) and Romanian speakers separates the region's two Slavic-speaking groups. Ethnic diversity is greatest in the southwest. In Yugoslavia, the federal system of government began to break down as ethnic-led rivalries between the constituent states came to a head in the 1990s.

Christianity and Islam (*above*) meet on common ground. This building in Pécs, Hungary, is now a church, but as the details of its architecture reveal, it was once a mosque. The Islamic crescent surmounting its dome has been topped by a Christian cross.

Kodály (1882–1967) in Hungary, Georges Enesco (1881–1955) in Romania and Bedřich Smetana (1824–84), Antonin Dvořak (1841–1904), Leoš Janáček (1854–1928) and Bohuslav Martinu (1890–1938) in Czechoslovakia.

Problematic boundaries

Although the present-day countries of Eastern Europe were created at the end of World War I on the supposed basis of national self-determination and common language, the linguistic and cultural demarcations were often far from clear and the new national boundaries were set with little or no regard for the inhabitants. Far from satisfying nationalist aspirations, the post-World War I territorial settlement in Eastern Europe created new tensions. Peoples of diverse, often unsympathetic, backgrounds were thrown together as one country. Yugoslavia stands out in this regard.

Linguistic groups were divided. Turks were settled throughout Bulgaria and Yugoslavia, and until 1939 Yugoslavia contained large numbers of Germans. Czechoslovakia embraced a large German minority, as well as Hungarians and Ukrainians. Romania still contains a large number of Germans, Yugoslavia a substantial Albanian population, and both of them large pockets of Hungarians. It was this situation of ethnic confusion that allowed Germany and Russia to move into Eastern Europe in 1939 to extend its protection to the German-speaking minorities in Czechoslovakia and elsewhere, and Russia to protect Ukrainians. The aftermath of the World War II conflict brought massive movements of people throughout the region. With the subsequent imposition of communist rule, ethnic and national differences were submerged in the attempt to create a uniform socialist culture.

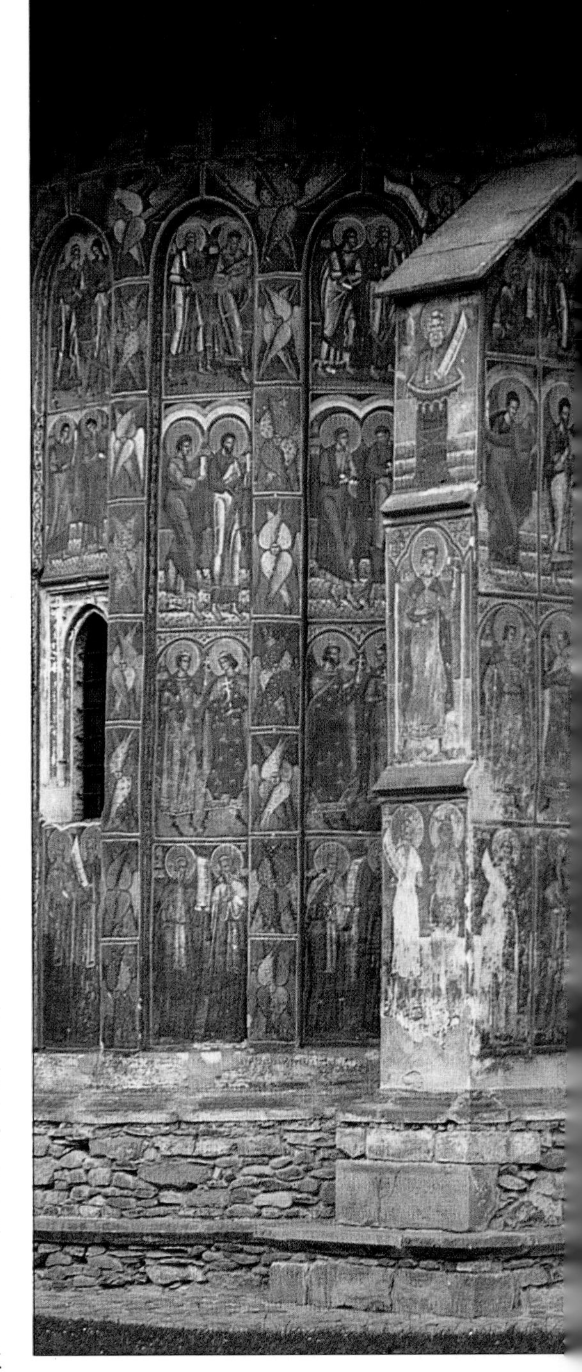

RESILIENT CULTURES

The distribution of ethnic groups and languages within the region today is the result of the refixing of boundaries and resettlement of populations that took place in 1945. The greatest changes were seen in Poland. In 1939, ethnic minorities accounted for one-third of Poland's population. They included Ukrainians, Germans and Jews. One in six Poles died in World War II. Forced deportations uprooted millions more, and the country's population dropped by 9 million. Most of Poland's large prewar Jewish community was exterminated. Virtually all that remained emigrated after the war.

In 1945, Poland's borders were moved westward into what had been Germany. Some 5 million Poles replaced 7 million Germans, and Poland's Ukrainian population was deported to the Soviet Union. Under communism, the population of Poland was officially of a single ethnic composition; since the collapse of communism in 1989 the minority communities of Germans (numbering about 500,000) and Ukrainians have been granted the right to their own language.

Other countries show much greater ethnic diversity. Czechoslovakia remains linguistically divided: two-thirds of the population speak Czech; most of the rest speak Slovak, and there are also Hungarian and German minorities. Slovaks still tend to regard themselves as being dominated by the Czech majority, and have called for greater autonomy.

Nearly all the region's ethnic minorities suffered from the policies of forced assimilation carried out by the communist regimes. In Romania President Ceausescu (1918–89) favored the traditional Romanian culture at the expense of minority cultures, and his increasingly repressive methods forced many of the country's 2 million Hungarians and 400,000 Germans to emigrate. In Bulgaria every effort was made to eradicate the influence of over 1 million Turkish inhabitants: Turkish customs, traditions and languages were forbidden; even Turkish names were outlawed. After 1989, however, these policies have been reversed.

Growing dissension

The granting of minority rights by postcommunist governments, however, cultivated public awareness of the ethnic differences, enflaming old resentments, renewing tensions, and kindling fears among leading ethnic groups of intervention by interested stronger neighbors on behalf of the minorities – Turkey in the case of Bulgaria, and Germany in the case of Poland. Nowhere did ethnic feeling run so high as in Yugoslavia, where the country has four official languages, all Slavic: Serb, Croat, Macedonian (considered a dialect of Bulgarian until 1944) and Slovene. Serb and Croat are almost identical, except that Serbian uses the Cyrillic alphabet and Croatian the Latin. However, Croats are Roman Catholic and Serbs are Serbian Orthodox, and the country also contains a large Serbo-Croat Muslim community.

Despite attempts to defuse the tensions inherent in Yugoslavia's diverse ethnic composition by organizing government on a federal basis, from the late 1980s on there was increasing ethnic conflict. A major factor in this was the political domination of Serbia, the largest and most powerful of the provinces. This was resented by the other provinces, particularly Slovenia and Coatia. The most prosperous parts of the country, Slovenia and Croatia also objected to subsidizing the poorer south. In Serbia, the communist party attempted to stay in power by taking a strong nationalist stance. This fueled the demand of both Slovenia and Croatia for independence, resulting in civil war.

Religious diversity

The spread of religions within Eastern Europe is no less varied than that of languages and ethnic groups, and for the same reason – within this border region – successive invaders have left the imprint of their religious beliefs. The countries in the north and center of the region – Poland, Czechoslovakia and Hungary – are predominantly Roman Catholic: their emergence as medieval kingdoms about 1,000 years ago coincided with their conversion to Christianity by missionaries from western Europe. All the countries contain Protestant minorities.

The conversion of the southern Slavs was carried out in the 8th and 9th centuries by monks of the Eastern Orthodox church sent from Constantinople: the Cyrillic script, based on the Greek alphabet, was devised in order to introduce them to the Bible. Orthodox Christianity remains the dominant faith in the south of the region. The religious picture in Romania is particularly complex: as well as the Romanian Orthodox church there are Protestant sects and 1.5 million Roman Catholics, most of them Hungarians or Germans. Some 1.5 million Eastern rite Catholics (Christians who use the Orthodox rite but recognize the authority of the Roman Catholic pope) were forcibly incorporated in 1948 into the Orthodox church.

Isolated mosques in Romania and a single Muslim community in northeastern Poland are reminders of the Tatar invasion; a larger minority of Muslims in Bulgaria reflects the centuries of Ottoman occupation. In Yugoslavia, the legacy of invasion is still more marked: in the south, some Serbs and Montenegrins converted to Islam under the Ottoman

Showing no emotion *(above)* The bride at this Albanian wedding fulfills her family's expectations and satisfies tradition by suppressing her feelings, at least until the ceremony is over. Practicing and passing on such rituals is intensely important to minority cultures surviving in pockets amid larger groups of people. Weddings in particular often provide excellent opportunities to assert cultural identity.

The Tree of Jesse *(left)* covers the exterior of this Orthodox church in Moldavita, Romania. According to the Bible, Christ was born into the house of Jesse, and the paintings represent Christ's genealogy as a huge family tree peopled with figures from scripture – a favorite theme of medieval religious art.

Gypsies in Romania *(right)* Estimates of Romania's Gypsy population vary between 100,000 and 250,000. Many speak Romany, which shows clear links with North Indian languages. With the collapse of communism, Gypsies once again found it easier to cross international borders, and many decided to travel west, particularly to Germany.

empire while others remained Orthodox. The Croats and Slovenes of the north, incorporated into the Austro-Hungarian empire from the 16th century until World War I, are mostly Roman Catholic. The Albanians, who had the longest unbroken experience of Ottoman rule, from the 14th century until 1912, are predominantly Muslim. Religion, however, was outlawed between 1967 and 1991.

Albania was an extreme example, but all the communist regimes attempted to secularize the population. Yet religion remained a potent symbol of national and cultural identity. For Poles, Hungarians and Slovaks in particular (though not to the same degree for Czechs) religious observance became a political statement. Only in Poland, however, was the church able to retain a voice in public life.

THE GYPSIES OF EASTERN EUROPE

Before 1939 there were as many as 1.5 million Gypsies in Eastern Europe, mainly in Bulgaria, Yugoslavia, Hungary and Romania. Some 400,000 were killed by the Nazis during World War II. Gypsies were subject to hostile treatment by the postwar communist parties who regarded them as potential sources of resistance. Official statistics tended to underestimate their numbers, making any accurate assessment difficult. Strenuous efforts were made to end their nomadic way of life, and travel within countries – let alone across borders – was severely restricted.

Gypsies have traditionally pursued a number of occupations on the margins of society, as horse traders and market sellers, violin makers, metal workers, tinkers and scrap dealers, and as itinerant circus entertainers, musicians and fortune tellers – roles that the communist rulers tried to eliminate. In some countries they also performed more specialized jobs, such as latrine cleaners in Hungary and undertakers and dog catchers in Romania.

Despite their history of persecution, Gypsies have had tremendous cultural impact on the region: Gypsy folk themes have entered the musical tradition of the West through the works of composers such as Liszt, Kodaly and Enesco. Because they did not move into cities, and resisted the pressure from communist regimes to assimilate them, they were able to retain once widespread customs.

Cultural pride *(above)* A Hungarian couple in Romania have decorated every inch of their home with traditional designs. The Hungarian minority was persecuted by Ceausescu, and the movement against him began in the predominantly Hungarian town, Timisoara.

A relic of the Habsburg empire *(right)* Swimmers relax in the splendid spa pool of the Gellerts Hotel in Budapest. Built in the Art Nouveau style of the 1890s, it is typical of the grandiose architecture that embellished the cities of the Austro-Hungarian empire.

COMMUNISM AND AFTER

With the imposition of communist rule throughout the region by 1948, all social organization and activity, for whatever purpose, came under Party and state control. Unofficial organizations were forbidden, and it became impossible to hold gatherings without the authority of the Party or the state. The consolidation of power was accompanied by police terror and purges: property owners, dissident politicians, and suspected or real opponents of the regime were imprisoned and often executed.

Party regimentation touched closely on all aspects of daily life, both within and outside the workplace. Travel was restricted, and newspapers, books, films and television were strictly censored. Religious worship was outlawed, or at best allowed to continue under severe restrictions. Art and literature had to conform to Party standards, and in every country an all-powerful bureaucracy, characterized by petty rules and paperwork, ensured the system was upheld. Party influence over daily life even extended to areas such as sport – international success was sought as a means of showing to the world the superiority of the communist system.

Within this general picture there were variations in the extent of state control over private life. Albania was particularly repressive. All places of worship were closed, and contact with foreigners was eliminated. In Romania, under Ceausescu's policy of systematization, whole villages were destroyed and Romania's cities rebuilt in monumental style, including in Bucharest a grandiose presidential palace, along with luxury apartment blocks for Party loyalists. In the last decade of his regime, Ceausescu – apparently believing his policies were in the country's best interests – deprived Romanians of food, clothing, heat and electricity, ostensibly to pay off the country's foreign debt, while accumulating a vast personal fortune.

As part of his policy of cultural assimilation, Ceausescu encouraged the performance of traditional Romanian music and the wearing of traditional dress. In other countries, too, traditional folk culture received official approval. In Poland, special efforts were made to preserve the traditional wooden architecture of the mountain areas in the south of the country, and peasant handicrafts were supported by the state. The government also allowed the nation's baroque cathedrals and castles to be restored and maintained, but its country houses of the 18th-century, symbolizing feudalism, were left to fall into disrepair.

The communist experience did not produce complete uniformity. Despite food shortages and agricultural collectivization, national preferences for particular spices and foods remained: paprika, goulash and strudel in Hungary; sauerkraut and kielbasa sausage in Poland. Czechoslovaks and Poles continued to eat cream cakes served with strong coffee. Czechoslovaks still drank beer, Poles vodka and Hungarians wine.

More importantly, even though driven underground, a dissident political culture also survived in some places. In Czechoslovakia, the temporary relaxation of Party control during the Prague spring of 1968 saw a revival of film, theater and literature. During this brief period Czech writers, artists and intellectuals regained a worldwide reputation. Persecuted and

SYMBOLS OF NATIONHOOD

When the region's communist governments fell from power in 1989, the new leaders lost no time in abandoning the emblems of socialism that had been adopted as national symbols to mark the advent of communist rule. In countries that had been under German occupation between 1939 and 1945 the communist rulers had deliberately made use of prewar national symbols in order to proclaim themselves to be the legitimate inheritors of national sovereignty. In Czechoslovakia, for example, a communist star was placed above the two-tailed Czechoslovak lion. In countries that had retained their independence new emblems were created: in Hungary the royal shield was replaced with one enclosed within sheaves of wheat and topped by a star.

The search for a symbol to mark the break with communism provoked a prolonged debate in Hungary. Some proposed the old royal arms incorporating a shield showing a double cross and surmounted by the distinctive crown of St Stephen, an 11th-century king and national saint. The other proposal, the Kossuth shield, takes it name from the leader of the 1848 revolt against Austrian rule. Its supporters argued that its themes of republicanism and revolution were to be preferred to the monarchical and religious symbols of the royal shield. However, a panel of scholars declared that the crown had come to signify statehood rather than kingship, and the double cross tradition rather than religion, so the royal shield was chosen.

imprisoned or forced into hiding when the Party took even firmer hold again, Czech intellectuals nevertheless maintained the tradition of dissent, and it was the outspoken writer Vaclav Havel who became the country's first president after the overthrow of communism.

Reacting to change

Some of the effects of change following the region-wide collapse of communist power in 1989 and 1990 were instantly obvious. Uniform Party newspapers were replaced by a wide range of publications. Political posters and graffiti rapidly appeared on walls and hoardings everywhere. Billboard advertisements sprang up. Statues of Lenin and Marx were toppled, churches repaired, and new ones built. Foreign goods, formerly unobtainable, began to fill the shops, and private trading became widespread.

Inevitably, these enormous cultural and social upheavals brought confusion and division. Anticapitalist views began to be reflected in a growing feeling against foreigners, who were frequently seen as buying national assets too cheaply and profiteering at the expense of the work force. The absence of established political parties allowed extremists to seize the political initiative and try to capture votes by appealing to resurgent nationalist sentiments; the communist parties themselves split on nationalist lines; and long-buried political rivalries and disputes, neither forgotten nor resolved under 40 years of communist rule, reappeared in the vacuum left by its disappearance.

The revival of religion

Under the monolithic rule of the communist parties secularization became the official doctrine, and many people – even those retaining a private religious belief – became unfamiliar with the practice and tenets of religion. With the return of political pluralism and religious freedom, many long-concealed religious minorities reappeared, and the churches assumed new authority. This compelled many believers to reevaluate their attitudes to such moral issues as abortion – the most widely used method of birth control – and divorce. Those who sought to give the church a greater voice in national and political life urged that the laws dealing with these questions, as well as with profanity and morality, should be altered to conform to church views.

The Jews: a people dispersed

The oldest synagogue in Europe (*above*) is in Prague's Old Town. The synagogue is the traditional center of the Jewish community – a place of assembly and study as well as worship – but the community that once supported this building has now vanished.

Silent witness (*right*) The experience of Nazi persecution and communist rule means there are few family survivors to tend the burial plots in this overgrown Jewish cemetery. Today wealthy emigrés in the West help to restore monuments like these.

The Jews are members of a worldwide religious group that traces its roots back to the ancient Hebrew people of the Old Testament books of the Bible. The Jewish belief in a single God laid the foundations for the world's other two major monotheistic religions, Christianity and Islam. The complex set of beliefs and laws that comprise Judaism emerged between 3,000 and 4,000 years ago in the area of the Middle East now located in the state of Israel. They are contained within the Torah, consisting of the sacred books of scripture and their interpretation.

Fundamental to these beliefs is the conviction that the Jews are a chosen people, bound with God through a covenant of mutual loyalty and guided by formal rules of behavior that set them apart from other peoples. The Jews have retained this sense of cultural identity throughout a long history of dispersal and migration. Those who consider themselves Jews by descent (usually through the maternal line) are generally accepted as Jews, even if they do not observe the rituals of the Jewish religion.

The Jews of Eastern Europe

By about 500 AD the Jews were scattered throughout the Mediterranean world, and parts of Asia. Here they were, compelled to live in ghettos, and subjected to periodic persecution by the dominant Christian community. Between the 13th and the 16th centuries forced expulsions from these countries led many Jews to settle in Eastern Europe, particularly Poland. They were also welcomed in parts of the region then under Turkish rule.

As a result, the number of Jews in Eastern Europe before 1939 was larger than that of any other region of the world, despite a rising tide of emigration to the United States in the 19th and early 20th centuries. In Romania there were 750,000 (4 percent of the population); in Hungary 450,000 (5 percent); and in Poland just over 3 million (10 percent). Most Jews were urbanized, working in trade, commerce and the professions, dominating them numerically. Even Warsaw, the capital of Poland, was over 30 percent Jewish at the start of World War II.

Yiddish, a language based on German, was the lingua franca of the region's Jewish community, but was by no means universal. There were small communities of Spanish-speaking Sephardic Jews in Yugoslavia and Bulgaria, who had reached Eastern Europe from Spain in 1492. The Jewish cultures of Poland and Romania were particularly varied. Poland, in particular, contained many important centers of Jewish learning.

Before 1918, the Jews of Eastern Europe had few civil rights and were the victims of frequent violence. The collapse of power in the region after 1918 brought political and legal rights, and a rich and vibrant Jewish culture flourished in the interwar period. This embraced a wide range of political, cultural and religious views, including orthodox and secular, conservative and radical positions. Some Jews advocated greater cultural assimilation by promoting use of Yiddish; others supported the revival of Hebrew and emigration to Palestine.

Before World War II antisemitism had been particularly virulent in Poland and

Romania, though it existed throughout the region. During the war, most of Eastern Europe's Jews were exterminated under Nazi persecution, including nearly all Poland's Jews. In Bulgaria, however, virtually the entire Jewish population was protected from the Nazis, and even in Romania half the nation's Jews survived.

Although antisemitism was officially outlawed by the postwar communist regimes, religious practice was discouraged and those Jews who stayed in Eastern Europe became cut off from the mainstream of Jewish culture. Moreover, the commercial activities that had provided their livelihood were taken over by the state, and their property was expropriated. After the founding of the state of Israel, they were treated with suspicion by the authorities. Not surprisingly, most chose to emigrate. Today, all that remains are empty synagogues and deserted cemeteries. Only Hungary, with over 80,000 Jews, has a sizable Jewish population. Romania has less than 20,000, Czechoslovakia less than 10,000, while Poland has even fewer than 5,000.

The post-1989 reforms that brought greater freedom to Jews were accompanied by a resurgence of traditional anti-Jewish feeling among some Christians, many of whom were airing old grievances after decades of suppression. The reopening of international relations, particularly with Israel, offered the surviving Jewish communities increased financial support from contact with the descendants of emigrants anxious to rediscover their cultural heritage.

Folk art

Before the spread of cheap factory-produced goods and modern building techniques people in the countryside produced their own furniture and implements and built their own dwellings. Usually they decorated them, using local styles and materials, creating regional traditions of folk art. Although most such traditions have now disappeared in Europe, some still remain in Poland.

Unlike most of Eastern Europe, the Polish peasantry resisted the attempts of the communist government to collectivize their land and locate them in new settlements. This left a large class of private farmers, generally with small amounts of land, who were able to continue their past ways of life more or less unchanged until the present day. Horse-drawn plowing and harvesting by sickle are still common in parts of Poland, while local folk art traditions flourish in places such as Zalipe, known throughout the country for its painted wooden interiors and exteriors.

Folk traditions inspired many of the painters of the 19th century as they attempted to forge distinctive national schools of art and, under the communist government, the authorities tried to organize folk art competitions, scholarships and marketing systems. All this wrought a change on traditional crafts, orienting them to the urban and tourist markets. Such changes inevitably took artistic activity out of its original context and geared it toward commercial markets. Ceramic-making, textile-making, and embroidery, once skills used to enliven everyday household objects, are now turned out to please the eyes of foreigners and townspeople.

Peacocks and flowers Traditional motifs still decorate these farm buildings in Poland.

A SOCIETY IN CHANGE

A PATTERN OF PEOPLES · CREATING A NEW SOCIETY · LIVING WITH THE SYSTEM

Most of the numerous ethnic groups within the Soviet Union as it existed before 1991 came together through a gradual process of conquest by the Russian people, who expanded southward and eastward from their power base in Moscow between the 15th and the 19th centuries. These disparate peoples were held together by the autocratic power of the tsar. In 1917 this was exchanged for the rule of the Moscow-dominated Communist Party, which attempted to create a unifying Soviet character that would transcend linguistic, religious and cultural differences. National groups were effectively denied self-determination, cultural traditions were forbidden, and opposition suppressed. With the collapse of communism, differences of nationality assumed even greater significance.

COUNTRIES IN THE REGION

Estonia, Latvia, Luthuania, Mongolia, Union of Soviet Socialist Republics

POPULATION

USSR	278.7 million
Mongolia	2 million
Lithuania	3.7 million
Latvia	2.7 million
Estonia	1.6 million

LANGUAGE

Countries with one official language Estonian (Estonia); (Khalka Mongolian) Mongolia; Latvian (Latvia), Lithuanian (Lithuania); (Russian) USSR

Over 200 languages are spoken in the Soviet Republics. The principal nationality groups are Russian (48%), Ukrainian (15%), Belorussian (3%), Azerbaijanian, Kazakh, Tatar (each 2%), Armenian, Georgian, Moldavian, Tadzhik (each 1%), Bashkir, Chuvash, German, Jew, Kirghiz, Mordvian, Polish, Turkmenian. Chinese, Kazakh and Russian are spoken by minorities in Mongolia.

RELIGION

Mongolia Tibetan Buddhist (95%), Muslim (4%)

USSR Nonreligious and atheist (70%), Russian Orthodox (18%), Muslim (9%), Jewish (3%), plus Protestant, Georgian Orthodox and others

A PATTERN OF PEOPLES

The multi-ethnic character of the peoples of the region is immediately apparent in their enormous variety of languages, cultural traditions and religions. A diversity of traditional lifestyles reflects the broad range of physical environments; small groups within the Arctic Circle still practice a hunting–gathering way of life, and nomadic pastoralists range the grasslands of the central steppes. But since earliest times these Central Asian plains have acted as a corridor for successive migrations of people westward and southward, and the Soviet Union's cultural complexity was also the legacy of its long history of invasion and conquest.

Cultural diversity

The former Soviet government officially recognized the existence of almost 100 different nationalities, distinguished by language and locality, and further divided into hundreds of subgroups. The nationalities ranged in size of population from the 500 Aleuts living on the Pacific fringe of Siberia to 140 million Russians. Of the 93 nationalities listed in the 1979 census, only 23 were attributed with more than 1 million people.

Slavs are the largest linguistic and cultural group of nationalities within the region, both in numbers (they make up over 70 percent of the total population) and in political and economic influence. Originating in the Carpathian Mountains of Eastern Europe, they had moved into the western part of the region by the 7th century AD, and include, within the Soviet Union, Russians, Belorussians (White Russians) and Ukrainians. In spite of close linguistic, cultural and historical ties, deep-seated rivalries continue to exist between them.

The Russians, the largest of the Slavic nationalities, began to move eastward to fill a vacuum left by the collapse of the Mongol empire in the 14th century, into what came to be the Russian Republic, by far the largest of the constituent republics of the Soviet Union. During the early years of communist rule, under the Five-Year Plans inaugurated by Joseph Stalin (1879–1953), millions more Russians (as well as Belorussians and Ukrainians) were transplanted to the new industrial cities east of the Urals. As a result, the Russian population today is found concentrated in cities rather than rural areas throughout the region.

Central Asia is populated by a number of Turkic-speaking peoples, originally from Mongolia, who spread into the area they occupy today during the 6th century. They are the second largest group in the Soviet Union, though coming some way behind the Slavs. They include Kazakhs, Kirghiz, Turkmen and Uzbeks.

The Caucasus Mountains, between the Black and Caspian Seas, are particularly rich in contrasting cultures and languages. Protected by the remoteness of the mountain terrain they inhabit, these different peoples have resisted cultural assimilation by their neighbors. Some of them, such as the Georgians and the Armenians, have been settled in this area for thousands of years, and their written history can be traced back at least as far as Greek and Roman times. Soviet Armenia is the original homeland of a nation that once extended into large areas of Turkey and Iran, and the Armenian Orthodox church claims to be the oldest organized Christian church in the world. Other groups in this area belong to larger nations now divided by international boundaries. The Azerbaijanis are also found in northeastern Iran, and the Kurds in Iran, Iraq, Syria and Turkey. The people of Tadzhikistan speak Tadzhik, a language closely related to Farsi, which is the language of neighboring Iran and Afghanistan.

Recent migrations

Very few of the territories that comprise the Soviet national republics have homogeneous populations – in most of them Russians have come to form a substantial minority. This was partly the result of a deliberate policy at the end of World War II to weaken the ethnic minorities who were felt to have been disloyal to the Soviet Union during the German occupation – thus Germans in central Russia and Tatars in Crimea, remnants of the Mongol horde that overran Asia from China to the Black Sea in the 13th century, were transplanted to Central Asia. Among those who suffered most were the people of Latvia, Estonia and Lithuania who were only incorporated into the Soviet Union in 1940. Thousands of people were deported to labor camps or dispersed throughout the Soviet Union, and were replaced by Russian emigrants.

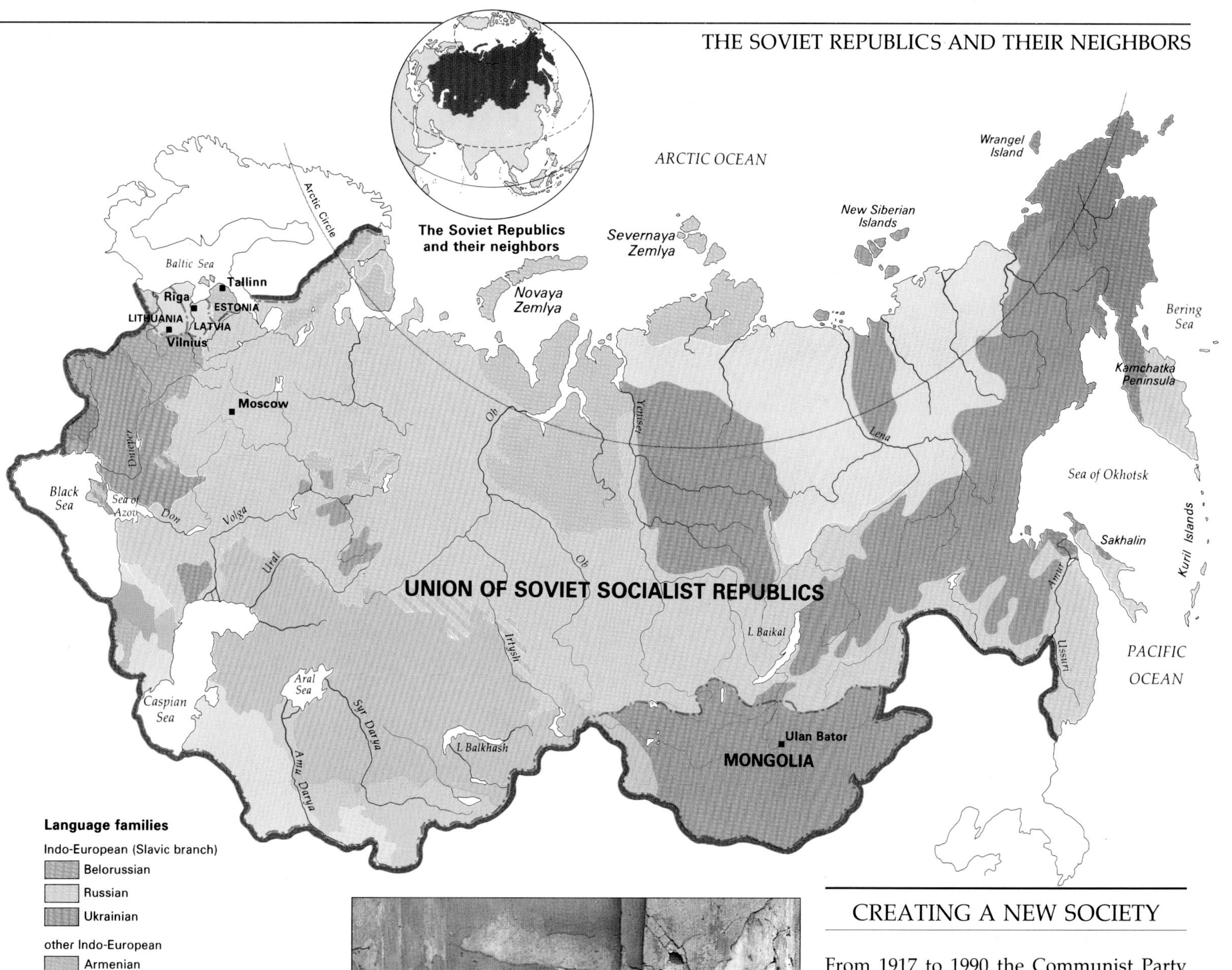

Language families

Indo-European (Slavic branch)
- Belorussian
- Russian
- Ukrainian

other Indo-European
- Armenian
- Latvian, Lithuanian
- Moldavian
- Tadzhik

Altaic (Turkic branch)
- Azeri, Turkmen, Yakut
- Bashkir, Tatar
- Karakalpak, Kirghiz, Uzbek
- Kazakh

other Altaic
- Manchu, Mongol

Caucasian
- Abkhazo-Adyghian
- Georgian
- Nakho-Dagestanian

Uralic (Finno-Ugric and Samoyedi branches)
- Estonian, Karelian, Komi, Lapp, Nenets

Paleo-Siberian
- Chukchi, Gilyak, Koryak

Language diversity *(above)* The numerous languages that are spoken throughout the vast region of the Soviet Union fall into several families, the largest of which are the Slavic and the Turkic.

Strong traditions *(right)* Two elderly Muslim inhabitants of Bukhara in traditional costume. Bukhara was, with neighboring Samarkand, once a leading intellectual and cultural center of the Islamic world.

CREATING A NEW SOCIETY

From 1917 to 1990 the Communist Party was the Soviet Union's sole political party. According to communist ideology, individuals left on their own cannot be trusted to act in the best interests of society. To be most productive, they need to be organized and controlled. The goal of the Party was to create a new Soviet citizen who favored the collective over the individual good, and worked in the interest of the state. Members of the party (among whom Russians were in the vast majority) were not supposed just to provide leadership, but were expected to embody Party standards of morality and behavior, setting an example to their fellow citizens.

Creating a Soviet society was thought to necessitate making the non-Russian nationalities, particularly those of Central Asia, catch up with the developed areas of western and central Russia by ensuring they become secularized, urbanized and Russified. Nomads were settled, peasants were collectivized and people were moved into the cities. Each nationality was to become part of this new Soviet culture, discarding customs that did not fit the Soviet way of life. To this

end, literature, theater and the arts were rigidly controlled and national traditions and motifs eliminated.

A policy of Russianization

For the Party authorities, advancement of the nationalities often meant cultural assimilation by the dominant group. Russian, the language of the dominant Slavic group, was the lingua franca and became the language of higher education and research. Although many nationalities developed a written language for the first time only under Soviet rule, and literacy levels rose (before 1917 only one in 500 Kirghiz could read or write), the non-Russian languages were reformed by introducing Russian words. In Uzbekistan, for example, the number of Russian words used in newspapers rose from 2 to 15 percent between 1923 and 1940. The Cyrillic alphabet of the Slavs replaced Latin and Arabic script in most regions. However, Georgia and Armenia were able to retain their distinctive alphabets, and Latin script is still used in the Baltic republics and, since 1990, in Moldavia in the southwest.

Many smaller minorities have been completely Russianized by these policies, and are separate nationalities in name only. They use the Russian language and have adopted Russian traditions. For many non-Russians, the only means of personal advancement was to join the Communist Party and conform to Russian cultural standards.

Russians and other Slavs were encouraged to migrate to the non-Russian republics. Today over one-third of the people of Latvia are Russian. Only one half are Latvians. In Kazakhstan, Russians account for two-fifths of the population, which also contains substantial numbers of Germans and Ukrainians. Kazakhs comprise less than 40 percent.

The religious dimension

Before 1917, in the tsarist empire, church and state were closely intertwined. In the Soviet state, religion was considered reactionary, and was actively discouraged. Religious ceremonies for marriages and funerals were replaced by new Soviet rituals. Religious holidays were replaced by secular ones. Religious groups were persecuted and driven underground. As attitudes changed with *glasnost*, the state came to permit recognized religions to function, but determined what church buildings they could occupy and how many priests they could maintain.

There are today at least 40 million Russian Orthodox faithful. They hold the beliefs and rituals of the eastern (Orthodox) branch of Christianity, which became the accepted faith of the Russian Slavs in the 10th century. Moldavians, Armenians and Georgians are also Orthodox Christians. Like the Russians, each of these peoples has its own self-governing national church, and jealously guards its distinct ceremonies.

Most Latvians and Estonians are Protestant Christians, reflecting past links with their Scandinavian neighbors. Lithuania and the western Ukraine, both incorporated within neighboring Poland for a long period of their history, share that country's predominantly Roman Catholic faith. In both Lithuania and Latvia the expression of nationalist feeling frequently led to antagonism with the Russian Orthodox church.

Among the Turkic-speaking peoples of Central Asia, Islam is the majority religion, having entered the region in the 7th century. Azerbaijanis are also Muslims. Although all outward signs of Islam were made illegal – traditional Muslim wedding ceremonies, for example, were forbidden – most Muslims continue to practice their faith, which is an inseparable part of their cultural tradition. Distinctive forms of dress have been retained, along with traditional customs.

The birth rate is high among these predominantly rural peoples, and the population growth rate is higher still, due

A SECULAR FESTIVAL

May Day, celebrated on the first day of May every year, commemorates the role of the workers in the struggle for communism, in addition to international brotherhood and peace. It has been one of the two most important holidays in the Soviet calendar, the other being 7 November, honoring the creation of the Soviet state and the Communist Party and marked by a military parade in Moscow's Red Square. The May Day celebrations were not so solemn, though in the past they were orchestrated by the authorities and the police. As well as the official Moscow celebrations, parades were held in the capitals of all the republics and in every provincial and district center.

Although it became so firmly associated with the Soviet system, May Day was one of the few Soviet holidays that predated the 1917 revolution – it originated in the United States in the 19th century – and one of the few that is also celebrated outside the Soviet Union. It was established as an official holiday to help to create a Soviet culture that would override religious and ethnic differences. Other public holidays commemorated New Year's Day, a Woman's Day, and an Army Day. In the past, May Day parades were carefully organized, but in the years of *glasnost* they became more informal. Dissenting political groups, such as those calling for greater national autonomy, were able to march freely with placards openly expressing their views.

to reduced infant mortality and longer life-spans. Divorce is rare, in contrast to other nationalities. The strength of Islam can be seen in the frequent outbreaks of violence that took place, for example, between Muslim Azerbaijanis and Christian Armenians, and also between Muslim South Ossetians and Christian Georgians, as the demand for greater national autonomy strengthened more and more from the late 1980s onward.

The religion that suffered most from Soviet repression was Judaism. Jews were recognized as a nationality but not as a religious group. They were allowed to have their own language – Yiddish – and their own alphabet, but synagogues and rabbinic schools were closed and religious observances proscribed. Added to this was a pervasive antisemitism, often supported by state policies, that created quotas for Jews in universities and the professions. Millions of Jews were exterminated by the Nazi occupation forces during World War II. Today most people classified as Jewish have no religious experience. However, it has become comparatively easier for those who wish to do so to emigrate to the West, particularly to Israel and the United States.

A living faith *(above)* Young girls in colorful dresses and headscarves are given instruction in a Qu'ranic school. Attempts by the Communist Party to suppress Islam were largely ineffective; most Soviet Muslims continued to practice their faith.

Reemergence of religion *(above right)* Russian Orthodox monks at the historic monastery of Saint Sergius in Zagorsk. The Orthodox church suffered greatly after the Russian revolution, but the ending of state restraints on religion revived its fortunes.

Sign of identity *(left)* A Moldavian activist affixes a placard announcing a political protest meeting in both Latin and Cyrillic script. The reintroduction of Moldavia's Latin script in 1990 reflected the upsurge of nationalist feeling taking place within the Soviet Union.

LIVING WITH THE SYSTEM

In the drive to create a Soviet culture, an independent legal system – the foundation of a society apart from the state – vanished. Laws, seen as just one tool in the development of a socialist culture and morality, were enforced selectively, ignored, or reinterpreted, officially for the benefit of the Party and the state, but more usually to suit the purposes of individual bureaucrats and local leaders. Under Stalin, the autocratic methods used to eliminate opposition drifted into arbitrary terror.

The Party and the state had not only the right, but also the obligation, to make sure that all art, sculpture, music, and literature embodied a properly socialist attitude. At times, national leaders intervened personally in cultural affairs. Stalin, for example, told the world-renowned composer Dmitry Shostakovich (1906–75) to make radical changes to his style of music.

In the words of Mikhail Kalinin (1875–1946), first president of the Soviet Union, the aim of the Soviet leadership was "to teach the people of the Kirghiz steppe, the small Uzbek cotton grower, and the Turkmenian gardener to accept the ideals of the Leningrad worker". It was in this spirit that the major nationalist epics of Azerbaijan, Kazakhstan, Uzbekistan and Turkmenistan were denounced by local communist leaders as works of religious fanaticism, Mongol epics were attacked as pro-feudal, and a Kirghiz epic was rewritten to remove all passages considered reactionary.

The histories of the nationalities were rewritten to show Russia in a favorable light. For example, a history of the Lithuanian Soviet Republic was denounced in 1961 for, among other faults, minimizing the struggle of the Russian people against the Mongols and for failing to stress Lithuania's pro-Soviet feelings between 1919 and 1940, when it was then an independent state. Non-Russian writers were expected to produce literary works that expressed the leading role of the Russian people: to do otherwise was unpatriotic to the Soviet Union.

As a consequence of these policies literature, music and the arts became very bland, and conformity became more important than originality. Artistic life was organized on the basis of professional unions, membership of which was mandatory. Artists and writers who did not conform to the offical line were unable to have their work produced.

A Soviet outlook

Although the Soviet leadership largely failed in its attempt to replace ethnic and cultural identities with a new Soviet culture, it succeeded in creating a distinctive Soviet outlook. The terror imposed by Stalin's methods of coercion left lasting scars on the national consciousness. People learned to mistrust the state and the organs of the state, particularly the police. Their mistrust was sharpened by chronic food shortages, exacerbated by laws forbidding private trading.

As a result, virtually every Soviet citizen, from a housewife to a factory manager, had to break the law. A culture with a double set of standards thus developed: one consisting of the public display of officially sanctioned opinions and attitudes, and another – shared among close friends and family – consisting of personal beliefs. Others found a

A more open society (*above*) Young Muscovites, wearing fashionable jeans and leather jackets like their peers in Europe and the United States, joke with a soldier. Before *glasnost*, it would have been impossible to imagine this scene taking place.

Unruffled by change (*left*) Turkish influences are clear to see as community elders in Tadzhikistan discuss the future of local schools while drinking tea. All the vicissitudes of Soviet politics failed to alter this traditional way of life.

Catching up on the news (*right*) Three supporters eagerly scan a sports newspaper for the latest results. Sports reporting was always an area of the news that was politically uncontroversial enough to escape heavy censorship by the authorities.

different escape route – alcoholism became widespread in the Soviet Union.

Throughout the Soviet Union a way of life prevailed that to some extent grew out of and adapted the methods already established by the all-embracing civil service of the tsarist state. All aspects of daily life came to be controlled by a bureaucratic system, characterized by paperwork, that created a network reaching to every district, blanketing town and country. Forms and passes became necessary to accomplish even the simplest tasks, and everyone had to carry an internal passport on their person at all times giving exhaustive details about their personal history.

Reform and collapse

After 1956, when Stalin's methods were publicly denounced by the Soviet leadership, the Soviet Union tried to come to terms with its recent past. However, it was not until the late 1980s that the Soviet Union began a move toward the creation of a civil society based on rule by law and an independent judicial system; in the new climate of *glasnost*, free speech was legalized, restrictions on religions eased and, in 1991, free market trading allowed.

These changes released a surge of popular feeling. New movements ranging from antisemitism to ecology sprang up, communist authority was widely discredited and there was much questioning of formerly accepted values and standards. The shift to free market trading and private ownership, for example, angered many people who saw it not only as a betrayal of socialist principles, but as the legalization of criminal activity. Others were led to reevaluate their lifestyles in the light of the growing influence of the church. When the church had no public voice, response to moral issues such as abortion, the widest form of birth control, did not matter. Today it does.

The final collapse of the communist state in 1991 made it possible for many nationalities to rediscover their past and to negotiate with each other independently of Moscow. Almost all the republics demanded greater autonomy, while Leningrad's citizens voted to return to the city's former name, St Petersburg. The newly independent Baltic states restored cultural links with the Nordic Countries, but remained economically tied to the Soviet Republics. They faced demands for autonomy from enclaves of Russians and Poles within their own borders, particularly in Lithuania.

The lifting of one threat – that of hardline communism – exposed another – interethnic strife. Tatarstan, for example, an enclave within Russia of 3.6 million people, many of them nonRussian Muslims, was typical of the areas that now demanded greater cultural freedom. As yesterday's minorities became today's majorities, the nationality question in the collapsing Soviet empire began to resemble a series of Russian dolls, nestling one inside the other.

A HIGH RATE OF DIVORCE

More than one in three marriages in the region ends in divorce – one of the highest rates in the world. There are variations in the pattern of divorce – the rate is highest in Latvia and among urbanized Slavs in the larger cities, where as many as one in two marriages fails. Divorce tends to occur early in the marriage – in Moscow, for example, one in ten marriages lasts under a year. Rates are lowest in Lithuania, where divorce is frowned upon by the majority Roman Catholic church, among the Christian Orthodox communities of Armenia and Georgia, and the rural peoples of Central Asia.

Alcoholism is blamed in almost half the cases of divorce, but stress is also a factor. Most women work full time and are financially independent. They are also usually responsible for domestic chores and childcare. Housing shortages cause strain since couples must often live in a hostel or share a small apartment with their parents until they find a home of their own.

People wishing to move to Moscow and other major cities require a permit. Often, temporary marriage to a resident provides a short cut to obtaining one. Similarly, since single people can rarely obtain an apartment, a marriage of convenience may be the best way of finding one. Finally, although no official stigma attaches to illegitimacy, or to abortion, which is the main form of birth control available, many women who are unable to obtain an abortion may get married for the birth of the child, and divorce later.

Art in the service of the state

Massive buildings in a neoclassical style, sculptures of men and women in heroic poses, and larger than life paintings glorifying the achievements of socialism sum up the official art of the Soviet Union. However, this monumentalist approach also existed in pre-Revolutionary Russia – nowhere were the imperialist aspirations of the tsarist state better symbolized than in the ornate palaces and classical buildings of St Petersburg, renamed Leningrad by the communists.

After the Revolution, Soviet architects and artists began to search for styles that could represent the life and achievements of the new state. Initially they turned to the avant-garde ideas of the Futurists and Constructivists, then in vogue among Russian artists, especially those that had trained and worked in Paris. Although they used a modern idiom, they thought in monumental terms, hoping to surpass the creativity of other countries.

They argued that in the ferment of new ideas and enthusiasms that the Revolution excited, the stale, outmoded images of the old order should be replaced by new socialist forms. Soviet architects began to formulate new, sometimes impracticable schemes, of which the most notorious was the tower designed by Vladimir Tatlin (1885–1953) as a monument to the Third Communist International. A model was exhibited in 1919, showing a striking design consisting of a leaning iron framework supporting a glass cylinder, a glass cone and a glass cube, each of which could be rotated at different speeds. The final structure, never built, was to be more than 480 m (1,300 ft) high. Frequently the state lacked the technology to build the projects. More often it lacked the money. Despite this, many international architects offered their services to the Soviet Union, planning factories, flats, offices and entire towns. They included the Swiss-born Le Corbusier (1887–1965), then just beginning to formulate his influential ideas on architecture, who designed the Tsentrosoyuz building in Moscow. In turn the innovative ideas of Soviet architects and artists such as the Vesnin brothers, Moses Ginsberg and Tatlin, won international renown, helping to establish the credentials of the new state.

A change of direction

In the late 1920s Stalin achieved a position of unopposed power. His taste in art and sculpture was conventional, and he also had a preference for neoclassical architecture. Modernism and experimentation were consequently declared anti-Soviet. They were replaced by socialist realism, approved by Stalin as the proper style to express the heroic goals and aspirations of socialism. Works of art and architecture were to be built as inspirational monuments to the socialist future, not just replacing, but dwarfing and outlasting, monuments from the capitalist past.

Moscow, the Soviet capital, became a showcase for the achievements of Stalin's rule. It was rebuilt in an imperial style with new, broad streets lined with monumental buildings. When Moscow's metro, or subway, was first opened in 1935, each station was an underground palace: corridors were lined with marble and mosaics, and platforms were paved with polished stone.

Excitement of the new *(above)* An architect's fantasy of domed palaces captures the post-Revolutionary spirit of innovation and experimentation.

Tatlin's Tower *(left)* A model of the giant iron and glass tower designed by the internationally renowned architect Vladimir Tatlin to commemorate the Third Communist International. The tower – like the International's dream of world communist revolution – was destined never to become a reality.

Monument to a dictator *(right)* Soviet architecture returned to more traditional monumentalist forms as Stalin's power increased. Moscow University, for example, was the largest of seven huge buildings built in the city during his lifetime. The connection between monumentalist architecture and dictatorship is well known; in the late 20th century Kim Il Sung in North Korea, Nicolae Ceausescu (1918–89) in Romania and Saddam Hussein in Iraq all conceived grandiose schemes that were intended to glorify and legitimize their power.

As Stalin's power grew, the scale of Soviet architecture increased. After World War II the rebuilding of the country's major cities, destroyed by the Germans, was planned on monumental lines, to symbolize the extent and nature of the Soviet victory. Prison labor was used extensively. Seven massive tower blocks were built in Moscow, the largest of which – Moscow University – is 240 m (800 ft) high and 480 m (1,600 ft) long.

Plans were made for a new Palace of the Soviets, which it was hoped would be the world's tallest and largest building, and would be topped by a colossal statue of Lenin, the Soviet Union's founding father. A cathedral was demolished to provide a site for the building, but the scheme was dropped after Stalin's death, when it was discovered that the site was too near a river for the deep foundations required. The excavated foundations were transformed instead into a vast outdoor swimming pool.

Monumentalist architecture came to typify Soviet life, its grandiose details and often gimcrack construction repeated in Party headquarters, regional congress buildings and factories throughout the country. Although such schemes are now a thing of the past, in many ways they remain Stalin's most enduring legacy, now that his statues have been toppled.

LIVING BY THE BOOK

BIRTHPLACE OF BELIEF · SHADES OF DIFFERENCE · SOCIETY IN FLUX

The cultural history of the Middle East has been one of innovation: its peoples were among the first to domesticate plants and animals and to develop urban settlements, and the world's three great monotheistic religions – Judaism, Christianity, and Islam – which take their authority from the divinely revealed books of the Bible and the Qu'ran, all originated here. Over the centuries civilizations and empires have risen to power and declined: the region's strategic importance as a route between east and west means that it has been fought over by outside powers. In the 20th century population growth and the discovery of vast oil resources have transformed what was a mainly agricultural society and exacerbated the ethnic, cultural and religious rivalries that frequently plunge the region into bitter dispute.

COUNTRIES IN THE REGION

Afghanistan, Bahrain, Iran, Iraq, Israel, Jordan, Kuwait, Lebanon, Oman, Qatar, Saudi Arabia, Syria, Turkey, United Arab Emirates, Yemen

POPULATION

Over 50 million Iran, Turkey

10 million–20 million Afghanistan, Iraq, Saudi Arabia, Syria, Yemen

1 million–10 million Israel, Jordan, Kuwait, Lebanon, Oman, UAE

Under 1 million Bahrain, Qatar

LANGUAGE

Countries with one official language (Arabic) Bahrain, Iraq, Kuwait, Lebanon, Oman, Qatar, Saudi Arabia, Syria, UAE, Yemen; (Farsi) Iran; (Turkish) Turkey

Countries with two official languages (Arabic, Hebrew) Israel; (Dari, Pushtu) Afghanistan

Other languages in the region include Armenian, Baluchi, Kurdish and Luri

RELIGION

Countries with one major religion (M) Afghanistan, Bahrain, Iran, Iraq, Jordan, Kuwait, Oman, Qatar, Saudi Arabia, Syria, Turkey, UAE, Yemen

Countries with more than one major religion (C,J,M) Israel; (C,D,M and other) Lebanon

Key: C–Various Christian, D–Druze, J–Jewish, M–Muslim

BIRTHPLACE OF BELIEF

More than 12,000 years ago the region was generally less arid than it is today, and was populated by hunter–gatherers. It is thought that a sudden climatic change that took place about 10,000 BC provided the impetus for people to domesticate plants and animals: a rise in temperature dried up the land, forcing people, animals and plants to congregate in places where water was available. Archaeological remains – mostly seeds and bones – reveal that settled agriculture in the region developed in well-watered valleys between 9000 and 8000 BC.

The gradual mastery of techniques for irrigating the land led to profound changes in social organization: the construction and maintenance of largescale irrigation works made labor specialization necessary; agricultural productivity increased; food surpluses made it possible for some sections of the population to take up other occupations, such as being artisans, traders, soldiers and priests; and cities began to develop.

The first urban settlements probably originated about 3500 BC in the fertile Mesopotamia area of modern-day Iraq; subsequently over the next millennia powerful civilizations, such as the Sumerian, Babylonian, Assyrian, and Phoenician, developed in the strip of territory (the "Fertile Crescent") running from the mouth of the rivers Euphrates and Tigris across to the Mediterranean, and bordered by mountains to the north and the Arabian desert to the south. Archaeological evidence indicates that various religious cults existed among these peoples: it was the Jewish people (Israelites or Hebrews), nomadic pastoralists who migrated from northern Mesopotamia to the area along the Mediterranean coast (Lebanon and Israel today) about 1200 BC, who first developed a belief in the one creator God.

The Persians, an Indo-European people, moved into the region of present-day Iran, to the east, about 1000 BC: the vast empire they established stretched at its height in the 6th century BC across the center and north of the region to challenge the Greeks who had established colonies in Asia Minor (today the coast of Turkey). Thereafter the Middle East became for centuries a battleground between the Greek (later the Graeco-

Roman), civilization in the west and the Persians in the east. Christianity, the religion founded by Jesus Christ in Roman-occupied Judaea (today Israel) in the 1st century AD, which began as a movement within the Jewish religion, spread over the next three centuries throughout the whole of the Middle East and the Roman world. The Jewish people revolted twice against Roman rule: after the second occasion, in 135 AD, the population was scattered.

The expansion of Islam

In the Arabian peninsula nomadic pastoralism predominated, with some settled agriculture around oases. Urban culture was possible only in Yemen, in the south, where the development of sophisticated irrigation techniques supported a series of trading kingdoms. During the 7th century AD, Islam, the religion founded

Islam in the modern world Muslim faithful gather for communal prayer in one of Saudi Arabia's impressive new mosques. Much of the wealth generated by the development of the oil industry has been directed toward religious and charitable goals.

by the Prophet Muhammad (c.570–632), a member of a nomadic desert people that controlled the trade routes around Mecca, united the Arab-speaking peoples of the peninsula and spread rapidly into the whole of the Middle East.

In the century that followed an Arab empire was created that carried Islamic (or Muslim) culture into Northern Africa, Sicily, the Iberian peninsula and large areas of central and southern Asia. Its influence was challenged in the 10th century by the incursion of the Turks, an Islamicized people originally from Central Asia, into the Middle East. This division within the Islamic world encouraged the Greek-speaking Byzantines, heirs to the eastern Roman empire, to attempt to retrieve their former influence in the region; help in regaining the holy places of Christendom was enlisted from Christians in Western Europe.

The age of the Ottomans

The fall of Constantinople – the principal city of the Byzantines – to the Turks (who called it Istanbul) in 1453 marked the end of Greek influence in the Middle East. Except for Persia and Afghanistan, the entire region fell within the vast empire of the Ottoman Turks, who continued to rule it until the 20th century. The four centuries of Turkish domination are regarded by many Middle Eastern Arabs today as a period of stagnation and loss of greatness. However, Arab Muslim subjects of the empire, as coreligionists of the Turks, managed to retain a significant amount of independence. Turkish never replaced Arabic, and a large share of political power was devolved into the hands of the local Muslim elite. Subjects and rulers were united in the desire to protect their faith and culture against encroachment from the growing power of the Christian West. It was only as the Ottoman empire fell into decay that hostility began to mark the relations between Turks and Arabs.

The collapse of the Ottoman empire at the end of World War I and the subsequent division of the region by the League of Nations into mandated territories controlled by Britain and France unleashed a surge of Arab nationalism. This grew still stronger with the creation in 1948 of the state of Israel – the realization of long-standing Jewish aspirations for a national homeland. Formed from parts of Palestine, then under British mandate, it left Palestinian Arabs without a nation state of their own.

SHADES OF DIFFERENCE

Although Arab Muslims vastly outnumber other groups, the Middle East's long and complex history of migration and invasion means that there is a much greater ethnic and religious variety than is often supposed. Significant numbers of people are neither Arab nor Muslim, such as the Israelis; are non-Arab Muslims, such as the Kurds, Turks, Iranians and Afghanistanis; or are non-Muslim Arabs, such as the Druze and Christians of Lebanon and Syria.

Arabs and non-Arabs

Arab peoples have a strong sense of common identity, which is rooted in their shared use of the Arabic language. This means that a Syrian will feel closer in spirit to a Yemeni, or even to a North African Arab, from several thousand kilometers' distance, than to a Turk, living in much closer proximity. Arabs are predominant in the countries of the Arabian peninsula, Iraq, Syria, Jordan, Lebanon and the province of Khuzestan in southwestern Iraq. The Palestinians who live in Israel and the occupied territories of the West Bank and Gaza, or in exile (very many of them in Jordan), are also Arabs.

The main areas of non-Arab population are Israel and a northern belt of territory stretching across Turkey, northern Syria and Iraq, Iran and Afghanistan. Some of these peoples have inhabited the lands they occupy for thousands of years, and are distinguished by a range of religious, ethnic, linguistic or cultural differences. Turks and Farsi-speaking Iranians are the major non-Arab ethnic groups in the region; other significant minorities include the Kurds (living in mountain areas of Turkey, Iraq and Iran) and the Armenians of northeastern Turkey (a greater number of whom live across the border in the Soviet Union).

Israel is the only Middle East state where Muslims do not form a majority. More than half the Jewish population of Israel (accounting for over 80 percent of the total) were born there since the state was created in 1948, about a quarter were born in Europe or the United States, and 19 percent in Asia and Africa. In the early 1990s most new arrivals were from the Soviet Union. Most Israeli Jews speak Hebrew, while the non-Jewish minority groups are mainly Arab-speaking.

The Muslim world of the Middle East is divided between Sunnis and Shi'as. This division stretches back to the very early years of Islam – the Sunnis recognize the first four caliphs as Muhammad's rightful successors, whereas the Shi'as believe that Muslim leadership belongs to the descendants of Muhammad's son-in-law alone. The Sunnis are by far the most numerous; Shi'as are restricted almost entirely to Iran and to Iraq, where they have suffered disadvantage at the hands of the ruling Sunni minority. Sunni control of nearly all Islam's most holy places, including the cities of Mecca and Medina, has been a cause of bitter conflict between the two communities. The Druze of Syria and Lebanon are a small, but closeknit Islamic sect of uncertain origin, who have fiercely preserved their traditions and independence for hundreds of years.

Friday qat market *(above)* Women in Yemen buying the week's supply of qat. This shrub is cultivated for its leaves which, when smoked, chewed in wads or infused as tea, have a stimulant narcotic effect. The shrub is widely grown in the region but is particularly common in Yemen.

Christianity is the region's other major religion. There are significant minorities in several countries such as Israel, Syria, Turkey and Iraq, and in Lebanon Christians account for about 32 percent of the population. The Maronites, the largest sect, trace their origins back as far as the 4th century; there are also Greek Orthodox, Greek Catholics and Armenians among other sects.

Veiled from view *(below)* An Omani woman in traditional costume with the full face and hair covering that satisfies the requirement for female modesty.

Mountain people of Afghanistan *(above)* A crowd eagerly watches as two owners release their fighting quails for battle. Isolated communities in Afghanistan's mountainous countryside have retained many distinctive customs and styles of dress.

WOMEN AND ISLAM

When a Muslim woman chooses to appear unveiled in public and to wear Western-style clothes, many observers will interpret it as a triumph for female emancipation. Others will regard it as a sign of the insidious Western corruption of Islamic morals. Ever since the Egyptian feminist Huda Sha'rawi symbolically dropped her veil into the Mediterranean in 1923, the veiling of women has been a contentious issue. Since the rise of Islamic fundamentalism in the 1970s many women have returned to veiling their faces, and to wearing the traditional dress that emphasizes female modesty. In Iran, this is virtually universal.

The veil is one manifestation of Islamic custom that requires the seclusion of women. Outside their family circle, women live a highly segregated existence and are prevented from taking any active part in public life. In Saudi Arabia, women are forbidden to drive cars or to work with men. However, while women still have inferior rights to those of men in many states, every country in the region accepts the right of girls to be educated at all levels – a view that is not necessarily shared by everyone within Islamic society. Heads of rural families often allow only primary education for their daughters. Nevertheless, women's increased access to education is already leading many to exercise greater social and political freedom, particularly in the cities, and may eventually erode the stark differences in opportunity currently available to the sexes.

Nomads, farmers and city-dwellers

Differences in ways of life within the traditional farming societies of the Middle East were largely determined by the environment, with settled agriculture taking place wherever there was easy access to water, and nomadic pastoralism predominating over the large tracts of desert and inhospitable mountain ranges. Although the number of nomads was always relatively small, they had the advantage of mobile military strength, which enabled them to exert great influence over the settled populations of the region, whose "softer" lifestyle they despised. The difficulty of surviving in a harsh landscape molded the nomads' traditional values of intense group solidarity. A whole family, clan or tribe could be held responsible for an individual offense, and vengeance was exacted in kind. These values of fierce loyalty to family and sense of personal and group honor have become an integral part of Middle Eastern culture.

Today only about 1 percent of the population are nomads, and are concentrated in areas of Afghanistan, Iran and Saudi Arabia. Most are herders – the caravan trade across the desert has almost entirely disappeared. Several factors have contributed to the decline of nomadism. Overgrazing has led to the deterioration of pastures; many nomads have taken advantage of the employment opportunities offered by the oil industry; others have been forced into a sedentary life by governments seeking to exercise greater political control over them.

Although the numbers engaged in farming everywhere are falling rapidly, the traditional village community still has great importance in the region. Often distinctive local customs and styles of architecture and dress are preserved here. For example, unlike other parts of Saudi Arabia in the agricultural southwest of the country, the women traditionally do not veil their faces, and wear colorful and decorative clothing.

Small farmers are the most conservative element in Middle Eastern society. Less exposed to outside influences than people living in towns, they accept change only with caution. Family allegiances remain strong and, in some communities, village elders still have the authority to settle disputes and guide communal decision-making. Nevertheless, rural lifestyles are inevitably changing. Literacy levels are rising with the wider provision of schools, and newspapers and television keep even remote communities abreast of developments in the outside world.

More and more people are leaving the timeless way of life of the countryside to swell the ranks of workers in modern cities. The city has always played an important role in Islamic culture: it is regarded as the repository of civilization, and great emphasis is placed on communal Friday prayer. Islam's influence on law and society has affected the way city life is organized. Houses have few windows and little external decoration, and are frequently built around a courtyard, which provides insulation against the heat, and also privacy for the family. Many traditional houses have separate living quarters for men and women.

SOCIETY IN FLUX

The enormous political, economic and social transformation of the Middle East this century has had widespread impact on daily life. Although those in the oil-producing and exporting countries have been most affected, nowhere has escaped change. Ordinary people are the chief victims of the continuing severity of the region's political tensions: armed struggle has occured in virtually every country in the region since 1945.

Population change

Population growth in the Middle East (except for Iran and Afghanistan) is high – between 1980 and 1985 it rose by almost 3 percent each year, a rate of growth exceeded only by parts of Africa. One of the causes is a decrease in infant mortality – which nevertheless is still high by Western standards. Modern methods of birth control are rarely used: the fertility rate of 37 births per 1,000 of the population is well above world averages. In some of the region's small countries large families have been encouraged by government programs in order to support rapid economic expansion.

The social effects of a rapid population rise – unemployment and insufficient housing, educational and health facilities – are most acute in the towns and cities where natural growth has been augmented by the huge influx of migrants from the countryside. New arrivals in the cities, who retain a traditional rural outlook, find their values come into conflict with those of longer-established city dwellers. They consequently often form enclaves with other migrants from their own or neighboring villages in slum districts on the city outskirts. Although population movement has always been a feature of Middle Eastern life, with its long nomadic tradition, international migration only became important after the 1940s to meet labor shortages in the newly oil-rich Gulf monarchies and Iraq. The immigrant workers who fueled their economic development at first came mainly from neighboring Arab countries in the Middle East and Northern Africa: more recently workers have come from the Indian subcontinent (particularly Pakistan) and Southeast Asia. The need of the small oil states such as Kuwait, Qatar and the United Arab Emirates to import labor has been so great that the migrant population now exceeds the natural population. This is the cause of frequent tensions between the wealthy minority and the poorer majority. In Kuwait, following the withdrawal of Iraqi troops after the Gulf War in 1991, violent retribution broke out against the sizable Palestinian community who were accused of having collaborated with Saddam Hussein's occupying forces.

Cultures in conflict

The sources of international conflict and armed struggle that have torn the Middle East apart in recent decades can be traced beyond the immediate political policies of the governments involved, to be found in the region's underlying social, cultural and religious divisions. This affects even overtly political conflicts such as the war in Afghanistan, which was fought between adherents of conflicting political ideologies from the mid 1970s. The failure of the Muslim Muhajidin to oust the discredited Kabul government after the withdrawal of Soviet troops in 1989 was directly related to its inability to unite the diverse ethnic and religious groupings of the population. In Lebanon, the religious and cultural diversity of the population has been a cause of civil war since 1970, with Maronite Christian, Druze, Sunni and Shi'a factions fighting each other to gain political influence.

The dramatic increase in the popularity of Islamic political movements throughout the region – part of the process of

THE HOLY CITY

The ancient city of Jerusalem is a holy city for Muslims, Jews and Christians. Its complex history and symbolic importance for all three faiths have made possession of the city an issue of enormous cultural and political sensitivity. From 1949 to 1967, the city was divided between Israel and Jordan: the older, eastern part of the city belonged to Jordan, while the newer area to the south and the southwest became the capital of Israel. Since 1967 Israel has also occupied the Old City – which contains most of the holy sites of all three religions – though Palestinians still claim it as their capital.

Within the Old City the Church of the Holy Sepulcher, reputedly the site of Christ's crucifixion and burial, lies a short walk from the Wailing Wall, the western wall of the former temple and Judaism's most sacred place. Jews come here to pray and leave written messages for God in the crevices of the wall. The Islamic shrine of the Dome of the Rock is built over the place where Muhammad is believed to have ascended to heaven and where, according to both Muslim and Jewish tradition, the creation of the world began.

Throughout the ages, Jerusalem has evoked deep passions; piety has frequently been mixed with violence as competing faithful claim the city as their own. Religious fervor has today become inextricably mixed with nationalism, heightening the tension between Jews and Palestinians, which often results in bloodshed.

A people in flight *(above left)* Thousands of Kurds fled to the mountains with whatever they could carry when their uprising against the Iraqi government failed in the aftermath of the Gulf War in 1991. Many died from exhaustion and the bitter cold.

Where Europe meets the East *(above)* The exclusively male clientele of a terraced café in Turkey combines Islamic culture with influences from the southern European Mediterranean world.

Sacred ritual *(below)* Jews wearing prayer shawls and phylacteries – small black boxes containing passages of scripture that are tied to the forehead or arm – carry the scrolls of the Law to the Wailing Wall in Jerusalem.

reform and revitalization that has taken place within the religion of Islam in the second half of the 20th century and is characterized in the West as " fundamentalism" – has had wide reaching effect. Since its revolution in 1979 Iran has stood as a model to all those who argue for the closer integration of Islam within politics, and all the states in the region, with the exception of Turkey, Lebanon and Israel, have had in some way to accommodate Islam within their political systems.

Political tensions in the region have contributed to the growth in Arab nationalism, which found fresh inspiration and militancy after 1948 in the existence of a Jewish national state in Palestine and in the ideas of Gamal Abdel Nasser (1918–70), president of Egypt from 1956 to 1970, who was a powerful advocate of Arab unity. This political ideal is the official policy of the ruling Baathist parties of both Syria and Iraq, and nearly all the Arab leaders of the Middle East pay lip-service to the concept of a common Arab identity. Yet unity has proved very hard to achieve, never more so than during the Gulf War in 1991 when the Iraqi occupation of Kuwait aroused conflicting loyalties within the Arab world. One effect of the crisis was to highlight the distinct identities and aspirations that separate the individual Arab states.

Perhaps the most developed sense of a distinct national identity within the Arab world has been fostered by the one large community that is without a state, the Palestinians. The Intifida, or uprising, in the Israeli-occupied territories from 1987 reinvigorated a national spirit that had survived the creation of the state of Israel, itself a cultural and political expression of Jewish national and religious identity.

The plight of the Kurds

Like the Palestinians, the Kurds are also a distinct national grouping without a state. They appear to have occupied the same upland area of eastern Turkey, northern Iraq and northwestern Iran for millennia. Originally nomadic, they are now settled farmers or have moved to the towns, and their traditional tribal organization has been diluted by urbanization. Nevertheless, they have a strong awareness of their ethnic distinction, and demand for a separate state of Kurdistan has intensified as attempts to suppress them culturally have strengthened. Long subject to hostile treatment by the Turkish government, which outlawed the Kurdish language, forbade the wearing of traditional dress and encouraged migration to the industrial towns of western Turkey, they were, after 1988, subjected to increasing pressure from the Iraqi authorities. Their plight in the aftermath of the Gulf War in 1991, when many fled from Saddam Hussein's army to the refuge of the mountains but were forced to return to a "safe haven", drew worldwide attention to their cause.

The world of Islam

Islam shares with Judaism and Christianity its Middle Eastern origins, its belief in only one God, and its derivation from a sacred text. Of these three great world religions, Islam is the youngest and fastest growing. It is the dominant faith of the Middle East – where its principal shrines of Mecca and Medina are situated – as well as of many North African countries, and also Pakistan, Bangladesh, Malaysia, Brunei, the Comoro Islands and Indonesia. The religion was spread by military conquest, trade and missionary activity through these countries and as far as southern Europe, eastern Africa, central Asia, India, China and the Philippines, where Muslim minorities remain today. The migration of workers from Asia and Africa to Europe and the United States during the 20th century means that Islam is now practiced widely in these regions as well.

The origins of Islam lie in the 7th-century revelations of the Prophet Muhammad, collected in the Qu'ran or holy book. Muhammad believed that the Qu'ran was dictated to him by God; it would be blasphemous to suggest to a Muslim that Muhammad was the author of the Qu'ran. Muhammad was born in Mecca in 570 AD and received the call to proclaim the message of Allah (the one God) when he was 40. His emigration to Medina, or the *Hijra*, in 622 marks the beginning of the Islamic calendar.

A dispute over the leadership of the Muslim community caused a rift in the faith after Muhammad's death in 632. The followers of Abu Bakr (573–634), the father of Muhammad's wife Aisha (c. 618), became known as Sunnis, and are generally considered more traditionalist and orthodox. The Shi'as are followers of Ali (c. 600–661), Muhammad's cousin and son-in-law, and differ from the Sunnis in matters of law and ceremony. Shi'as form the majority of the Muslim population in Iraq, Iran, Bahrain and Lebanon, but are in the minority in other states.

The art of Islam (*left*) A Shi'a Muslim from Iran reading the Qu'ran in front of one of the intricately decorated surfaces that characterize Islamic art and architecture. As the Qu'ran forbids figural representation in religious works, Islamic artists evolved a unique style using complex abstract, geometric or floral designs and brilliant colors. Mosques are generally light and spacious with a dome and interior decoration designed to draw the eye heavenward.

Journey of a lifetime (*right*) Thousands of pilgrims waiting to enter the Great Mosque at Mecca during the *hajj*. All Muslims should make the pilgrimage at least once in their lifetime, provided they can afford it. The *hajj* consists of several ceremonies spread over a number of days. These include walking seven times around the Ka'ba, the shrine within the mosque; kissing and touching the Black Stone; and running seven times between two elevations 400 m (1,300 ft) apart beside the mosque. About 2 million people perform the *hajj* each year.

Pillars of faith

The Qu'ran is more than just a spiritual book. It gives precise guidance on the way a Muslim's life should be led in accordance with the "five pillars of the faith", which are the principal religious duties a Muslim is obliged to follow. The "five pillars" are: to recite the profession of faith (the *shahada*), "There is no God but the one God, and Muhammad is His Prophet", at least once in an individual's lifetime; to pray five times daily facing toward Mecca; to give alms; to fast during the daylight hours of Ramadan, the ninth month of the Muslim year; and to make the pilgrimage, or *hajj*, to Mecca – Islam's holiest site – at least once in a lifetime.

The advice in the Qu'ran also encompasses practical, legal and cultural commands. There are, for example, passages which concern marriage, divorce and inheritance. Muslims are prohibited from eating pork, drinking alcohol, gambling, committing fraud, usury, and making images. *Jihad*, or holy war, stresses the responsibility of the Muslim community for collective defense against external and internal enemies. It has been an important element in spreading and protecting the faith.

A second component of Islam is the body of the Prophet's sayings and deeds (*Sunnah*) that are not contained in the Qu'ran. The task of the Islamic law schools was to arbitrate in areas not explicitly covered by the Qu'ran by interpreting and applying these implicit commands, but their judgments were often contradictory, and divided Muslims by allegiance. However, the overriding importance of the explicit law of the Qu'ran is universally accepted. The main body of Islamic law (*Shari'a*) is followed in varying degrees by Islamic governments.

The mosque is the center of community worship, where the Friday services are held. In the past it was used for many functions – military, political, social and educational – but today these have mostly been taken over by other institutions. However, many mosques still have an attached *maktab*, or elementary school, where the Qu'ran is taught and classes are given in Islamic law and doctrine.

Although local architectural styles and building materials may be used, the layout of all mosques is unchanging, modeled on the place of worship used by the Prophet Muhammad – the courtyard of his house at Medina. A roofed-over open space, sometimes with a minaret attached to it from which the call to worship is proclaimed five times a day, contains the *mihrab*, or niche, reserved for the prayer leader (*imam*); this points the way to Mecca. The *minbar*, a seat at the top of steps to the right of the *mihrab*, is used as a pulpit. Worshippers are barefoot; they perform the ritual prayer by prostrating themselves on mats or carpets in the body of the mosque. Because of the Qu'ranic injunction against the representation of images, mosques were traditionally decorated in the intricate geometric and interlacing patterns, using flower, foliage and fruit motifs, that have become characteristic of Islamic art.

DESERT AND MOUNTAIN PEOPLES

RISE AND FALL OF EMPIRES · THE CULTURAL LANDSCAPE · MODERN CHALLENGES

Northern Africa falls into two distinct cultural areas. The peoples of the north, living along the Mediterranean and Atlantic coast from Egypt to Morocco, are close to the peoples of Europe and the Middle East; to the south they face the heartland of Bantu-speaking Africa. In between lies the Sahara where – in contrast to the settled rural and urban cultures of the two peripheries – nomadism is the traditional way of life, though threatened now by modernization. Across this vast desert, the peoples of Northern Africa have communicated and traded for centuries, and this exchange has allowed them to develop a common history and social environment. Most people in the region are Muslim; more recently they have shared a history of colonial (mainly French) rule, which has left a clear cultural imprint.

RISE AND FALL OF EMPIRES

The original inhabitants of Northern Africa entered the region tens of thousands of years ago from eastern Africa, continuing to spread northward while the Sahara was still temperate and fertile. The famous rock paintings of the Ahaggar mountains and Tassili-n'Ajjer in southern Algeria show pastoralists with cattle, and herds of other animals such as elephants and rhinoceroses. Other migrations in prehistoric times from the Mediterranean and the Middle East account for the Berber population of the Maghreb countries (Morocco, Algeria, Tunisia and Libya), and for the enormous variety of peoples living in the Horn of Africa.

Human civilization has some of its earliest roots in Africa north of the Sahara. Some 5,000 years ago farmers working the alluvial plains of the Nile delta, enriched annually by the river's flooding, produced sufficient agricultural surpluses to support the rise of the highly centralized Egyptian empire with its urban elite, massive armies and sophisticated religion. To the west, separated by extensive intervening semidesert, settlement was patchy: the Berbers lived as pastoralists and settled farmers in scattered groups in the northern mountains

COUNTRIES IN THE REGION

Algeria, Chad, Djibouti, Egypt, Ethiopia, Libya, Mali, Mauritania, Morocco, Niger, Somalia, Sudan, Tunisia

POPULATION

Over 40 million Egypt, Ethiopia

20 million–30 million Algeria, Morocco, Sudan

1 million–10 million Chad, Libya, Mali, Mauritania, Niger, Somalia, Tunisia

Under 1 million Djibouti

LANGUAGE

Countries with one official language (Amharic) Ethiopia; (Arabic) Algeria, Egypt, Libya, Morocco, Sudan, Tunisia; (French) Mali, Niger

Countries with two official languages (Arabic, French) Mauritania; (Arabic, Somali) Somalia

Other significant languages spoken in the region include Bamabara, Bedawie, Berber, Djerma, Fulani, Hausa, Nubian, Oromo, Sara, Sarakole, Songo, Tigrinya, Toucouleur

RELIGION

Countries with one major religion (M) Algeria, Djibouti, Libya, Mauritania, Morocco, Niger, Somalia, Tunisia

Countries with more than one major religion (M,C) Egypt; (M,EO,I) Ethiopia; (M,I,C) Chad, Mali, Sudan

Key: C–Various Christian, EO–Ethiopian Orthodox, I–Indigenous religions, M–Muslim

Looking two ways *(below)* The architecture of the *medina*, or old town, of Tangier, is characteristic of Islamic influence in Northern Africa. However, for several centuries Tangier was closely tied to Europe, just a short distance away across the Strait of Gibraltar, and was only joined with the Kingdom of Morocco in 1956. There is still a small European community.

The Grand Malam (*above*), leading communal prayer in the desert, is the religious leader of the Kanimbo people of northern Chad. They are a relic of the Kanem-Bornu empire, the powerful nomadic dynasty that once controlled trade in the area. It was largely this nomadic influence that helped establish Islam among the people of the Sahel.

and along the coastal plain. Similar conditions prevailed in the Ethiopian Highlands and in well-watered areas of Sudan.

On the margins of the desert nomadism took over, particularly after the introduction of the camel into Northern Africa 4,000 years ago. These nomads controlled the routes across the Sahara, exacting tribute from the urban-based traders of the Maghreb, as they did from the settled agricultural populations of the oases. In this way a number of powerful states, such as the Garamantes 2,000 years ago and the Zaghawa 1,000 years later, came to brief dominance in the area.

Farther south, the savannas of the Sahel (in the countries of Mali, Niger, Chad and Sudan) supported similar extensive trading empires. The most famous of these was the Kanem-Bornu, which controlled a large area around Lake Chad from about the 9th to the 19th century. Ruled for most of this time by the Sef dynasty, its economic success depended on the ability of a mobile warrior-class to extract taxes from the local populations of semi-nomadic and settled farmers.

The long, narrow coastal plain of the Mediterranean has been subject to frequent invasion. The Phoenicians, traders from the eastern Mediterranean, set up a number of trading posts here in the 8th century BC, the largest of which was Carthage in what is now Tunisia. In 142 BC the Romans captured Carthage and for the next 500 years the Mediterranean littoral became part of the Roman empire, supplying it with grain. It was later to become an important center for Christianity – St Augustine (354–430 AD), one of the greatest of the early fathers of the church, was a Berber from Tunisia.

The impact of Islam

In the 7th and 8th centuries Arabs from the Middle East swept through Egypt and the Maghreb – an invasion that has had a longlasting effect throughout Northern Africa – providing it with a common religion, Islam. This was quickly established in the north of the region and was carried southward into the Sahel along the well-established trade routes: the Kanem-Bornu were Muslim by the 11th century. From the beginning, the Berber people of Morocco rebelled against Arab rule, and in the 11th century the Almovarid clan conquered the whole of the country and extended their influence into Muslim Spain; they were eventually succeeded by the Almohads.

Large numbers of Arabs from Syria were brought into the region to tighten control over the peoples of Libya and Tunisia. As a result, the Maghreb acquired the Arab-speaking character that still predominates today. Sizable pockets of Berber-speaking people, however, remain in Morocco and Algeria, where they form 20 percent of the population.

The European chapter

The latest influence to have welded the disparate cultures of Northern Africa into a loosely definable entity has been European colonization. With the exceptions of Egypt and Sudan (Britain), Libya, Somalia and, from the late 1930s, Ethiopia (Italy), and northern Morocco and the Western Sahara (Spain), the region fell within the orbit of France's African empire. Between 1830, when the French first moved into Algeria, and 1934, when the last rebellious tribes in northern Chad and western Morocco were finally subdued, French civilization extended a widespread linguistic and cultural uniformity, but its most enduring legacy may be the artificial national boundaries imposed by the colonial administrators. The struggle for political independence, born in the Arab nationalist movements after World War II, may prove to have had more effect in giving common identity to the peoples of Northern Africa.

THE CULTURAL LANDSCAPE

Africa is the world's most linguistically diverse continent, and Northern Africa is no exception. In Sudan, for example, there are over 400 local languages, and in Chad over 100. To some extent the colonial languages provide a common means of communication. French is still the official language of Mali, Mauritania and Chad, and is also widely spoken in Algeria, Tunisia and Morocco.

French television is received in these three Maghreb countries, while French remains an important literary language. Algerian writers such as Katib Yacine have established international reputations, and Tahar Ben Jelloun, a Moroccan writer, is a winner of the Prix Goncourt, the prestigious French literary prize. English remains widely spoken in Sudan, Egypt and Ethiopia, although mostly among the educated classes.

In an attempt to create a homogeneous Arab–Islamic culture, some governments (including Sudan, Algeria, Morocco, Libya and Egypt) have declared Arabic the official national language. Although Arabic is widely spoken, it has many regional dialects (*magrhibiyya* or *darija*), which are barely intelligible to each other; these governments, therefore, have tried to propagate a standard form of Arabic, *fusha*. This has, however, been resisted in Algeria and, to a lesser extent, in Morocco, by Berber-speakers (forming about a fifth of their populations), who have campaigned to retain French alongside Berber and Arabic dialects. An attempt by the Algerian government in 1980 to ban lectures on Berber literature given by the well-known literary critic, Mouloud Mammeeri, in the city of Tizi-Ouzou led to three days of rioting. Resistance to Arabic has also come from the southern peoples of Sudan, who speak a wide range of languages. Many of these – such as Dinka and Nuer – belong to the Nilotic family. Only in Somalia and Ethiopia have ethnic languages achieved official status, though in the latter, Amharic, the language of the dominant ruling group, is not spoken by either of the large Tigrean or Oromos populations.

Survival of indigenous culture (*above*) A Dogon blacksmith in Mali working on a flintlock gun; this will be used during death ceremonies to fire a volley of shots to see off the dead. The Dogon are primarily an agricultural people, and their few craftsmen – metal and leather workers – form distinct castes. Each Dogon district has its own *hogon*, or spiritual leader.

First lessons (*right*) A Qu'ranic school in Mauritania has set up for the time being on a dusty street corner; the boys sit in a semicircle round their teacher while a small girl – excluded from the class – listens in a doorway. Makeshift schools such as this are likely to be the only form of primary education available to many children.

The African face of Islam

With the exception of Ethiopia and the countries in the southern parts of the Sahel, particularly Sudan and Chad, where Christianity and animism also flourish, Islam is the dominant religion of the region. In the countries of the Maghreb more than 95 percent of the population belong to the Sunni sect – one of the two main divisions of Islam. Islam is also the offical religion of Egypt (where four-fifths of the population are Sunni), and of Somalia, Sudan and Mauritania.

Although Islam played a significant role in the hard struggle for independence, nowhere in the region did a true Islamic state emerge. Libya, under Colonel Muammar Qaddafi, came closest, but even there Qaddafi has wrested power away from the *ulema* or guardians of the law. In Sudan the introduction of the Islamic law (*Shar'ia*) in 1983, involving among other

Holy journey *(above)* Christians in Ethiopia make a pilgrimage to the shrine of a local saint. Pilgrimages to sacred places associated with particular events and people feature in many religions, including Christianity and Islam. The pilgrim gains spiritual benefit from undertaking the journey, especially if it is made over a great distance and in conditions of discomfort.

things a ban on alcohol and the introduction of strict Qu'ranic penalties for crime, has met with armed resistance in the nonMuslim south.

By contrast, in Tunisia, the most Westernized of the Maghreb countries, polygamy (permitted by the Qu'ran) has been banned. Restaurants remain open during the month-long fast of Ramadan to service the large tourist trade and, until recently, the wearing of headscarves and the public call to prayer were forbidden. In most countries the *hajj* – the pilgrimage to Mecca made by devout Muslims at least once in their lifetime – is still observed (though it was banned under French rule). Ramadan and local shrine festivals (*moussems*) are celebrated. Alcohol and smoking, however, are common.

Despite religious injunctions against it, birth control is widely practiced – for example, 60 percent of Egyptian women use contraception. In Egypt, too, limited experiments with fixed interest rate banking have begun, despite strict Islamic constraints against usury. In general, religious observance is stronger in rural areas, where Islamic family law upholds the traditional authority of the husband in the household and of the eldest son in the line of descent. In the cities, the increasing number of women who work, the reduction in family size among the middle classes, and the migration of many males to work abroad, have made it more difficult for religious codes to govern family life.

NonIslamic religions

The Coptic Christian church – a sect that dates back to the 4th century AD – has 4 million adherents in Egypt and Orthodox Christianity is the national religion of Ethiopia. It is followed by over half the population, though the Marxist state tried to suppress it after 1974. Cut off from contact with other Christian churches by the Arab conquests, the Ethiopian church embraces nonChristian beliefs in benevolent and malevolent spirits. About one-third of the Ethiopian population is Muslim, among whom there is also widespread animist belief. The Omoros, for example, nominally Muslim, have a strong spiritual connection with the land, given symbolic expression in the practice of burying a new-born child's navel cord. Separation from their territories, and from sacred trees and rocks, destroys their sense of cultural belonging.

Until recently there was also a small Jewish community in Ethiopia known as the Falashas. Their origins are uncertain, and their marriage and circumcision rites differ from those of Jews elsewhere. Two mass airlifts, one in 1984 and the other in 1991, which were organized by the government of Israel, evacuated almost the entire population.

There is religious diversity in other parts of the region. In Mauritania, for example, the Black Peul – settled agriculturalists living along the banks of the Senegal river – are Christians who blend their faith with belief in animist spirits and rituals. Elsewhere in the western Sahel, animist *bory* cults are well established, while Somalia is the home of the *zar* spirit cults.

THE SANUSI ORDER

The Sanusiyya was one of a number of prophetic religious movements that developed in the early 19th century in reaction against Western culture and to ensure that Islam would remain the dominant religion in the Middle East and Northern Africa.

The order, based in Cyrenaica, the eastern province of Libya, assumed control of the great nomadic peoples of the area – normally in a state of perpetual strife. Because these nomads dominated the trade routes across the Sahara, the Sanusiyya were able to penetrate Libya's Saharan hinterland and later came into conflict with French colonizing ambitions in Chad. From their fortresslike *zawiyas*, which combined a mosque, a university and a residential complex, they spread Islam among the nomadic Tubu peoples.

In the early part of the 20th century the Sanusiyya organized resistance to the Italian occupation of Libya. Idris I (1890–1983), Libya's first king after independence in 1951, was the head of the Sanusiyya. However, the revolutionary regime of Colonel Muammar Qaddafi that replaced him in 1961 banned the order and all reference to it. Yet its memory still persists in the film *Omar Mukhtar*, financed by the Libyan government to commemorate the country's struggle against colonialism.

MODERN CHALLENGES

Religious, racial and clan divisions combine to create conflicts across the region, many of which spill over the political boundaries left by the colonial powers. While some are the consequences of the immense social changes resulting from high population growth and rapid urbanization, others are rooted in age-old divisions that exist between nomadic and settled peoples.

In Mauritania, for example, the Badayane or "white" Moors of Arab-Berber descent – former camel herders – battle with the Harratin or "black" Moors and the black Fulani and Wolof farmers. Another nomadic people, the Tuareg, are engaged in a guerrilla war against black Africans in Mali and Niger, which began when they returned from their seasonal pastures in Algeria and Libya in the 1980s. In both countries nomads used to dominate the slave trade; although it was officially abolished in 1980 in Mauritania, it persists in some places and serves as a source of resentment among black Africans. The nomadic Sahrawis of Western Sahara are engaged in a war of independence with Morocco, which has left some stranded in refugee camps in Algeria, where they are obliged to adopt a more settled agrarian lifestyle.

Women in a changing society A group of students at Cairo University wearing either Western or traditional Islamic clothing. The feminist movement has a strong foothold in Egypt, leading many educated women to question traditional attitudes toward their role in society. The rise of Islamic fundamentalism is a conflicting pressure that has prompted some women to resume the veil, and to be less ambitious about careers outside the home.

Religious and ethnic divisions

In the troubled and desperately impoverished south of the region, religious and ethnic divisions lie at the root of the fighting that has left millions of people homeless. Sudan's civil war between a mainly Muslim and Arabic north and a mainly Christian and animist south has forced over 400,000 refugees – most of them young males from the Dinka and Nuer peoples – to move to neighboring Ethiopia. In 1988 250,000 people died of war-related famine and disease inside Sudan, and starvation threatened the lives of as many as 7 million people after prolonged drought in the early 1990s. Sudan also received three-quarters of a million refugees from other countries, the majority from Ethiopia.

Ethiopia's divisions are more ethnic than religious. In 1991 the predominantly Christian Tigreans overthrew the government, which was run by Amharas, also mostly Christian. The country's single largest ethnic group, the Oromos, however, are Muslim and animist. Conflict here was intensified by the attempts of the Marxist Amharic government to force peasants into villages and resettle many of them in more fertile parts of the countryside. Used to living in scattered homesteads connected by kinship ties, the Oromos objected to having to live with non-kin, and to severing their close links with the land.

Two-thirds of the population of Somalia, almost entirely Muslim, are nomads. Clan-based loyalties lay behind the alliance of the larger Isaak and Hawiye clans that deposed the minority Marehan clan government in 1991 after a bitterly fought civil war that left much of the country totally devastated.

Islamist movements

In the north of the region conflict has centered around the efforts of various Islamist groups, ranging from moderates to militants, to replace existing civil laws with the *shar'ia*. Such movements have been fueled both by poverty and by opposition to Westernization. The populations of the Maghreb countries, including Egypt, have increased rapidly – by tenfold in 90 years in Egypt, for example. In most, about 40 percent of the population are less than 15 years old. Migration to the cities has created severe problems of over-crowding.

Economic failure has driven many out of the region to seek work. One million Moroccans are employed overseas; there are 750,000 Algerians in France alone;

People in crisis Displaced from their land by fighting in the civil war, refugees in one of Somalia's camps wait desperately for food to arrive. Persistent fighting between different ethnic or religious groups – which often prevents aid from reaching those who need it – coupled with frequent drought, has left millions of the region's people homeless and close to starvation.

between 2 and 4 million Egyptians are migrant laborers, mostly in the Gulf (though the Iraq–Kuwait war forced many to return home in 1991). Those who remain behind in the Maghreb have become dissatisfied and frustrated with their Western-style governments, and their desire for change attracts them to the Islamist movements. Such groups include the Ennahdha (meaning "renaissance") in Tunisia, the Islamic Salvation Front (FIS) in Algeria, and the Muslim Brotherhood in Egypt and Sudan.

The Islamist groups of the different countries have no common political program or organized structure. Many of their demands are similar, however, including the enforcement of Islamic marriage and family law, the exclusion of women from work, and the covering of women in public places, as well as bans on the consumption of alcohol and some types of popular entertainment. They also call for stricter adherence to obligatory religious duties, such as the *zakat*, or distribution of money to the poor at the end of Ramadan. Many women have begun to adopt the veil and the full dress required by Islamic law, some through religious devotion, but others as a gesture against the invidious Western commercialization of women's bodies and appearance by the wearing of cosmetics and immodest clothing.

In Algeria, the FIS has organized itself through hundreds of small mosques, rather than through the grand state-built mosques, using them as welfare and education centers. In the 1990s activists began to occupy public spaces in Algiers, filling the streets with slogans relayed over loudspeakers and invading some of the university classes. Although women were involved in these protests as well as men, their actions were segregated from those of the men.

Women are also well represented among the opposition to the Islamist groups. Egypt, for example, has a strong feminist movement among the educated middle classes, which stretches back to the 1920s. The struggle to make sense of the rapid social changes that the countries of the Maghreb are experiencing has led to many conflicts, particularly among its urban population, between people with Western-derived cultural ideas and aspirations and those who live by the code of Islamic fundamentalism.

Traditional ways of life and patterns of behavior are more persistent outside the cities. Traditional social values remain strong, and concepts of honor and shame continue to govern social relations. However, the continuing urban drift, and a steady rise in tourism, means that modern economic values and Western cultural ideas filter back even into the most remote rural areas, and it cannot be long before they, too, are affected by the change sweeping the region.

POP-RAI: THE MUSIC OF REVOLT

One of the chief ways that Algerians have expressed their alienation from official ideology has been through music. Berber singers found initial favor among dissident groups, particularly with the songs of Idir and Ait Menguillat. Later rai, a traditional form of music from the city of Oran, became a vehicle for disaffection. In rai – which means "opinion" – the singer, with chorus, expresses his views: he is accompanied by traditional instruments such as flutes, lutes, violins and percussion. The impact of rai was spread, however, by the introduction of Western instruments and amplified sound in the 1970s, creating pop-rai.

Pop-rai took sex, drugs and fast living as its themes. The improvised songs often commented on topical events and politics. Discouraged by the government and by Islamist groups, it found an audience among Algerian communities overseas. Pop-rai was closely associated with a spate of riots in every major Algerian city in 1988 that forced the government to abandon its censorship laws. However, the success of Islamic fundamentalists in municipal elections 20 months later brought a return to censorship, and a nightclub in Oran that had become a shrine to pop-rai was forced to close.

Most Chebs, as pop-rai singers are called, are working-class boys who began singing in clubs and at weddings. In the 1980s they were taken up by professional record companies, and their records have been widely distributed outside Algeria. They have had a particularly strong influence on contemporary groups in France.

Disappearing nomads

At home in the desert *(above left)* A Tuareg family enclosure in the central Sahara. Adult males traditionally wear veils in the presence of women, strangers or in laws, though this has largely been abandoned by Tuareg who have become urbanized, either by choice or through compulsory settlement.

Overtaken by progress *(above right)* The romantic sight of a camel caravan in stately procession through the desert is rapidly becoming a thing of the past. Motorization has replaced the camel as the nomad's main form of transportation; now it is more common to see a truck parked beside the tent.

Under the pressure of modern life, and often discouraged by government resettlement schemes, nomadism as a way of life is slowly dying out in virtually all the regions of the world where it has traditionally been practiced as a response to environmental constraints. Nevertheless, the Sahara remains one of the great strongholds of nomadism – about a third of its 1.5 million population still lays claim to such a lifestyle. Today, however, nomadism is often a recreational activity, carried on with the use of motorized vehicles, by people who also have a permanent place of residence.

Saharan nomads belonged traditionally to one of four ethnic groups: the Tubua and the Zaghawa in the east, the Tuareg in central Sahara, and the Moors in the west. Some Arab nomadic peoples also lived along the desert edge. For survival, nomads moved freely from one sparse pasture to another in a repetitive annual cycle in order to feed their major source of wealth – livestock. For most nomads, the most important animals were camels, which provided milk and meat, were ridden in raids and carried goods along the trade routes across the Sahara. In small desert market centers such as the towns of Ouargla, Al-Oued and Gardaia in Algeria, nomads traded animals, skins and dates for manufactured goods.

As well as practicing animal pastoralism, raiding and trading, the nomads also made a living by extracting expensive transit fees from traders for protection while passing through their territory in the Sahara. Most nomads were also in

control of chains of oases running through the desert. There they held the settled populations – the Kamadja in Borkou or the Harratin in southern Morocco – in a state of vassalage, offering protection in return for taxes paid in cereals, vegetables and, above all, dates.

Vanishing lifestyle

Much of this traditional way of life has now disappeared. Roads across the desert have completed the destruction of the trans-Saharan trade, which had already been severely damaged at the beginning of the 20th century by the growth of the railroads. Modern developments, reinforced by urban drift, have persuaded many nomads that the rigors of nomadic life are no longer justified – a decision that was hastened by the droughts that occurred throughout the Sahel in the 1960s and 1970s.

In addition, successive governments have tightened up the border controls, thereby rupturing traditional migration routes. Previously nomads had crossed modern international borders with a fine disregard for the niceties of passport and customs regulations. One reason for this has been the border disputes and civil wars that have occurred in many parts of the region, such as the hostilities in Western Sahara between the Polisario Front and Morocco, and between Chad and Libyan forces after Libya had come to the support of Frolinat rebels who were opposing the Chad government.

Frolinat found its basis of support in the Tubua and Zaghawa nomadic groups, who successfully adapted their traditional military skills to modern warfare – just one example of the way that traditional nomadic values have adjusted to the modern world. Many dispossessed Tuareg, ruined by the Sahelian drought, have moved into truck-driving, the modern equivalent of the trans-Saharan camel caravan. In these and other ways, the people of the Sahara continue the age-long struggle to survive in the inhospitable environment of the desert.

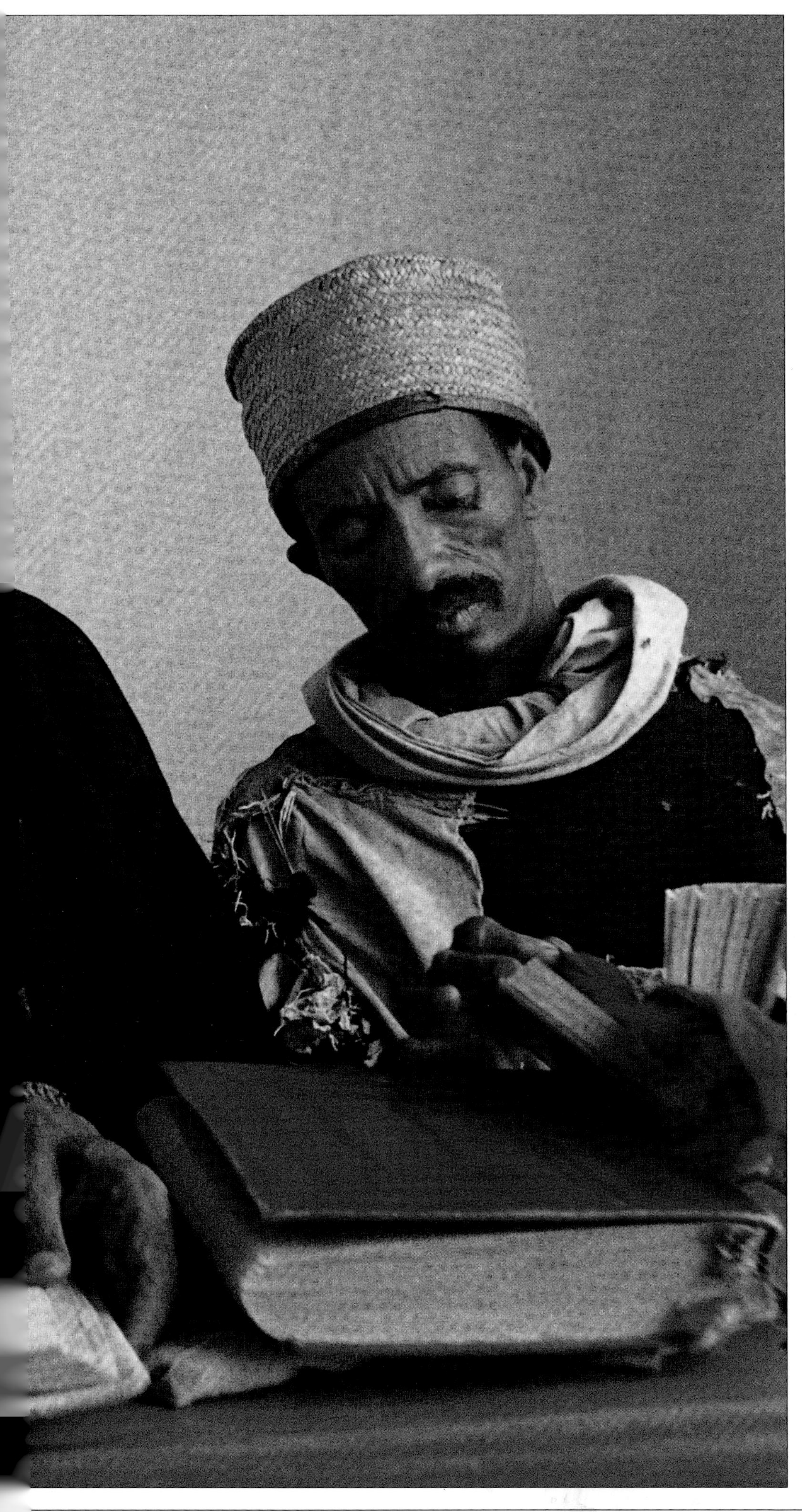

Christianity in Ethiopia

Ethiopia was among the first parts of Africa to be converted to Christianity. Its distinctive form of the religion, known as Monophysitism, was introduced by missionaries from Syria between the 4th and 6th centuries. It holds that Jesus Christ had a single divine nature, and not – as the Roman Catholic church believes – two natures, human and divine. This doctrine was declared a heresy in 451, so the Ethiopian Orthodox church split with Rome, and came under the control of the Egyptian Coptic church.

The spread of Christianity in Ethiopia was long and sporadic, competing with Islam, and over the years it gradually became fused with many older beliefs to produce a unique, syncretic version. Some of its elements are animist, particularly the belief in spirits and magic, while other aspects owe much to a Hebraic inheritance. Like Jews, Ethiopian Orthodox Christians practice circumcision. For a time, the Jewish sabbath was observed as well as Christian Sunday, and the Ark of the Covenant, the container holding the tablets bearing the Ten Commandments that God gave to Moses, is kept inside every church, as in the Hebrew temple. There is a legend that the Ethiopians, and not the Jews, are God's chosen people, since they claim to be descended from King Solomon and possess the original Ark of the Covenant. This legend was described in the epic *Kebra Nagast* (The Glory of Kings), written down by monks in the 13th century.

Before the revolution of 1974, which deposed the emperor, the church had an extensive clergy, which was most numerous in rural areas. It was composed of priests and deacons, who conducted the services, and clerics known as *dabtaras*, who specialized in learning, writing and magic formulae. The church's close association with the emperors, however, placed it in opposition to the Derg revolutionary regime, and its clergy were consequently subjected to persecution.

A unique tradition Two priests and a deacon study sacred texts in a monastery school.

SETTLERS AND NOMADS

THE PEOPLES OF CENTRAL AFRICA · A KALEIDOSCOPE OF CULTURES · ETHNIC DIVISION AND SOCIAL CHANGE

The predominantly rural region of Central Africa is occupied by numerous ethnic groups with distinct territories, languages and customs. Differences in their traditional ways of life and social organization – notably, the distinction between settled and nomadic peoples – were shaped by the physical environment, which ranges from desert to parched grasslands and equatorial rainforest. Superimposed on them have been a variety of values and beliefs that were imported from outside the region. In the early medieval period Arab traders were responsible for introducing Islam along major trade routes into the region, and the more recent experience of European colonial rule has left a pervasive influence in the form of Christianity and the prevalence of Western poltitical, economic and cultural values.

COUNTRIES IN THE REGION

Benin, Burkina, Burundi, Cameroon, Cape Verde, Central African Republic, Congo, Equatorial Guinea, Gabon, Gambia, Ghana, Guinea, Guinea-Bissau, Ivory Coast, Kenya, Liberia, Nigeria, Rwanda, São Tomé and Príncipe, Senegal, Seychelles, Sierra Leone, Tanzania, Togo, Uganda, Zaire

POPULATION

Over 118 million Nigeria

10 million–35 million Cameroon, Ghana, Ivory Coast, Kenya, Tanzania, Uganda, Zaire

1 million–10 million Benin, Burkina, Burundi, Central African Republic, Congo, Gabon, Guinea, Liberia, Rwanda, Senegal, Sierra Leone, Togo

Under 1 million Cape Verde, Equatorial Guinea, Gambia, Guinea-Bissau, São Tomé and Príncipe, Seychelles

LANGUAGE

Countries with one official language (E) Gambia, Ghana, Liberia, Nigeria, Sierra Leone, Uganda; (F) Benin, Burkina, Central African Republic, Congo, Gabon, Guinea, Ivory Coast, Senegal, Togo, Zaire; (P) Cape Verde, Guinea-Bissau, São Tomé and Príncipe; (S) Equatorial Guinea

Countries with two official languages (E,F) Cameroon; (E,Sw) Kenya, Tanzania; (F,K) Burundi; (F,R) Rwanda

Country with three official languages (C,E,F) Seychelles

Key: C–Creole, E–English, F–French, K–Kirundi, P–Portuguese, R–Rwandan, S–Spanish, Sw–Swahili,

Numerous indigenous languages are spoken in the region

RELIGION

Countries with one major religion (M) Gambia; (RC) Cape Verde, Equatorial Guinea

Countries with two major religions (P,RC) São Tomé and Príncipe, Seychelles

Countries with three or more major religions (I,M,RC) Benin, Burkina, Gabon, Guinea, Guinea-Bissau, Ivory Coast, Liberia, Senegal; (C,I,M,P,RC) Kenya, Zaire; (I,M,P,RC) Cameroon, Burundi, Central African Republic, Congo, Ghana, Rwanda, Sierra Leone, Tanzania, Togo, Uganda

Key: C–Various Christian, I–Indigenous religions, M–Muslim, P–Protestant, RC–Roman Catholic

THE PEOPLES OF CENTRAL AFRICA

Archaeological evidence shows that today's countries of Kenya, Uganda and Tanzania are the birthplace of the human species. The Rift Valley of eastern Africa has yielded an enormous number of fossil remains, enabling archaeologists to retrace the evolutionary trail of our earliest ancestors, and providing an unsurpassed record of Stone Age culture. The region's prehistory is still being pieced together from a variety of sources, including linguistics, botanical studies and oral history: archaeological remains in the Central African Republic, for example, indicate that a civilization existed here earlier than that of ancient Egypt, which dates from about 3000 BC.

Among the earliest known migran groups peopling the region were Bantu-speaking peoples who spread out of the area that is now Cameroon to establish themselves across much of the region by the end of the 1st millennium AD. Their culture, based on ironworking and settled agriculture, largely displaced that of the indigenous hunter–gatherer groups of Pygmy peoples. By the 10th century a system of chieftainships and empires had developed, some of them – particularly those in areas where trading contacts outside the region were strongest – with their own sophisticated cultures and systems of commerce.

Spices from eastern Africa were being traded with the Mediterranean region as long ago as Greek and Roman times, but it was under the influence of Arab traders in the Indian Ocean that coastal cities such as Zanzibar, Mombasa, Lamu and Malindi, still important cosmopolitan centers today, were founded. Arabic has had a major influence on Swahili, the dominant Bantu language of eastern Africa and Zaire, and a regional lingua franca. After the Arab conquest of Northern Africa in the 7th century, Arab merchants and migrants acted as unofficial missionaries for the Islamic religion, helping to spread its influence and culture across the Sahara and into the west and center of the region. A number of cities and empires, such as Yorubaland, which is now part of Nigeria, developed along the major trading routes that linked sub-Saharan Africa with Mediterranean Europe in precolonial times.

The impact of Europe

In the area that is now the Congo, various Bantu groups, including the Teke and the Kongo, formed themselves into powerful kingdoms; they habitually captured men from other areas, and a trade in slaves had already been established through eastern Africa to Arabia when Portuguese colonists arrived in the west of the region in the late 15th century. They were offered slaves in exchange for European goods, and over the following few centuries Europeans fully exploited this woeful trade, exporting millions of Africans to provide labor on plantations in North and South America. More slaves were taken from the area that is now occupied by the Congo, Zaire and Angola than from any other part of Africa – more than 13.5 million people over three centuries – and

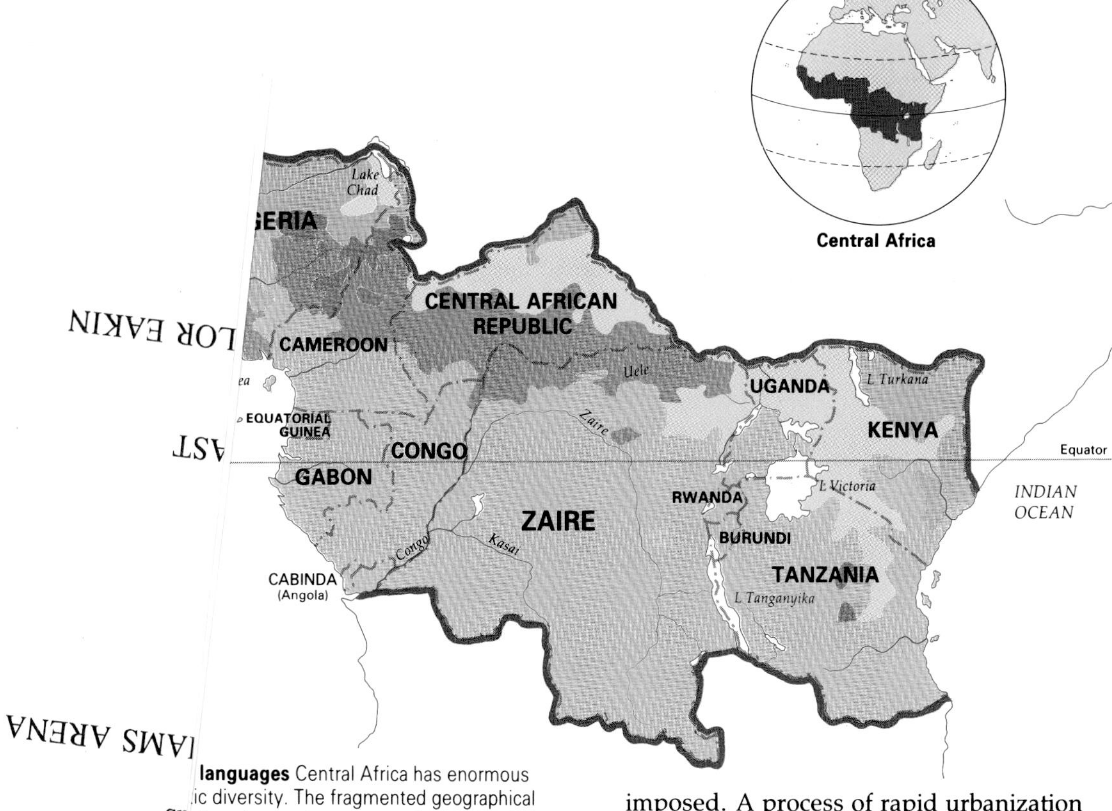

languages Central Africa has enormous ...ic diversity. The fragmented geographical ...ution of some languages reflects historical ...vals and migration within the region. Very many ...e are bilingual, with French or English acting as ...gua franca.

A trading history *(above)* The wealth that Ashanti kingdoms derived from the slave trade is symbolized by the elders' traditional golden headdresses. The Ashanti still have influence in trade and commerce and have played a major role in Ghana's economic development.

The call of Mecca *(left)* The Islamic religion was first spread to Africa by Arab traders and merchants. There are Muslim populations throughout the region; they form the majority in most of western Africa from Nigeria to Senegal.

a number of small, increasingly powerful coastal kingdoms came into being to control the trade.

These vanished with the ending of the slave trade in the 19th century. European exploitation of Africa's human resources was now transformed into the direct plunder of all its natural wealth. By the end of the century the colonialists exerted wholesale political, economic and social control throughout the region; missionaries sought to convert even the most remote groups to Christianity, and the use of European languages was generally imposed. A process of rapid urbanization began to meet the needs of European-owned trade and industry.

Cultural artefacts were among the products appropriated by the colonizers. Traditional works of art, especially those from Benin, Burkina and Niger, became greatly sought after by European and American collectors, and continue to fetch high prices on world art markets today. These wealthy collectors encourage the excavation of settlement sites or burial mounds that date from before the 7th century, in the hope that burial pots and bronze and copper objects will be unearthed. Although this trade is now illegal under national and international law, the countries involved are ill-equipped to prevent the looting of their heritage.

Many of the national boundaries drawn by the colonizing powers cut across the natural divisions between the territories of different ethnic groups: every country is now inhabited by at least four ethnic groups; some contain more than 50. As a result, although most of the languages spoken in the region belong to the Bantu group of languages – one of Africa's five basic linguistic groups – some countries have enormous linguistic diversity. For example, more than 43 dialects are spoken by Gabon's population of little more than 1 million. Consequently, European languages – particularly English and French – often still remain as official languages, and Africanized versions, or pidgins, are the unofficial lingua franca of the cities.

Village communities The family is the basis of village life; settlements are arranged to reflect kinship patterns. Women share the task of caring for the children and do much of the work in the fields; communal decisions are usually made by the men.

A KALEIDOSCOPE OF CULTURES

Despite the near-universality of the colonial experience, there is still enormous cultural diversity among Central Africa's hundreds of ethnic groups. These vary considerably in size of population and territory. Most groups, such as the Anga and Birom, two of Nigeria's roughly 200 ethnic groups, and the Elgeyo and the Tugen of Kenya, number less than 150,000 and the area of land they occupy – sometimes referred to as village states – may be only a few hundred square kilometers. But there are also larger groups whose members are counted in millions and are spread over hundreds of thousands of square kilometers. These include the Yoruba (12 million people), the Hausa (10 million) and the Ibo (8 million) of west Africa and the Kikuyu (2.2 million) in the east of the region.

African society is traditionally organized within kinship groups known as lineages. Each lineage consists of all the descendants of one common ancestor, either real or mythical; the great majority of ethnic groups are related and inherit property through the male line (patrilineage), though some do so through the female line (matrilineage). Occasionally small matrilineal groups can be found among predominantly patrilineal people, such as Nigeria's eastern Ibo clans of Ohafia and Abiriba.

In much of the region, most groups live as settled farmers. The smallest living unit is the homestead, a cluster of houses or huts that contains all the members of the extended family. The arrangement of individual homesteads in a village, or of villages within a group, is determined not – as in many societies – by occupation, but by lineage. Village groups or clusters are usually united by the name of a common mythical ancestor, and the spatial layout within them reflects the different branches of descendancy, so that they can be read in a similar way to a family tree.

Within patrilineal groups, the extended family consists of a male head (the patriarch) who wields absolute social and economic power over his wife or wives, their sons' families, and occasionally the families of younger brothers who have yet to set up homesteads of their own. In common with many other aspects of traditional life, however, the extended family is changing. Rapid urbanization and the spread of Western ideas and education, combined with the new economic forces of the colonial and post-colonial period, have tended to make the younger generation less dependent upon their elders. In cities particularly, the extended family is now being replaced by the nuclear family.

The nomadic way of life

In contrast to the settled life of these agriculturalists are a number of nomadic peoples, living in areas where rainfall is sparse, who move their livestock to distant grazing lands every year in the dry season. The Cattle Fulani, for example, range across the dry areas of the Sahel, south of the Sahara, from Senegal to Cameroon, while the Masai and Turkana graze their livestock across the east African plateau.

Cattle form the basis of the nomadic culture. The number of animals a herder possesses establishes his social standing and wealth, and determines his rights and obligations in respect of others. Land is only important to the pastoralists as pasture for their animals. Although the Masai live in areas of the region that support cultivation, they scorn agriculture and buy millet and maize from settled farmers to supplement their diet of milk, blood and meat. When local supplies of grazing and water have been exhausted, they abandon their temporary settlements – *manyattas* – of huts made of mud and cow dung and move on.

There has been a tendency in some official quarters to regard nomadic pastoralism as a primitive form of animal rearing, and some of the region's governments have made various attempts to settle these groups. Neither the Masai nor the Cattle Fulani have shown much interest in these settlement schemes, and there is now considerable evidence to show that the periodic use of pasture in arid areas makes sound ecological sense, allowing vegetation time to recover and avoiding the risk of desertification.

A MIXING OF RELIGIONS

The oldest independent republic in Africa, Liberia, was established in the early 19th century as a homeland for slaves that had been freed, mostly from North America. The repatriated slaves brought with them the religions they had adopted in servitude, and in 1843 the capital, Monrovia, boasted 23 Christian churches: 13 Methodist, 8 Baptist, and 2 Presbyterian. Today there are more than 40 church buildings – representing over 20 different Christian denominations – and 4 Muslim mosques to cater for all the varied religious persuasions in the capital.

Christian and Islamic places of worship have not completely displaced traditional beliefs and religious sites. Many Liberian villages, old and new, are distinguished by giant cotton trees. Towering to 50 m (165 ft), they have been planted in the belief that they will ward off evil spirits and adverse magical forces. Several bundles of dried palm leaves, hung from the top of a pole in the village center and usually populated by flocks of weaver birds, serve the same purpose.

Some Liberians, particularly in the north, regard kola trees as holy, and they are found growing in or near most settlements. In the past kola nuts, which contain caffeine and are chewed, were given as peace offerings. Today they are served as a token of friendship to visitors.

The power of magic The medicine man traditionally occupies an influential place in rural society. The fetishes strung around his waist are endowed with magical powers derived from ancestral spirits, which he is able to use on behalf of those seeking his help to effect cures and avert harm.

Traditional beliefs

Continuity between people and nature, and between the living and the dead, is at the center of indigenous religious beliefs. Because all of nature is a manifestation of a god or gods, natural objects are attributed with a living soul. Ancestors are an extremely important part of African religions. They are regarded as the real owners of family land; living descendants are allowed to enjoy its benefits, but they would incur the wrath of their ancestors if they sold it. The primary role of ancestors, however, is to defend their descendants. These traditional beliefs are handed down from generation to generation by word of mouth at the fireside.

Magic is another key element in indigenous religions: good magic keeps evil spirits at bay. Medicine men or sorcerers dispense charms, give advice on how to avoid danger, tell fortunes and carry bags of fetishes (small figures embodying good or evil spirits). The power of a fetish depends on the magical substance buried in the cavities of its head or abdomen. Each fetish represents a specific ancestor, and they are believed to cure disease, inhibit sorcery or ensure success in hunting, depending on the substance.

Despite the prevalence of Christianity and Islam (which has grown in strength in some areas following the worldwide Islamic revival), many people continue to practice traditional religions. Moreover, the popular following of the syncretic churches – which blend Christianity with traditional music, dance and healing rituals – is expanding rapidly.

ETHNIC DIVISION AND SOCIAL CHANGE

The national boundaries, arbitrarily dividing ethnic territories, that were inherited by many of the region's states when they gained independence have been the source of much friction and conflict. In Nigeria, for example, the difficulties of establishing a national consciousness in a country composed of no less than 200 ethnic groups proved enormous, and simmering tensions between the three main groups – the Hausa in the north, the Yoruba in the west and the Ibo in the east – erupted into a bloody civil war in 1967 when the Ibo declared the independent state of Biafra. Since then, one practical solution to Nigeria's long-term ethnic problems was the building of a new capital, Ajuba, in the center of the country, at the common peripheries between Hausa, Yoruba and Ibo territories. This replaced the former colonial capital of Lagos in the territory of Yoruba.

Politics and cultural values

A major factor in the decision of a number of post-independence states, such as Ghana, Guinea and Tanzania, to reject the forms of government inherited from the European colonial powers in favor of the socialist models of Eastern Europe and the Soviet Union was that they claimed to be based on broad mass movements and seemed to offer a way of controlling influential minority groups. A further argument in favor of socialism was that it was seen to be more compatible with African culture than capitalism. The socialist system of cooperative production appeared to blend well with the traditional African concept of the extended family. Julius Nyerere, president of Tanzania from 1962 to 1985, claimed that African economic and social organization was based on the socialist principles of

Living in a dry land (*above*) A girl from one of Kenya's nomadic ethnic groups collects water from a dried-up river bed. In times of drought, it may take a day's journey to find a few jars of dirty water.

Transition in the cities (*right*) Traditional values meet Western influences in the towns and cities. Most immigrants are young males who have left their home villages to earn enough money to marry or to support their family at home.

TRADITIONAL RIVALS

Three major ethnic groups, the Hutu, the Tutsi and the Twa, make up the populations of Burundi and Rwanda, two of the smallest countries in the region, straddling the high plateau that forms the watershed between the Congo and Nile river systems. Each group maintains a distinct cultural identity, but the Hutu and the Tutsi, who speak closely related Bantu languages, compete keenly for political power.

Traditionally, the three groups all performed very different functions in society. The Twa, the traditional Pygmy inhabitants of the area, were specialists in hunting and pottery, and the Hutu were agriculturalists who had moved into the area, probably from western Africa, well before the arrival of the pastoralist Tutsi from the north in the 14th century. Although the Tutsi form a minority in both populations – about 10 percent in Rwanda and 15 percent in Burundi – they came to dominate the others, deriving their power from their control over the supply of cattle, which all three groups regarded as the basis of economic and social status.

Colonial rule exacerbated the rivalry between the Hutu and Tutsi. In Rwanda the Hutu had attained the upper hand shortly before independence in 1961, and the next 10 years were characterized by repeated raids from the Tutsi who had been driven from the country. Tutsi minority rule continued in Burundi, but the deep distrust between the two groups was the cause of similar turmoil, most notably in 1972 when, following an organized uprising by the Hutu, the Tutsi systematically massacred about 100,000 people.

communal ownership in which kinship and family groups took part jointly in economic activity, welfare and security.

Western export-oriented economies had created social imbalances in many countries. The rapid growth of major cities fueled largescale migration from rural areas; this had serious social implications in the countryside. Socialist ideas were harnessed in an attempt to reverse this process. Decentralization policies were introduced in both Kenya and Tanzania; in Tanzania these moves were complemented by the establishment of traditional communal village production units, *ujamaa*, though these were later abandoned.

The city and the extended family

Although Africa is the world's least urbanized continent, present rates of urban growth are among the highest in the world. Immigration, both from rural areas and from smaller urban centers, has to a great extent been responsible for the phenomenal expansion of tropical Africa's major cities, from Lagos in the west to Nairobi in the east. The extended family system contributes to this development, since many new arrivals in both older precolonial cities such as Lagos and Ibadan (both in Nigeria) and in newer colonial cities such as Dakar (Senegal), Accra (Ghana), Brazzaville (Congo) and Kinshasa (Zaire) stay with members of the extended family until they can afford to pay rent for a place of their own. But as people become established in the cities, the influence of the extended family is being simultaneously reduced since the increasing number of urban residents discourages the continuation of a family system that places heavy demands on small incomes.

This mass movement of people to the cities has affected the age and sex composition of African society in all parts of the region. Most of the migrants congregating in urban areas are young male workers who are either single or have left their wives behind in the rural home until they can secure employment and accommodation for their families. As a result, women and elderly men are left in the villages to take on the arduous tasks of crop cultivation that were traditionally performed by the younger men. Consequently, a good deal of village land has ceased to be farmed. These traits are most pronounced among Muslim groups, such as the Hausa, that frown on the free movement of unmarried women.

The modern influence of the city has also resulted in a general desire for fewer children among urban families. In a rural setting, the contribution children make to communal labor adds to the prosperity of the kinship group; in cities, they are another mouth to feed. The decline in urban fertility levels compared to those in rural areas is also influenced by easier access to birth control clinics and by the availability of contraception. Urban women, who generally have access to better education, also tend to marry later. Zaire is the exception to this rule. Fertility rates in the capital city of Kinshasa are higher than those in the countryside, where they have traditionally been low. This is probably due to the better health care and higher living standards that are available here.

The sounds of Africa

Tune in to a radio station playing African pop music anywhere in Central Africa and you could be forgiven for thinking you are in South America or the Caribbean. Latin rhythms are just one of the influences that give a typical African pop band its unique flavor and exuberant sound – Western jazz and rock, Black American music, and the scales and rhythms of traditional African music are all likely to be present as well.

These wide influences can be traced to three major sources. Africans who became members of colonial regimental bands learned typical European rhythms, such as marches and polkas, and became familiar with regimental brass, wind and stringed instruments. Then there were the songs, hymns and melodies of various Christian religious groups. Finally, in the ports that grew up to carry the products of colonial Africa around the world, sailors' shanties and ballads played on a variety of instruments such as harmonicas, accordions and guitars added another sound. All these were blended with traditional music and instruments, particularly with drums and stringed instruments made from gourds, animal skins and horns, to produce the extraordinary rhythms of African pop music.

Highlife and Congo

Outside influences were perhaps most keenly felt in Ghana, where a Westernized music called "highlife" emerged

Live performance *(above)* A Cameroonian television team records a performance of a ritual dance by the Bantu-speaking Ibibio people of the country's southwest. In traditional African dance the performers interpret the percussive patterns of the music through their movements. These complex rhythms, taken to America by slaves, have influenced jazz and blues music, which are in turn an ingredient of African pop.

Regimental marches *(right)* An army band at Gabon's national day celebrations on 12 March. European military music, with its range of brass, wind and percussion instruments, its brisk rhythms and cheerful marching tunes, has added its lively sounds to today's blend of African music. Many musicians acquire their musical training in army bands like this one.

World Music *(below)* The Senegalese musician Youssou N'Dour on a European tour in 1989 – one of many African performers who have found a large and appreciative audience on the world scene. A number of international stars such as Paul Simon have made successful recordings with African musicians.

among the dance bands of Accra and Kumasi before and during World War II. Troops stationed in western Africa injected the newest sounds of Big Band and American Swing into highlife. After the war, touring western African bands continued to assimilate new sounds and rhythms from across the Atlantic: Black American music, jazz, soul and more recently, reggae.

Meanwhile another African pop sensation was being created in the Congo. "Congo-bars" sprouted in Leopoldville (now Kinshasa) and Brazzaville during and after World War II, offering dancing and refreshments. The music performed there had a strong Cuban flavor as radio stations had been popularizing early rumba stars. Solo guitars were accompanied by small brass ensembles and a rhythmic background provided by the tinkle of bottles played like gongs. Congo later developed into Africa Jazz and OK Jazz when electric guitars and amplifiers arrived on the scene and were taken up by large orchestras.

The Congo music of Zaire and the Congo remains very popular in western Africa, but competes with Cameroonian Makossa music, a mixture of highlife and soul, that originated in the nightclubs of Douala. In Nigeria Afro-Beat, heavily influenced by the jazz of the American James Brown, vies for popularity with Juju music, which has its origins in the traditional music of Yoruba.

African pop, which has absorbed so many different musical influences, is itself now influencing Western music. Its unique sound and exotic rhythms attracted an ever-growing international audience during the 1980s and 1990s. "World Music", which combines ethnic music with Western pop, became increasingly popular in Europe, and the worldwide televising of huge benefit concerts such as *Live Aid* won many new fans for African music and dance.

MIGRATIONS AND INVASIONS

THE PEOPLE OF SOUTHERN AFRICA · FROM FARMERS TO TOWN DWELLERS · THE EUROPEAN INFLUENCE

Farmers and pastoralists had been living on southern Africa's plateaus and coastal plains for thousands of years before the arrival of Europeans in the 15th century. The impact of colonialism on these groups brought about enormous cultural and social change. The migration of thousands of Europeans to farm the fertile lands of the region and to exploit its mineral resources undermined local societies and destroyed their economies by depriving them of their lands and drawing adult males away to work in the mines and cities. South Africa's history of racial conflict between whites and blacks has distorted outside perceptions of the region and obscured its cultural diversity. Today southern Africa is being drawn firmly into the modern world, but the transition is not easy, and produces inevitable tensions.

COUNTRIES IN THE REGION

Angola, Botswana, Comoros, Lesotho, Madagascar, Malawi, Mauritius, Mozambique, Namibia, South Africa, Swaziland, Zambia, Zimbabwe

POPULATION

Over 30 million South Africa

10 million–20 million Madagascar, Mozambique

1 million–10 million Angola, Botswana, Lesotho, Malawi, Mauritius, Namibia, Zambia, Zimbabwe

Under 1 million Comoros, Swaziland

LANGUAGE

Countries with one official language (E) Botswana, Mauritius, Zambia, Zimbabwe; (M) Madagascar; (P) Angola, Mozambique

Countries with two official languages (A,F) Comoros; (Af,E) South Africa, Namibia; (C,E) Malawi; (E,Se) Lesotho, (E,Si) Swaziland

Other significant languages in the region include Comorian (the majority language of the Comoros), ChiSona, Kimbundu, Lunda, Makua, Setwana, Si Ndebele, Tombuka, Umbundu and numerous other indigenous languages

Key: A–Arabic, Af–Afrikaans, C–Chichewa, E–English, F–French, M–Malagasy, P–Portuguese, Se–Sesotho, Si–siSwati

RELIGION

Countries with one major religion (C) Lesotho, Namibia; (M) Comoros

Countries with two major religions (C,I) Angola, Botswana, Malawi, Swaziland, Zambia, Zimbabwe

Countries with three major religions (C,I,M) Madagascar, Mozambique

Country with more than three major religions (C,DR,H,I,M,RC) South Africa

Key: C–Various Christian, DR–Dutch Reformed, H–Hindu, I–Indigenous religions, M–Muslim, RC–Roman Catholic

THE PEOPLE OF SOUTHERN AFRICA

Although there is evidence of humans having lived in southern Africa as long ago as 250,000 years, the first signs of organized social groups appear at the end of the Stone Age, about 10,000 years ago. The change from a hunting–gathering existence to farming seems to have taken place during the 1st millennium AD as iron-using peoples moved out of the Congo Basin into the region.

The presence of minerals in the east of the region – iron, copper and gold – gave rise to a number of powerful trading kingdoms, such as Zimbabwe, whose massive stone ruins, built in the 14th century, are still visible today, and the Muenemutapa empire of Mozambique. Existing alongside these well-organized states were a great number of smaller, less developed groups of hunter–gatherers, such as the San (Bushmen) of the Kalahari Desert, pastoralists and settled farmers, such as the Zulu on the well-watered plateau of eastern South Africa. The latter's military expansion to gain land and cattle at the beginning of the 19th century, the period known as the Mfecane, sparked off a complex series of migrations across the region as one group fled after another. This still affects the distribution of peoples today.

The colonial impact

By this time European settlers were already well established in southern Africa. The first to arrive were the Portuguese at the end of the 15th century. From coastal trading stations in Angola and Mozambique they organized, with the cooperation of some tribal leaders, a largescale export of slaves to their plantations in South America. Slavery devastated tribes psychologically and undermined the rural economy by removing the most active members of the population. Nevertheless, the military strength and organization of the inland tribes helped prevent the Portuguese from extending their influence beyond the coast for four centuries.

Other European nations soon followed the Portuguese intervention. The Dutch were settled in the Cape as farmers from the 17th century onward, but the interior was depicted as barren and hostile, and little attempt was made to penetrate it. However, after the British took over control of the Cape Colony in 1814, many Dutch (Afrikaner) colonists grew dissatisfied with their rule and in 1831 migrated northward in the Great Trek to found the independent Boer republics of the Orange Free State and Transvaal.

The discovery of diamonds at Kimberley in northern Cape Province in 1866 and of gold in the Witwatersrand, Transvaal, in 1886 changed European perceptions of the interior, bringing rapid economic development and a scramble for a share in the region's resources. A flourishing mining industry quickly sprang up, using migrant black African laborers from as far north as Malawi, who were forced en masse into the modern economy in order to pay colonial taxes. Indentured laborers from the Indian subcontinent were introduced to work the large sugar plantations in Natal on the east coast.

Colonialism had varied impact. Some areas were little affected, others profoundly so, especially where European policies of divide-and-rule encouraged separatism among ethnic groups. Many of them were cheated of their lands by the representatives of mining companies. The destabilizing effects of migrant labor on ethnic integrity presented missionaries with the opportunity to convert "pagan" Africans to Christianity. In the new urban areas Africans gradually absorbed the values of a capitalist money economy. Elsewhere, horrific acts of subjugation were carried out; thousands of Herero and Nama were massacred in the haste to colonize South West Africa (now called Namibia).

The amibitions of the colonists were often couched in rhetoric that masked cultural imperialism and stressed the "civilizing mission" of Europeans to expunge barbarism from "Darkest Africa". Yet acute rivalries also existed among the Europeans, especially between the British and Afrikaners. Conflict between the two came to a climax in the Boer wars (1896, and 1899–1902) when Britain defeated the Boer republics to seize control of the Transvaal gold mines.

Island peoples

The history of the islands off the coast of Mozambique is very different, giving rise to a mix of peoples and influences from Africa, Arabia, India, Southeast Asia and Europe. Arab traders had established

Cultural imperialism *(above)* Shanty towns like this one outside Cape Town are a grim reminder of the impact of European settlement on southern Africa's indigenous peoples. Their lands were seized, their cultures destroyed and people were uprooted from their homes, all in the name of progress.

A mix of influences *(right)* Madagascan dancers wearing costumes derived from the fashions of 19th-century French colonists. Of African and Indonesian origin, the island's population is split between Christians and those following traditional beliefs.

bases in all the islands before the arrival of European traders and colonizers in the 16th and 17th centuries. Dutch, and then French, settlers established sugar cane plantations in Mauritius; in 1815, the island was taken over by the British, who brought in indentured Asian labor. The French exercised a strong influence in the Comoro Islands and Madagascar (first settled between 2,000 and 1,500 years ago by people from Africa and Indonesia) for many years before they formally became part of the French overseas empire in the second half of the 19th century.

FROM FARMERS TO TOWN DWELLERS

Cultural differences between the large numbers of groups that make up the black population of southern Africa tend to go unnoticed from the outside because of the amount of international media attention given to South Africa's racial conflict between whites and blacks. Among these millions of people, differences in ways of life and language, and opposing elements of traditional and modern, have threaded together to create a complex tapestry of cultural diversity. Although modernization has been widespread throughout the region, many traditional forms of social organization still exist among the ethnic groups in many parts of southern Africa, especially in rural areas.

In such societies, men are still the

dominant sex: the local male chief has the power to decide on the use and ownership of communal land and on legal and family matters. In Lesotho, for example, personal disputes are settled by a tribunal consisting of the chief and the warring parties, but he has the final word. Ethnic integrity is maintained by the performance of certain rituals. For example, the Umlangha ceremony, a mass dance performed by young girls from all over Swaziland, venerates the queen mother and celebrates womanly virtues, thus reinforcing the role of women in traditional Swazi life.

Cultural identities

One way in which distinctive cultural identities are maintained is through linguistic diversity. Many different languages and dialects are spoken, falling within several major language groups, such as the Tswana and Xhosa. These may straddle international boundaries. In Botswana, for example, there are nine ethnic groups speaking Tswana, but each has its own dialect and colloquialisms. These linguistic differences persist even in the cities, although most people are bilingual. English is the lingua franca used in most of the region with (to a lesser extent) Afrikaans in South Africa and Namibia, and Portuguese in Angola and Mozambique.

For some, symbolic references from the past may bolster cultural identity. The Zulus draw on the embellished histories of victorious campaigns during the Mfecane period to establish physical and mental agility as cultural characteristics that separate them from other groups. Somewhat differently, the Indian community of Natal expresses its distinctive culture in its architecture, which reflects the styles of the subcontinent, and in the continued wearing of traditional, brightly colored dress.

Different uses and perceptions of the natural environment influence everyday life substantially. For the millions of smallscale farmers across southern Africa, the ownership of land is vital. They have a strong feeling of attachment to their home area and an awareness of what makes it unique from other places. For the few remaining nomadic pastoralists of the Kalahari, however, a sense of place is less important than an empathy for their animals and for the desert environment on which they depend.

The journey to work (*above*) A truck carrying men from outlying villages to their work on one of Malawi's largescale commercial farms, or in the city, is already overflowing at the seams, and stragglers have to hitch a lift from the next one following behind. Many young men, single and married, have to leave their homes and families for long periods to find work. They keep in touch with their home villages only through the remittances they are able to send back.

A cure for everything (*right*) A contemporary version of the village medicine man, an urban healer in Zambia waits for custom beside a sign declaring that he can provide cures for afflictions ranging from backache to bilharzia as well as luck charms and protection against being bewitched. Belief in magic and the power of good and bad spirits is an important part of indigenous religious beliefs.

A NATIONAL ICON

The Zimbabwe Ruins – consisting of an acropolis of huge stone walls several meters thick, ingeniously constructed without mortar around great natural bolders at the summit of a 100 m (330 ft) high granite hill – were probably built in the 14th century by a powerful and flourishing people. Their sheer size and craftsmanship speak of a very sophisticated social structure able to undertake such a largescale construction.

Since Zimbabwe's independence in 1980, the Ruins have acquired great cultural significance for Zimbabweans. They have given their name (Zimbabwe means "place of stones") to the new state, just as in the colonial era the country was named after the industrialist and pioneer Cecil Rhodes (1853–1902) who established British rule over the region. Although the Ruins had been excavated and a museum established there before independence (they were made a National Monument in 1960), they had not attained the status of national icon that they have today. For white Rhodesians, Cecil Rhodes' grave in the Matopo Hills near Bulawayo, and the flagstaff in Salisbury (today Harare) where the first British flag was raised in 1890, assumed that role. Today, the Zimbabwe Ruins have been used to undermine the myth developed in the colonial period that African peoples were primitive and backward, and are celebrated as a symbol of Zimbabwe's illustrious past.

Urbanization and nationalism

Colonialism and its after-effects have thrust southern Africa firmly into the modern world, diluting and transforming the traditional ways of life of the majority of people. A very large proportion of Africans are now urban dwellers – if not permanently, at least temporarily. They work mostly in low-rank industrial and service jobs. City dwellers are exposed to a Western money economy, to a consumer ethic and to the material benefits of a technologically sophisticated society, substantially altering their cultural values and aspirations.

Nationalism has become increasingly important as a focal point for people's group loyalty, replacing the ethnic group. Newly post-independent states such as Angola, Mozambique, Namibia, Zambia and Zimbabwe have as yet little in their history to draw on to display a distinctive national character. Consequently they have tried to foster a national loyalty toward principles such as democracy and equality, with mixed success. The failure of policies to improve the standard of living, such as those pursued by the ruling Frelimo party in Mozambique, gives the phrase, political principles, a hollow ring: little incentive exists for people to abandon older loyalties.

As a result, ethnic jealousies and rivalries can still have divisive force, and one-party rule, as is emerging in Zimbabwe, may represent the domination of one ethnic group over another or several others. Although the socialist program of the African National Congress (ANC) offered an agenda to unite the culturally divided blacks of South Africa in common cause against apartheid, traditional conflict between the Xhosa and the Zulu still erupts in bitter violence.

Missionaries during the colonial period introduced a wide range of Christian sects and faiths among black Africans. The Swazi are one of the few peoples still to practice traditional religion, consulting the spirits of their ancestors. However, for many Christians, elements of animist ritual and ancestor worship may exist comfortably beside their formal religious allegiance. These religious beliefs have had to adapt to contemporary political and economic conditions. Regular church worship and weekly attendance at the local ANC rally may have equal weight in the day-to-day routine of many black South African families. For some, religious belief has been replaced by secular political ideologies. The communist governments of Angola and Mozambique, for example, officially discourage religious practice, and religious teaching is forbidden in schools.

Race and sport *(above)* Years of segregation produced different sporting preferences among South Africa's ethnic groups. Although the country is now pursuing a policy of integrated sport – resulting in its readmission to the 1992 Olympic Games – the training facilities and equipment available to blacks lag far behind those of their white compatriots.

THE EUROPEAN INFLUENCE

Although South Africa at the beginning of the 1990s was the only country in the region where whites still rule, the European influence remains pervasive. Most whites are urban dwellers, employed mainly as skilled workers and professionals; a few are still largescale commercial farmers. In countries such as Zambia their lifestyles contrast strongly with those of the black majority, reflecting their elevated socioeconomic position: they live in large suburban houses with swimming pools, drive imported cars, and send their children to all-white private schools; in South Africa, this separatism was institutionalized by the apartheid system.

In contrast to the British, Dutch and Germans in the rest of southern Africa, the Portuguese in Angola and Mozambique mixed more freely in the past and there was considerable intermarriage. As a result, European and African cultures have hybridized to an extent not met elsewhere. The Portuguese that have remained in Angola since independence, however, often have a very narrow outlook and do not identify with the much wider black culture.

Throughout the region, the cultural preferences of large sections of the black population strongly reflect those of whites. Architecture and urban design are a case in point: rather than design buildings and city complexes to create a distinctive African style, modern planned towns are based on Western models; for example, Botswana's capital of Gaborone is built in the precinct style favored by British architects in the 1960s.

At a popular level soccer has become an almost universal sport, with groups of men and boys kicking a ball among themselves whenever opportunity allows. In the South African homelands there are thriving professional and semiprofessional soccer leagues, and Western fashion items, such as brand-name training shoes and track suits, are popular with some of the teams.

Although international sporting links were resumed with South Africa, following the restructuring of the sporting organizations to allow mixed participation

in the early 1990s, segregation still existed between white and black on a day-to-day level. Many games remained the exclusive preserve of whites (rugby football – the national sport of white South Africa – is not popular among blacks, for example). There is, however, greater mixing in other areas; for example, black students attend some of the "white" English-speaking universities, such as Johannesburg University.

Divisions among the whites

It would be erroneous, however, to imagine that uniformity exists among the white minority. In South Africa there is a very deep cultural divide between Afrikaners and British-descended whites that stretches back to the Boer wars. The farmers of the *platteland* (flat or plateau land) of the Orange Free State and the Transvaal are the descendants of the Voortrekkers who took part in the Great Trek, and who are regarded as the personification of the spirit of Afrikanerdom.

In celebration of nationhood (*below*) Afrikaner schoolchildren watch the annual recreation of the Great Trek. Afrikaners regard the 19th-c[illegible]ury migration away from British influence as a ce[illegible]t in their history and the origin of their dis[illegible]ntity.

This distinctive culture is given expression in the Dutch-derived Afrikaans language (the Afrikaners are the only European group in Africa to have evolved a language of their own), in ritual ceremonies such as the annual recreation of the Great Trek, when 19th-century Boer dress is worn and bullock-drawn waggons formed into a *laager* or circle, and in landscape icons such as the imposing Voortrekker Monument near Pretoria, South Africa's capital city. There are food distinctions too: the *braai* (barbecue) is a favorite social event, when a particular sausage, *boerewors*, is eaten. Afrikaners are one of the most religious groups in southern Africa, most belonging to the Dutch Reformed Church. Religious observance among them is strong, attendance at the Sunday service being a regular occasion to reaffirm group identity.

Whites of British descent, who can be found in most of the countries of the region, are a more disparate group. In South Africa they are to some extent distinguished by their liberal politics, particularly among those who have been settled there for several generations: new arrivals tend to feel less secure and therefore more racially prejudiced. Unlike the Afrikaners, who are culturally isolated by the Afrikaans language, British whites are drawn into a wider international culture by their use of English: their lifestyles and cultural preferences in literature, music and entertainment are similar to those enjoyed in Britain and the United States. Religious observance is [illegible]w, and predominantly Protestant.

The world of the islands

The diverse religions, languages and cultures practiced by the peoples of the islands reflect their rich mix of origins. On the Comoro Islands, Arabic influences are shown in the dominant role of Islam, with its strict religious and moral code. Swahili, the trading language of eastern Africa, is the main language, and French and Arabic are also spoken. Christianity and Hinduism are the main faiths of Mauritius, the latter being practiced by descendants of the Indian labor force imported there in the 19th century. English is the first language, but Hindi, Urdu and creole French are also widely spoken by the inhabitants. On Madagascar and Réunion the predominance of the Roman Catholic faith is a reflection of the strong French influence.

In the islands, daily life is a curious mixture of the new and the traditional. On Mauritius soccer is the national sport, as it is on the mainland, while the aping of pop idols reflects the strong appeal of Western music. In Madagascar, on the other hand, folk music is very lively, based on traditional dance rhythms, such as the African *watsa watsa* and using traditional instruments like the *valiha*, which consists of strings strung around a tubular wooden sounding-box. In rural parts of the island ancestors, or *razana*, directly influence the way people behave. This is manifested in a complex system of taboos, known as *farly*. It may, for example, be *farly* to whistle in one village, while in a neighboring settlement it may be *farly* to eat pork.

THE GREAT TREK

[illegible]frikaners' strong sense of a [illegible] cultural identity is rooted in [illegible]icular faith of the Dutch [illegible]Church and forged around a [illegible]rpretation of their history. [illegible] language of the Old [illegible]ks of the Bible, this [illegible]ners as God's Chosen [illegible]ica.

[illegible]his history is the [illegible]s since acquired [illegible]ymbol of Af[illegible]rd migration [illegible]. the Cape took [illegible] beset with great [illegible]ships. They were at[illegible]e Matabele and Zulu [illegible]who stole their cattle and [illegible]heir leader. Eventually they triumphed at the famous Battle of Blood River, and settled down to enjoy their Promised Land.

Many Afrikaners today regard the Great Trek as proof that they belong in South Africa: God chose His people (*volk*), brought them into the wilderness, and saved them at Blood river. This powerful message is reaffirmed each year when the Great Trek is recreated by hundreds of Afrikaners, thus perpetuating its position at the heart of Afrikaner cultural consciousness. Opposition to the dismantling of apartheid, which institutionalizes the right of the *volk* to live as they will in the land God entrusted to them, is consequently strongest among small farming communities of Afrikaners.

Women coping with change

The changing role and status of women has been one of the most significant social transformations that colonialism has made in southern Africa. In traditional African societies women acted as wives and mothers, and were largely responsible for cultivating tribal land. Men customarily looked after the livestock, controlled rural affairs and had ultimate authority in household and tribal matters.

The widespread use of migrant labor has radically altered this. African men may be away from their rural homes for periods lasting between one and two years. This means that women form a majority in the villages, and for the first time are heading the households. Married women are still junior to their absent husbands under tribal law, but in practice they make all the decisions. But the absence of men is affecting social structures still more profoundly: many women cannot find husbands, and illegitimate births are consequently rising, producing a significant number of single mothers who have no traditional decision-makers to act for them and their families.

These changing circumstances have disadvantages as well as advantages for women. There are physical problems in coping with specialist male tasks such as livestock rearing and plowing. Even

Female enterprise *(above)* By running a small roadside food stall, a woman hopes to be able to earn enough money to achieve some kind of financial independence. Many women have chosen to become self-employed rather than to accept the poor wages and lowly work that is generally all that is available to them in the cities.

In domestic service *(left)* A maid to an affluent white family carries their baby on her back in traditional African style while doing the housework. Most white families employ black domestic labor; for many urban women, this offers the opportunity for reasonable job security and relatively high wages compared to most jobs.

Wives and mothers *(right)* In traditional African society, women raise children and cultivate the land, but men make all household or communal decisions. Largescale male migration is altering the nature of rural life, forcing women to tackle new responsibilities and become more independent in the handling of their affairs.

where technology exists to compensate for the absence of male labor, women often do not receive the necessary technical training because they lack the time, the education and the money for it. Their difficulties increase if their men neglect to send back a regular part of their wages from their work in the mines or the cities. Lesotho miners, for example, earn over 3,000 Rand ($1,000) a year. Even a quarter of this amount makes life significantly easier for their wives and families, but many men fail to send money home – the demoralizing effects of mine labor and of being far from home and family lead all too easily to neglect of family responsibility. As a result, women are left with the daunting task of raising children and attending to the farm work without the benefit of any additional income.

Slow progress

The increased independence of women is taking place in the context of strengthening tribal loyalties, which reaffirms traditional male seniority. The tensions that this situation creates make it clear that for many women the changes are only skin-deep, and progress toward real independence will inevitably be slow. In Lesotho and Swaziland, for example, unmarried mothers are still not allowed to own and cultivate their own land, which can place them in an economically desperate situation. In socialist Angola and Mozambique, although the principle of male–female equality holds out hope of a better future, change has been slow.

Nevertheless, the new status of rural African women gives them the beginning of an independent voice in southern African society. This independence is still more noticeable among the single women who have migrated to urban areas in search of work. The traditional attitude to women means that they tend to command menial jobs at the lowest wages. To escape from this, many have opted to become self-employed; for example, by making and selling food and clothing at roadsides. Although the returns are small, they allow women to gain a foothold in the urban economy, offering them the prospect of better employment in the future, and the financial independence this will bring. Women raised in urban areas already enjoy this independence. In Johannesburg, for example, large corporations employ black secretaries at relatively good wages.

MANY PEOPLES, MANY TRADITIONS

A SUCCESSION OF CULTURAL INFLUENCES · ETHNIC AND RELIGIOUS DIVERSITY · CULTURES IN CONFLICT

The enormous cultural diversity of the people of the Indian subcontinent is reflected in the vast numbers of languages spoken, and in the variety of their beliefs, customs, diet and dress. The diversity is both a response to the region's range of natural environments, and the result of successive migrations of people over the centuries who have spread new cultural influences and modified others – a process that has continued into relatively modern times. Conversely, a number of indigenous religious cultures now have worldwide distribution, including Hinduism and Buddhism. In spite of the region's cultural complexity, several unifying features can be identified, most particularly the caste system, which affects a great many aspects of life, and the common experience of British imperialism.

COUNTRIES IN THE REGION

Bangladesh, Bhutan, India, Maldives, Nepal, Pakistan, Sri Lanka

POPULATION

India	801 million
Pakistan	109.43 million
Bangladesh	107.75 million
Nepal	18 million
Sri Lanka	16.6 million
Bhutan	1.36 million
Maldives	202,000

LANGUAGE

Countries with one official language (Bengali) Bangladesh; (Divehi) Maldives; (Nepali) Nepal; (Sinhalese) Sri Lanka; (Urdu) Pakistan

Country with two official languages (English, Hindi) India

Country with three official languages (Dzongkha, English, Lhotsam) Bhutan

India has 14 officially recognized languages. As well as Hindi and Urdu, the most significant languages in the region include Gujarati, Malayalam, Marathi, Punjabi, Tamil and Telugu. There are hundreds of local languages and dialects

RELIGION

Countries with one major religion (M) Maldives, Pakistan

Countries with two major religions (B,H) Bhutan; (H,M) Bangladesh

Countries with three or more major religions (B,H,M) Nepal; (B,C,H,M) Sri Lanka; (B,C,H,J,M,S) India

Key: B–Buddhist, C–Various Christian, H–Hindu, J–Jain, M–Muslim, S–Sikh

A SUCCESSION OF CULTURAL INFLUENCES

From earliest times, the fertile flood plain of the Ganges, in the north of the region, provided an attractive habitat for people, and archaeological evidence suggests that farmers were settled here 8,000 years ago. An early urban civilization developed in the Indus valley, in the northwest, in the 3rd millennium BC, and extended over the whole of present-day Pakistan and well into northwestern India. Excavation of its scattered sites, the best known of which are Mohenjo-daro and Harappa, reveals a distinctive culture, with characteristically decorated pottery and sculpture, a developed town planning, and a binary method of counting.

People from the north

About 1500 BC, groups of seminomadic people, known as Aryans and probably originating in central Asia, moved into the Ganges basin from the north, forcing the original inhabitants, the darker skinned Dravidians, to move south. These newcomers spoke Sanskrit, an early Indo-European language related to modern Hindi, widely spoken in northern India today. Their oral traditions are preserved in the books of knowledge – the Vedas, the Brahmanas and the Upanishads – and the classic epic poems, the *Mahabharata* and the *Ramayana,* that still make up the sacred (Vedic) literature of Hinduism.

Gradually the Aryans became scattered in a number of states ruled over by kings. Society was divided into four classes, or castes – priests (*brahmins*), warriors, farmers and traders, and servants – and it was at this time that the hierarchical caste system, which ranks individuals by birth, began to develop. Buddhism, which originated in northern India in about 500 BC, inspired by the teachings of Siddhartha Gautama, the Buddha (Enlightened One – c. 563–483 BC), came about as a reaction to such aspects of Hinduism. Both Buddhism and Jainism, which had developed a century earlier and is still practiced in parts of northern India, believed in the basic holiness of individuals (in the case of the latter, of all life) and their ability, through their own actions, to attain enlightenment, irrespective of their social status.

The monastic structure of Buddhism, with its emphasis on alms-giving and preaching, gave it a missionary character that helped it spread throughout the subcontinent and beyond. It reached Sri Lanka from southern India about the 3rd century BC and was later carried to the Maldives by the earliest settlers of the islands. Buddhism continued to be a strong cultural and social force in India until 600 AD, but the decline of the Mauryan empire (321–185 BC), which had unified the entire subcontinent under its rule, was followed by political fragmentation and the emergence of independent states including the Himalayan kingdoms of Nepal, Kashmir, Bhutan and Assam. Hinduism gradually took over from Buddhism as the dominant ideology, its caste system bolstering the feudal government of local noble families. Regional loyalties supplanted the political unity of the previous empires and local cultural traditions flourished.

External influences were also woven into the subcontinent's cultural tapestry. In the south, trade with Europe and Asia influenced historical development. The introduction of Christianity is attributed to St Thomas, one of Christ's twelve apostles, but it is more certain that traders from the Middle East founded churches on the southeast coast at the beginning of the Christian era. Reference is also made in some 5th-century records to a white Jewish community at Kerala. The Parsis, practicing Zoroasterianism, a religion that originated in Iran, migrated to India in about the 8th century to escape Muslim persecution; today they are concentrated in the cities of Karachi (Pakistan) and Bombay. In the north, links were developed through Nepal and across the Himalayas with China and Tibet.

Invasions of the recent past

About 1000, Muslim invaders from central Asia spread across the plains of the Indus and Ganges, establishing full control there by the 13th century. They had a profound effect on cultural development, particularly in architecture and the arts, and through their distinctive system of Islamic laws and style of government. Under the powerful Mogul empire, from the mid 16th century to the mid 18th century, present-day Pakistan, Bangladesh, and much of India were consolidated under one Muslim ruler and one bureaucracy. Only Nepal, Bhutan, Assam, India south of the Deccan and Sri Lanka remained outside Islamic influence.

Life beside the tracks *(above)* A family makes a home alongside the railroad tracks at Patna in Bihar state, one of the poorest areas of India. Rural poverty forces many people to migrate to the cities in search of work, and the railroads are their main means of transportation. Most of the railroad system was built by the British during their imperial rule in the subcontinent, a useful legacy of colonialism.

A Sri Lankan devotee *(left)* gazes up at a statue of the Buddha. Images of the "Enlightened One" are important features of Buddhist holy places, used as a focus for meditation and to inspire reverence for his teachings and example. Effigies are found in a variety of postures – meditating, teaching or reclining – that represent different aspects of the Buddha's life.

A final strand of cultural influence was introduced by the British, who held political sway in all the present-day countries of the region, except Bhutan, Nepal and the Maldives, from the 18th century until independence in 1947. They brought the Protestant tradition of Christianity and inaugurated widespread changes in government, administration, the legal system, transportation and communications, agriculture and trade. Public buildings, churches and railroad stations were built in the grandiose Gothic style favored by the 19th-century British. They even introduced cricket, which remains a popular national sport in India, Pakistan and Sri Lanka. They also encouraged the migration of large numbers of people from the subcontinent to the Caribbean, eastern and southern Africa, Fiji and other parts of the British empire, to provide labor on plantations.

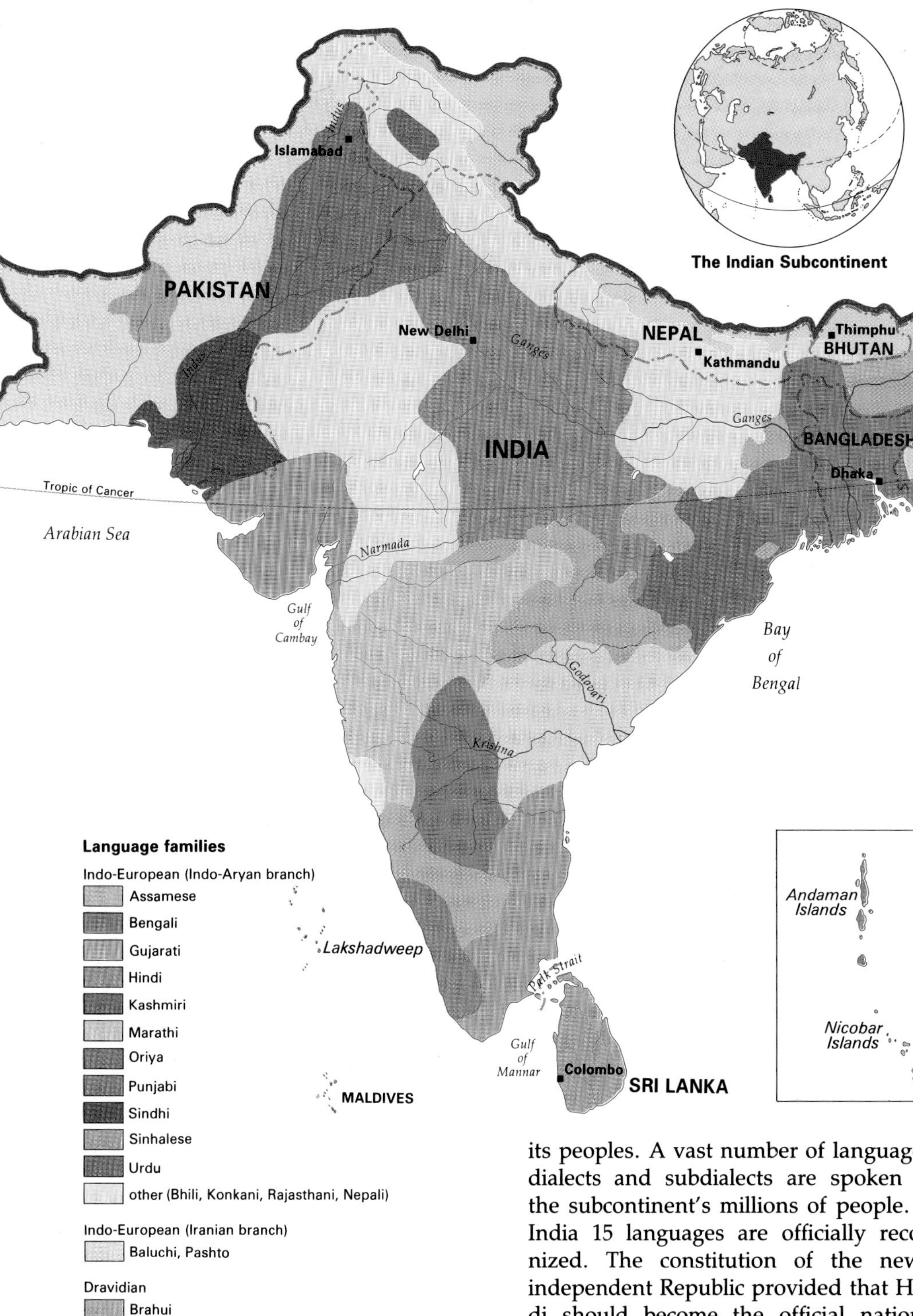

Indian women (*right*) gather to worship Lakshmi, the goddess of wealth and beauty. Festivals occur frequently in the Hindu calendar and are woven into the fabric of daily life.

Linguistic diversity (*left*) Most of the hundreds of languages spoken in the region belong either to the Indo-European family (in the north) or the Dravidian family (in the south). Urdu (related to Hindi) is the official language of Pakistan, though it is the mother tongue of only a small percentage of the population.

ETHNIC AND RELIGIOUS DIVERSITY

The population of the Indian subcontinent has mixed to such a degree as a result of its long history of migration and invasion that language rather than ethnic origin is the primary distinction between its peoples. A vast number of languages, dialects and subdialects are spoken by the subcontinent's millions of people. In India 15 languages are officially recognized. The constitution of the newly independent Republic provided that Hindi should become the official national language. English, though not classified as an official language, would be retained in use until 1965, when it would be replaced by Hindi. Since it is claimed as a mother tongue by some 150 million speakers in northern India and is related to some other northern languages, such as Gujurati and Punjabi, Hindi has some entitlement to be regarded as the lingua franca of the north, but it has little in common with the Dravidian languages that are widespread in the south, including Tamil, Telugu, Kannada and Malayalam. Southern states have consequently resented attempts to promote Hindi as a national language and, to avoid sectarian strife, English continues to be used as the language of government.

Group cultural identification within the region is also determined by religious affiliation. All the religions of the subcontinent have been shaped by their cultural surroundings – Sikhism, for example, the predominant religion of the northwest state of Punjab, blends elements of Islam and Hinduism. As a result, forms of social organization, daily observance and custom associated with a particular religious group may extend beyond that group, or have become modified within it. For example, Ismaeli Muslims in the north of Pakistan do not observe the Ramadan month of fasting, and women travel unveiled in public. Hindus in the north of India have adopted the Muslim practice of *Purdah*, or female seclusion.

The caste system

The most pervasive cultural influence throughout the subcontinent is the caste system. Although Hindu in origin, it affects all people within society, despite the emphasis that Islam, Christianity, Buddhism and Sikhism place on equality between individuals. Caste determines the individual's economic and social standing from birth. Within its four broad categories, society is divided into thousands of hierarchical but mutually interdependent ranks with defining rules of behavior. In rural communities, it governs the network of obligations and services between landowners and tenant peasant farmers.

A central element in caste is the concept of pollution; this links caste ideology with social practice and presupposes a set of ideas concerning the relationship between spiritual and physical life. The lowest caste rank were the untouchables, or Harijans. Traditionally, they were placed outside society because their habits of life involved polluting activities, such as killing animals, treating the hides of cattle, sweeping and cleaning, and disposing of human excrement. Orthodox Hindus

regarded the hill tribes as untouchables because they ate beef. The untouchables were often segregated in hamlets outside towns and villages, and forbidden entry to temples and schools, or from drawing water from wells used by higher castes.

Although many caste taboos with regard to eating, drinking and dress have been relaxed, their legacy is apparent in a bewildering variety of behavioral patterns found throughout the subcontinent. By and large, marriage still takes place within caste categories, thus preserving these cultural traditions. The economic function of traditional caste occupations has diminished over time but the old jobs are often replaced with similar substitutes, and occupational mobility tends to be greater among the higher castes who have the advantage of better education and higher literacy.

A prescribed role for women

Although the rapid modernization of society has modified many traditional cultural practices, particularly among more wealthy urban groups, women of every class remain subject to the cultural stereotyping that dictates their social behavior and determines their expectations. For all religions and castes, marriage is a universal goal; it is therefore difficult for women to remain single or to combine marriage and motherhood with a career. Even in the cities there is continued reliance on the practice of arranged marriage; in the absence of a network of relatives and friends to find a bride of suitable family and caste, such as exists in small rural communities, advertisements are very often placed in newspapers and magazines.

Such marriages strengthen the authority of parents and elder relatives, and ensure conformity to traditional roles. Once married, a woman is expected to fulfill the part of dutiful wife, mother, and, very often, daughter-in-law: it is still common practice for married couples to live with the husband's family. Only when a woman becomes a mother-in-law herself is she likely to attain freedom and authority. The persistence of the traditional images of women conflicts with the ideas and expectations of the increasing numbers of women emerging from higher education and into the formal workplace today. However, a major factor in their perpetuation appears to be the conservatism of women themselves.

FOOD CUSTOMS

A Hindu proverb that "caste is only a question of food" provides an insight into the cultural significance of food and its preparation. Environmental factors probably played a part in shaping differences of diet and food taboos originally, but subsequent patterns of migration, religious belief and caste behavior have had greater influence. The best known of the rules relating to diet are the prohibitions that prevent Hindus from eating beef, out of respect for the cow's sacred status, and Muslims from eating pork. Vegetarianism, which is widespread throughout the region, and is particularly strong in southern India, reflects the doctrine of *ahimsa*, or reverence for all life, that is fundamental to Buddhists and Jains as well as many orthodox Hindus.

Notions of purity and pollution affect many of the customs surrounding the eating of food. Traditionally, Hindus would not accept food prepared by someone belonging to an inferior caste. Only the right hand is used to carry food to the mouth, the left hand being considered ritually impure. Menstruating women are still considered unclean in many households and are forbidden to cook at this time.

The distribution of food is also culturally determined. Women often eat after the men of the household have finished, a custom that has serious implications for their health.

Pakistani Shi'ite Muslims eating during a festival. Certain foods hold significance in different religious celebrations.

CULTURES IN CONFLICT

During the second half of the 20th century the Indian subcontinent has seen a considerable increase in communal violence. There are several underlying reasons. The ending of British rule in 1947–48 removed a common cause for cooperation between cultural and ethnic groups and – despite the efforts of the politicians and administrators – the new national and state boundaries did not correspond to the major religious and linguistic divisions. Furthermore, in the past the widespread caste system had maintained both the economic interdependence and social separation of different groups within society, ensuring their peaceful coexistence. In the course of the 20th century, however, society has been transformed by a rising level of education and reforms of the economic and social structure, particularly in the rapidly growing cities. Population growth means that more and more people are seeking to share in economic and political power. Clashes that appear to be based on ideological and cultural differences between ethnic groups or castes are fueled by economic and political disaffection. For example, the demands of the Indian southern states for the promotion of the Dravidian languages were linked to a belief that the government had been dominated by the north. As a result, it was asserted, the south had been prevented from sharing in industrial and economic development.

The role of religion

More than any other cause, religious differences provide a focal point for sectional conflict in the subcontinent. In India, where Muslims account for only 11 percent of the population, concentrated mainly in the northwest, and in the state of Jammu and Kashmir in particular, violence between Muslims and Hindus is increasing. Some 60 intercommunal riots were recorded in 1960; in 1986 it had risen to 500, and tension has intensified since then. At the center of clashes in the early 1990s were claims that a disused Muslim mosque lay on the site of an ancient Hindu temple.

The regional strength of Sikhism also lies in the northwest, in the Punjab, where it was founded in the 15th century. Its adherents observe strict laws of obedience, set down in the Adi Granth, the Sikh sacred book, written in Punjabi. At independence, the Punjab was partitioned between Pakistan and India. Amid extreme violence and loss of life, a mass exchange of population took place, with about 2.5 million Sikhs moving into Indian East Punjab, and equal numbers of Muslims crossing to Pakistan. Sikhs now form a majority in East Punjab, though only about 2 percent of the total Indian population. The demand for a separate Sikh nation has led to frequent clashes with the national government.

Despite the ostensibly ethnic basis of their separatist claims, conflict in the

Untouchability *(above)* Although untouchables were renamed Harijans (children of God) by Mahatma Gandhi (1869–1948), who worked for their emancipation, they are still to be found doing the dirtiest and most menial jobs.

City life *(left)* A movie poster dominates the other signs mounted high above this crowded Bombay street. The Indian movie industry is the largest in the world, offering an easy escape route to romance and adventure.

Punjab must also be seen in its economic context. Living in one of the most affluent states in India, Punjabis resent the authoritarian rule of the national government in New Delhi and the seemingly one-way transfer of resources from their state to the rest of the nation. At the same time, the mechanization of agriculture and insufficient investment in industry has led to increasing unemployment among young Sikhs. For the poorly educated, unemployed farmer, a career in terrorism can reap rewards.

Such incidents are matched elsewhere in the region. Rioting has become endemic in Sindh province in Pakistan. Here the principal clashes take place between the rural Sindhi population and the Urdu-speaking Muslims who settled there after partition. Among the demands of the Sindhi nationalists are the promotion of the Sindhi written language and cultural heritage.

Race and mythology have been used to fuel conflict in Sri Lanka. After independence, the ruling Sinhalese majority stressed their Aryan ascent to promote their image as custodians of Buddhism; Sinhalese was established as the official language, and the mainly Hindu Tamil minority were excluded from a share in economic resources. After centuries of peaceful coexistence, relations between the two groups deteriorated to produce one of the most divisive and bitter of the ethnic separatist conflicts in the region.

Caste conflicts

Not all conflicts are regionally or territorially based. The "Scheduled Castes Movement" is concerned with the protection of the untouchables, a vulnerable minority in all regions. The label of untouchability has provided a sufficiently cohesive force that transcends cultural and economic differences between those that carry it. In recent years many untouchables have converted to Buddhism, theoretically a casteless religion but the label of untouchability continues to affect their quality of life. Attempts to introduce legislation to end discrimination against untouchables, such as the enforced use of quota systems to ensure places for lower caste members in education and government offices, have heightened the tide of feeling against them. The announcement in 1990 that 27 percent of government and public-sector jobs were to be set aside for "intermediate" or "backward" castes provoked riots in several states, with upper caste students protesting that merit rather than caste should determine employment. A number of students set fire to themselves to reveal their depth of feeling.

For a number of years individual states (particularly in the south of India) have been setting high quotas for backward caste places. The attempt to implement this policy on a nationwide basis has met with widespread protest from those who suggest that the backward castes do not suffer as much discrimination as the untouchables and tribespeople (for whom the quota system was first established) and that the reservation system enhances rather than defuses caste feeling.

ETHNIC DIVERSITY IN NEPAL

Nepal is a country of great geographical and ethnic variety. It comprises a myriad of ethnic groups all with their own languages, cultures and religions. Broadly, two external influences can be identified: the Tibeto-Burman from the north and the Indo-Aryan from the south. However, shamanist and animist practices are also strongly integrated in the Nepali culture. In the Himalayan villages, religious paintings, architecture, festivals and ritual ceremonies are more likely to be in the Tibetan Buddhist style. In the central hills, Hindu and Buddhist influences blend harmoniously. The Magars, for example, are of Tibeto-Burman origin, but many of them are nominally Hindu. Hinduism becomes increasingly more orthodox traveling down to the plains in the south.

For many years Nepal has been portrayed as a peaceful haven, free of external and internal conflict. As such, the country, and most particularly its capital, Kathmandu, became the destination for hippy pilgrims from the West on the trail of *nirvana* in the 1960s. For many Nepalis, the effects of mass tourism have been negative, and fears about hard drug usage contributed to much stricter visa controls in 1984.

The picture of Nepal as a harmonious kingdom was shattered in 1990 when strikes and demonstrations were met with curfews and killings. The popular movement for democracy in Nepal arose in protest against the partyless system of politics in the country and the power of the monarchy. While many Nepalis continue to recognize their king with reverence and affection (for many he is a descendant of the Hindu god Vishnu), calls for democracy were sufficiently strong to bring about a break in tradition.

Hinduism: an ancient religion

Hinduism is one of the world's oldest religions. It was developed by the Aryan invaders of the subcontinent between 1400 and 500 BC, and has been the major influence on the nature and organization of society, economy and culture within the region. Today it has more followers than any other religion in south Asia. The central concept of spiritual progression by reincarnation, or rebirth in a new form after death, provides an organizing principle for all aspects of Hindu life.

The caste system is a clear example of Hindu philosophy in action, each social caste indicating a person's spiritual status. At the top of the ladder the "twice born" castes – Brahmin, Kshatriya and Vaisya – are all closest to liberation from the wheel of birth, death and rebirth. Beneath all the castes are the untouchables who, because of their "polluted" status, have traditionally carried out the most menial and degrading tasks. The law of *karma* justifies this system of social differentiation, as an individual's position in life is considered to have been determined by his or her deeds and conduct in previous lifetimes. *Karma* also provides a strong incentive to conform to the laws and duties defined by an individual's *dharma* or moral code: not to do so may result in a lower reincarnation.

Traditional rites-of-passage are performed to prepare the incarnating soul. "Twice born" males, for example, receive a sacred thread made of cotton, wool or hemp (according to status) when undergoing their "spiritual birth". Marriage ceremonies are highly ritualized, often taking a number of days. In Bengal, among the Hindus of the upper class, the bride-to-be may take a ceremonial bath in water from the Ganges, the sacred river, several times a day to purify her body and her soul. During the actual wedding ceremony the girl makes the same offerings to her husband as she makes to the deities in the temples, symbolizing devotion to her spouse. Death rites are also important in Hindu society. Traditionally the dead must be carried to the riverside by their relatives, placed on a pyre and cremated. The mourners may bathe in the river for purification.

A living religion

The Hindu religion is essentially monotheistic, the many gods and goddesses symbolizing aspects of the one, omnipresent God. Nevertheless people choose to worship specific deities, the choice being governed by an individual's problems, caste or inclinations. Worship or *puja* is a very important part of everyday life and every orthodox Hindu's home, ranging from that of the urban elite to the slum dweller, has a domestic shrine where representations of the chosen household deity are placed, offerings of incense, food or flowers made, and daily prayers offered.

Pilgrimages to holy rivers such as the Ganges, the great number of religious festivals and the proliferation of temples and shrines bear witness to the importance of religious belief. Hinduism has no formal creed and rites and ceremonies vary enormously, though certain festivals such as Divali (the festival of lights) are held throughout India. During Holi, which is celebrated mainly in northern India and Nepal, devotees throw colored water and powder at each other.

Hindu mythology continues to be a major inspiration for culture and the arts, and for popular entertainment. A rich oral tradition of village storytellers kept the myths and legends alive. Today these myths are disseminated more widely than ever through the mass media; radio stations broadcast devotional songs, and bookstalls display brightly colored comic books retelling the epics, which have also been the subject of popular cinema and television movies.

Simple shrines (*above*) are found in many Hindu homes, providing a place where the daily *puja* can be observed. Their presence underlines the central role that religion plays in everyday Indian life.

Varanasi on the Ganges (*right*) is Hinduism's most sacred city. At dawn the riverside steps (*ghats*) are crowded with pilgrims who have come to worship and bathe in the holy river, and cremate their dead.

Hindu theater and the *Ram Lila*

After the monsoon rains have ended the festival of Dassara is celebrated across northern India. In villages, small towns and big cities, local communities get together to perform the *Ram Lila*, one of the major Hindu theatrical traditions. Each evening, for a period lasting from 10 to 30 days, casts of amateur actors present episodes from the life of Ram to large audiences in outdoor arenas.

The aim of the play is to give the audience a glimpse of a deity – in this case Visnu, of whom Ram was the seventh earthly manifestation – and so inspire feelings of personal love or devotion, known as *bhakti*. Visnu is the sustainer, who delivers the world from the threat of demonic forms. In the story, Ram is denied his rightful throne by his jealous mother, exiled, and obliged to battle with the evil Ravan to rescue his wife, Sita. Having overcome Ravan he reclaims his throne and restores order. Ram and Sita are therefore ideals of Hindu masculinity and femininity. Ram is just and self-possessed, while Sita is modest, shows fidelity, and is never heard to complain.

The performances of the *Ram Lila* are organized by local committees and financed by small subscriptions. Each year young boys from the Brahman caste are selected and trained for the major roles. They wear elaborate make-up and costumes, and once fully dressed, become the divine embodiments (*svarupas*) of the characters they play – and so must be carried to the playing arena. The audience is not limited to Hindus, though in recent years the nationalistic overtones of the *Ram Lila* story have been the cause of some friction within religiously mixed communities.

Epic performance A young boy wearing elaborate make-up and jewelry plays the part of Sita, the wife of Ram, at a staging of the play in Varanasi.

CONTINUITY AND CHANGE

AN INWARD LOOKING SOCIETY · HAN AND NON-HAN · MILLIONS OF PEOPLE

The peoples of China – who number about 1 billion, almost a fifth of the world's population – have great ethnic and cultural diversity. Dominated by the Han, speakers of Mandarin Chinese, its borders encompass some 55 recognized minority groups, each with their own language and cultural traditions. China has one of the oldest continual civilizations on Earth. For much of its history it developed in isolation from the rest of the world. One of the world's few remaining communist states, it has kept many elements of its distinct traditional culture. China's leaders face many problems, not least the feeding of its growing population, as well as the challenge of creating a modern, technological society that is able to preserve both its traditional Chinese roots and its communist ideals.

Outdoor recreation *(above)* Four men enjoy a game of *mahjong*, which is believed to have been played in China for at least 1,000 years. Originally a card game, it is now usually played with 144 small ivory tiles enscribed with Chinese characters.

Seeking guidance *(right)* Chinese temples are busy, crowded places where people go when decisions have to be made about personal, family or financial matters. Predictions are obtained on slips of paper, which are then discarded.

COUNTRIES IN THE REGION

China, Hong Kong*, Macao**, Taiwan

POPULATION

China	1,110 million
Taiwan	19.9 million
Hong Kong	5.76 million
Macao	479,000

LANGUAGE

Countries with one official language (Mandarin Chinese) China, Taiwan; (Portuguese) Macao

Countries with two official languages (Cantonese Chinese, English) Hong Kong

Other significant languages in the region include Manchu, Miao, Mongol, Tibetan, Uighur and Yi. There are numerous local languages and dialects

RELIGION

China Although religion is officially discouraged, many people practice a combination of Confucianist, Taoist and traditional folk belief. There are smaller groups of Buddhists, Muslims and Christians.

Hong Kong Buddhist-Confucianist-Taoist (92%), Christian (1.1%), Muslim (0.1%)

Macao Buddhist (45.1%), nonreligious (43.8%), Roman Catholic (7.4%), Protestant (1.3%)

Taiwan Confucianist-Taoist-traditional (48.5%), Buddhist (43%), Christian (7.4%), Muslim (0.5%)

* *Colony of UK; due to be returned to China in 1997*
** *Colony of Portugal; due to be returned to China in 1999*

AN INWARD LOOKING SOCIETY

The remains of some of humankind's earliest ancestors – known as Yuanmou Man, Lantian Man, and Peking Man – who lived between 1.7 million and 600,000 years ago, used tools and (in the latter case) had a knowledge of fire, have been found in China. There is a considerable archaeological record of human habitation of the region into neolithic times. A distinct Chinese civilization, one of the world's oldest, began to emerge about 5,000 years ago. A dynastic system, in which power passed through a family line of descent, gradually evolved, initiating a series of dynasties that lasted until the 20th century.

It was not until the third century BC, under the Qin (Chi'n) and Han dynasties, that China begin to approach its present shape and attention turned to keeping the "barbarians" – the nomadic herdsmen of the north and of Central Asia to the west – at bay. It was at this time that the Great Wall was built as a bulwark against invasion, and the boundaries established then were later regarded as embracing the indivisible area of China proper.

A culture based on writing

The development of an ideographic system of writing played an essential role in establishing China's cultural distinctiveness. It created a medium for spreading cultural ideas; furthermore, command of the written language distinguished the Chinese from their barbarian neighbors, and became a requisite skill for holding office. It was through written texts that the teachings of Confucius (551–479 BC), which continue to influence Chinese thinking and behavior, were transmitted: the continuity of the language means that they can still easily be read today.

Confucius believed that humans are innately good and taught that ethics and morality were the way to justice and peace. He believed in an ordered society of superiors and subordinates. Confucianism, with its insistence on respect for elders, came to form the basis of political and social life in China: mastery of the Confucian literature was the recognized way of entry into the civil service until the beginning of the 20th century.

Taoism, the other main tradition in Chinese culture, was founded in the 6th century BC by Lao Zi, who taught that there is a natural order to the universe – the *Tao*, or Way – comprising a continual dichotomy between *yin*, dark, female negative energy, and *yang*, light, male positive energy. Its emphasis on individual freedom and mystical experience contrasts with Confucianism's concern

with moral duties and public behavior. Over the centuries it has come to incorporate a large number of folk deities. Buddhism, which spread into China in the 1st century AD, provided many people with religious belief in an afterlife, and its ideas fitted comfortably with Taoism, and also with the Confucian ideal of moderation and self-control.

Throughout its long history China has experienced little external cultural influence. It was, however, briefly incorporated into the Mongol empire, which at its peak stretched across Asia and Europe from Korea to Germany and from the Arctic Ocean to the Gulf, when Genghis Khan, a tribal chief who had unified the Mongol-speaking peoples, crossed the Great Wall in 1211 and marched on Beijing. By 1276 the Mongol Yuan dynasty ruled the whole of China, though its domination lasted for less than a century. From an early date, however, elements of Chinese culture, such as its philosophical ideas, its script, its literature and its technological knowledge, were transmitted outside its boundaries, particularly to Korea and Japan and to Southeast Asia, especially Vietnam.

China's rulers were intensely suspicious of Western ideas, which began to penetrate the region with the opening up of trade with Europe from the 16th century, believing that their long-developed technologies were superior to Europe's industrial cultures. They attempted to control Western influence by confining European traders to certain designated ports and by establishing monopolies. However, they were unable to resist the twin pressures of Western imperialism and unrest at home. A disastrous series of trading wars from 1839 fueled the dissatisfaction of the people, many of whom were living in acute poverty, and brought about the downfall of the last Chinese dynasty in 1911.

Remaking Chinese society

The prolonged period of social upheaval and political confusion that followed led eventually to China's becoming a communist state in 1949. Under the chairmanship of Mao Zedong (1893–1976) every element of China's traditional society – agriculture, industry, social organization, art and culture – was remodeled on Marxist lines, and every aspect of daily life controlled: rest, recreation, food and work. Undoubtedly the principal architect of the new socialist China, during the last years of his life Mao became the subject of a personality cult, particularly during the bitter years of the Cultural Revolution (1966–76).

Mao's death in 1976 led to a reevaluation of socialist ideology with a view to modernizing China without destroying the fundamental aim of equality. Under his successors a market economy was introduced, agricultural communes were abolished, and farmers again permitted to farm their family plots. Trade and foreign investment were encouraged, but a new emphasis on materialism was not matched by freedom in political ideas.

A Miao girl wearing traditionally embroidered costume for a festival, in Guizhou province, southwest China. The Miao have a long history of resistance to domination by the Han, and have kept many features of their distinct culture, including their own language.

HAN AND NON-HAN

By far the largest group of people in China – some 93 percent of its billion population – call themselves "people of Han", a name they have used since the days of the illustrious Han dynasty (206 BC–220 AD). Although Mandarin Chinese in one of its three main variant forms is spoken by the great majority of Han, there are also a number of dialects, many of which are scarcely intelligible to other Han. However, all use the same system of writing, which is the major factor in defining their cultural identity. The Han are traditionally peasant farmers, and three-quarters of them still live in the countryside, mainly in the eastern half of the country.

On the margins

In the past, the attitude of the Han to other nationalities was colored by their agricultural way of life. This, they felt, marked them out from the nomadic people on their borders, whom they considered primitive, backward and abnormal. This sense of superiority is still reflected in official policy toward the non-Han peoples of China.

There are 55 recognized minority nationalities within China. Areas with significant concentrations of non-Han people have in theory been designated autonomous regions and granted self-administration. However, the reality has been that the Han-dominated Communist Party has determined what policies should apply to the minority nationalities, including the large-scale, government-sponsored settlement of many of those areas by Han people, and the exploitation of their natural resources to benefit the Han economy. A major source of conflict between the Han and the nomadic Uighurs and Mongols of China's vast western plains has been over the economic use of land. Although the Han need the cooperation of the minorities, in particular those that secure the western boundaries, recent decades have seen frequent clashes between the Han-dominated government and these minority groups.

The most troubled of the autonomous regions is Tibet, in the far west, which was only incorporated into the Republic of China by invasion in 1950. Small scattered groups of Tibetan-speakers are also found in neighboring provinces. They are predominantly Buddhist.

To the north of Tibet, occupying some of the most inhospitable places on Earth, with baking hot summers, freezing winters and sparse vegetation, are the Turkic-speaking Kirghiz, Kazakhs and Uighurs, descendants of the nomadic populations of Central Asia who ruled a substantial area of north and northwest China in the 8th and 9th centuries. Many Kirghiz and Kazakhs are still tent-dwelling nomads, but a large number of Uighurs have become settled farmers. Although they account for less than half the 11 million population of Xinjiang Autonomous Province, they are fiercely protective of their language, which is written in Arabic script, and of their Muslim religion. They are renowned for their silverwork, and make their own instruments to accompany their intricate traditional dances.

Uighur dissent has challenged the government on a number of occasions. Muslim separatists fought a long guerrilla war against incorporation into communist China – the last Uighur guerrilla leader was executed only in 1961. In 1986 the Uighurs protested vigorously against the use of their lands near the salt basins of Lop Nor for nuclear testing.

Their neighbors to the northeast, the seminomadic Hui (Chinese Muslims who adopted the religion when it penetrated China from Central Asia), now found mainly in Gansu and Ningxia, and the Mongols of the Autonomous Region of Inner Mongolia, have experienced similar treatment. The latter are vastly outnumbered by Han Chinese (2 million to 19 million) but have resisted government attempts to restrict their nomadic way of life, preferring to remain independent horse herders than to become settled farmers, and choosing the *yer* (their traditional tent) to the brick house.

Ethnic diversity is particularly high in the mountains of south China, on the borders of Vietnam, Laos and Thailand. Although many of these people have become more or less culturally assimilated during the long history of domination by the Han, they are differentiated from them by language. The Miao, a large minority, are divided into more than a hundred subgroups, each distinguished by dress, dialect and customs, but all sharing a common language. Traditionally the Miao practice *azhu*,

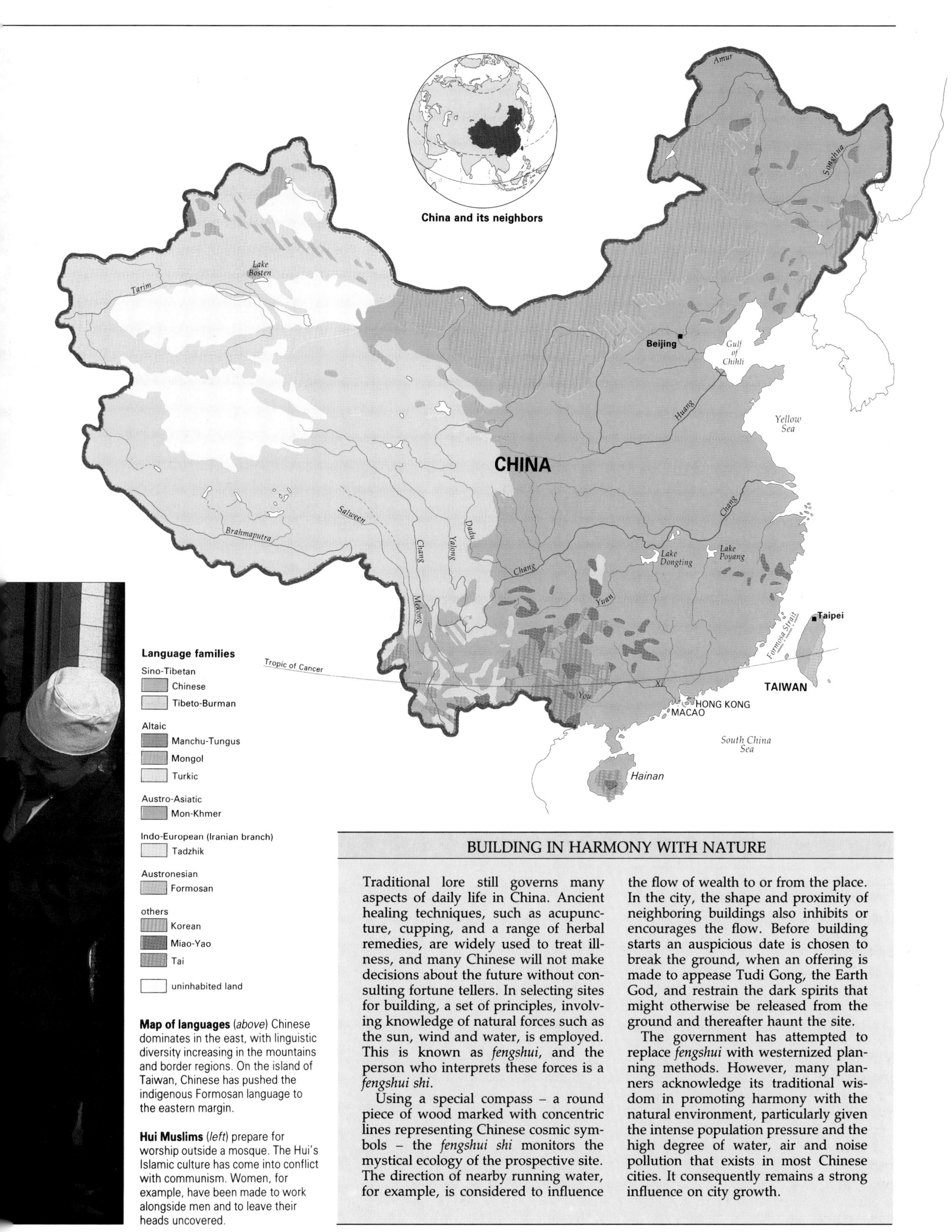

Map of languages *(above)* Chinese dominates in the east, with linguistic diversity increasing in the mountains and border regions. On the island of Taiwan, Chinese has pushed the indigenous Formosan language to the eastern margin.

Hui Muslims *(left)* prepare for worship outside a mosque. The Hui's Islamic culture has come into conflict with communism. Women, for example, have been made to work alongside men and to leave their heads uncovered.

BUILDING IN HARMONY WITH NATURE

Traditional lore still governs many aspects of daily life in China. Ancient healing techniques, such as acupuncture, cupping, and a range of herbal remedies, are widely used to treat illness, and many Chinese will not make decisions about the future without consulting fortune tellers. In selecting sites for building, a set of principles, involving knowledge of natural forces such as the sun, wind and water, is employed. This is known as *fengshui*, and the person who interprets these forces is a *fengshui shi*.

Using a special compass – a round piece of wood marked with concentric lines representing Chinese cosmic symbols – the *fengshui shi* monitors the mystical ecology of the prospective site. The direction of nearby running water, for example, is considered to influence the flow of wealth to or from the place. In the city, the shape and proximity of neighboring buildings also inhibits or encourages the flow. Before building starts an auspicious date is chosen to break the ground, when an offering is made to appease Tudi Gong, the Earth God, and restrain the dark spirits that might otherwise be released from the ground and thereafter haunt the site.

The government has attempted to replace *fengshui* with westernized planning methods. However, many planners acknowledge its traditional wisdom in promoting harmony with the natural environment, particularly given the intense population pressure and the high degree of water, air and noise pollution that exists in most Chinese cities. It consequently remains a strong influence on city growth.

Father and son together This army officer knows his son's family will care for him as he grows old, in traditional Han manner. But a consequence of the One-Child Policy is that couples with a daughter will be unsupported in their old age.

whereby husband and wife continue to reside in their mothers' houses, the husband visiting his wife each night. Another independently-minded minority, the Zhaung, are heavily influenced by Thai culture. Many still practice "levirate", the customary marriage of a widow to her deceased husband's brother.

The Chinese overseas

During the last half of the 19th century and the beginning of the 20th, large numbers of Chinese, spurred on by political turmoil and poverty at home, emigrated to many parts of the world in search of work, many of them as indentured laborers. Although some traveled as far as the United States, where they helped to build the railroads, by far the greater number ended up in Southeast Asia. Chinese overseas communities are closeknit, preserving their traditional customs, languages, festivals and foods. Although few emigrés today have any expectation of returning to live in China, many retain close emotional links with the country and send money to relatives in the Republic.

Taiwan, off the southeast coast, was first settled by the Han Chinese in the 17th century. Following the communist victory on the mainland in 1949, the nationalist government and its supporters fled to the island. Today 98 percent of the population are Han Chinese, 15 percent of whom are postwar arrivals from the mainland. The populations of the European trading colonies of Hong Kong and Macao, both due to be returned to the Republic of China before the year 2000, are also predominantly of Chinese.

MILLIONS OF PEOPLE

China is by far the most populous nation on Earth – over a billion people live there, one-fifth of the world's population. But some 90 percent of them live on only one-sixth of the land, concentrated in the fertile valleys of the east. Vast expanses in the west are virtually uninhabited.

Although China is still predominantly an agricultural country – more than two-thirds of the population are employed on the land – the cities are rapidly growing, and city life is marked by acute overcrowding. Three generations of one family often live together in a single room, sharing a bathroom and kitchen with several other families. Urban workers are mainly employed in large, state-run industries, though an increasing number of smaller, neighborhood workshops have been established since the 1980s, and lighter, modern industries developed.

Politics and the Communist Party dominate the lives of the urban Chinese. They will frequently attend political meetings, both at the neighborhood level and in the workplace. Party messages are relayed from street corners to the millions of cyclists that crowd the city at rush hour, and televised political broadcasts greet the worker on returning home.

Population control

As early as 1962, the government realized that China's population was outgrowing the country's ability to produce sufficient food to feed its millions of mouths. Mao Zedong, however, believed that China's strength would come from her people, and continued to encourage large families. It was not until the early 1970s that China began to develop a serious population policy. Late marriages, fewer births and longer intervals between children were advocated. For a brief period, marriage before 23 for women and 25 for men, was prohibited.

By 1980, the policy of "one child is best, two at most" was replaced by the much stricter One-Child Policy. Although men and women were now enabled to marry at 22 and 20 respectively, couples were required to restrict their families to only one child. Single-child families were given a range of economic incentives, including free education and housing allowances, which would be taken away on the birth of a second child. In addition, women had to obtain the permission of their workplace before becoming pregnant; those who failed to do so were often forced to have abortions. Due to their relatively small numbers, minorities were exempt from this policy.

The One-Child Policy did not win general acceptance, particularly in rural areas. Chinese agriculture is extremely labor intensive, and great emphasis is placed on family labor. Furthermore, the Han traditionally welcome sons as an assurance that they will be cared for in

CARING FOR THE ELDERLY

China is an aging nation. In 1990 old people accounted for just under one-tenth of the population, but this figure is expected to rise to one-fifth by the year 2025. This rapid growth in the proportion of the elderly is the result of relatively longer life-expectancies and the implementation of the government's population control policies.

Most elderly Han still live with their families. On marriage, the elder son remains with his parents, and his wife moves into their household. She then loses all obligations to her natural parents and takes on responsibilities for her in-laws, who become in effect her new parents. A couple without a son loses this support. Among younger people in the cities, moreover, the nuclear family is becoming common.

In 1982, the National Committee on Aging was set up to cope with the growing number of old people who were not being cared for within a family unit. A national program of homes for the elderly was instigated, and by the end of the 1980s nearly a third of a million old people were in residential homes. There are plans for every town in China to have an old people's home by the year 2000.

their old age. These sentiments were so strong that many rural families continued to have large families despite all the injunctions against doing so. In 1988 it was admitted that the One-Child Policy was not working fully, and rural families were permitted to have a second child, or even a third if the first two were girls.

Living through upheaval

The Cultural Revolution, instigated by Mao Zedong as a means of strengthening China's socialist revolution and protecting it from supposed capitalists within the Party, had enormous social and cultural impact. Schools and universities were closed, students compelled to join the Red Guards, and millions of young people sent from the cities to work in the countryside. Those considered to be the enemies of socialism were denounced.

The Chinese brand of communism draws on selective strands of Confucian teaching – in particular, the emphasis on a moderate lifestyle, respect for principle and rejection of superstition – to reinforce its position. In Confucianism the only

Lively debate fills a teahouse in Chengdu, Sechuan. The city is famed for its vigorous intellectual café society, and has been called "the Little Paris" of China. The center of a rich farming area, it is also renowned for the excellence of its cuisine.

sanctioned form of violence is righteous punishment. Under Mao Zedong, this principle was used to endorse the means of enforcing decisions whose absolute correctness is beyond question: power may come from the gun, but it is the Party that wields it. It sanctioned the use of violence during the Cultural Revolution, when the Party authorized the Red Guards to destroy symbols of "bourgeois influence". It also justified sending in the tanks against the dissenting crowds in Tiananmen Square in 1989.

Confucianism itself was the target of particularly vicious attack during the Cultural Revolution. Red Guards, for example, dug up Confucius' grave at Qufu in Shandong province. As faith in Mao and Marx waned, however, the Party became more conciliatory. Confucius was recognized once more as a "great ancient thinker of China".

Other aspects of Chinese culture suffered from the strict dictates of the Cultural Revolution. Artists, for example, were exhorted to depict socialist realist themes, and traditional Chinese subjects were banned. Although the worker and the soldier are still favored topics, the relaxing of codes under Deng Xiaoping allowed the introduction of western styles and subjects, particularly among younger artists, and a return to traditional forms, such as ornate calligraphy. Literature moved away from the principle that writers should serve the political cause, and saw a return to literary merit.

Some of the changes imposed by the Communist Party have brought undoubted benefits. Both the written script and spoken Mandarin have been simplified. *Baihua*, or colloquial Chinese, has replaced the elaborate, stilted classical Chinese that was the monopoly of educated people until the middle of the century. As a result, knowledge of only between 2,000 and 3,000 characters is now required to read most daily newspapers and general books.

Tibet – the forbidden country

The Tibetan Autonomous Region lies in the far west of China. It is a vast grassland plateau that covers an area of 1.2 million sq km (460,000 sq mi) and lies at an average height of 4,500 m (14,760 ft).The massive mountain ranges of the Himalayas create a daunting barrier to the south, and it is the inaccessibility of its situation, as well as its strong Buddhist tradition and ancient history, that account for Tibet's social and cultural isolation. More than 2 million Tibetans inhabit the region today.

In 625 AD, Strong-Tsan Gampo became the first king of a unified Tibet, a large empire that controlled part of northern India and stretched east to the kingdom of Nanzhou, now modern Yunnan in south China. The significance of his rule is still evident. In marrying two princesses – one Chinese, the other Nepalese, but both of them Buddhists – Strong-Tsan Gampo introduced the religion to Tibet and thus to China. The first Buddhist temple was built in Tibet in 651.

Although Strong-Tsan Gampo made Buddhism the state religion, Tantric, or Mantrayana, Buddhism did not emerge as the dominant form of the religion in Tibet until a century later. The Tantric tradition uses the power of the emotions to transfer the energies of the mind into wisdom and enlightenment and seeks to unify the two aspects of the human being – masculine and feminine, intellectual and intuitive, active and passive. This form of Buddhism was strengthened in Tibet by the Indian Padma Sambhava, "the Precious Guru", who founded the first community of Tibetan lamas, or monks, in 740 and by the Indian scholar monk Atisha, founder of the Khamdampa sect in 1042.

It was this sect that established the notion of religious succession and the belief that all spiritual leaders are reincarnations of their predecessors, which is applied to the Dalai Lama, the spiritual and temporal ruler of Tibet. Since the first Dalai Lama in the 15th century there have been 13 more, each believed to be reincarnations of the first. The 14th in the line was enthroned in 1940, but fled to exile in India in 1959, following the Tibetan people's unsuccessful revolt against the Chinese occupying forces.

Occupation and oppression

Until the Chinese invasion in 1950, the Tibetan people had lived beneath a form of feudal theocracy, under which arable

A market in Lhasa (*right*), Tibet's capital. These traders are selling juniper branches. The staple Tibetan diet is barley flour, supplemented with vegetables, meat and dairy products: rice is reserved for the well to do and for monks. Tea, flavored with salt and thickened with yak butter, is the most common drink. Butter is also the traditional fuel for lamps.

Regardless of their surroundings (*left*), religious devotees prostrate themselves at the entrance to the Johang monastery in Lhasa – one of many that have now reopened. Tantric Buddhism uses meditation and other techniques such as the repetition of a sacred phrase, or mantra, to channel the energies of the mind into enlightenment.

Return to tradition (*right*) A high-status monk in ceremonial dress. Behind him other monks wear the red robe of the Tibetan lama. After the Chinese invasion thousands of monks were thrown out of the monasteries and put to work. The invasion, based on a claim to sovereignty that stretches back 10 centuries, was described as a "liberation". Chinese troops easily overwhelmed the ill-equipped Tibetan defenses.

land was owned by nobles and the monasteries, and the peasants farmed it, working as tenants or hired laborers. Nomadic herdsmen pastured their flocks of goats and yaks on the high steppes. It is estimated that about 20 percent of the population were members of religious orders, and monasteries were the main seats of learning. Economic development was minimal.

Following the invasion, China began to modernize the country, and introduced a massive building program of roads, bridges, hospitals and schools (before 1950 most Tibetans traveled on foot or by animal). Thousands of monasteries were destroyed and nuns and monks put out to work. Its persistent undermining of religious authority provoked the popular uprising of 1959, which led to more repressive measures.

Thousands of Han were settled in Tibet, and society reorganized in line with China's own principles of government. Tibet's goat and yak herders were forced into collectives and arable farmers compelled to abandon growing barley in favor of other crops. Religious oppression was intensified during the Cultural Revolution: Red Guards stormed through the remaining monasteries, wrecking them and destroying precious Buddhist relics. The Tibetans, however, refused to abandon their religious beliefs.

Although permission is denied the Dalai Lama to return from exile, China has relaxed its control of Tibetan religion. Monasteries have opened once again, temples are rising, village prayer wheels spinning, and flags bearing passages from Buddhist scriptures are flying. Young men in the traditional red dress of lama monks are to be seen again.

Tibet, however, lags behind the rest of China in economic terms. Per capita income is half the national average, the illiteracy rate is 70 percent, and life expectancy 20 years less than the average. In 1980 the Chinese government, keen to promote tourism to Tibet, launched a new development program, constructing new hospitals, schools and housing projects, and encouraging still more Han workers to move to Tibet. Despite assurances, Tibetans remain fearful that Chinese policies will overwhelm their traditional culture and way of life.

MEETING OF THE WAYS

PEOPLES OF THE FORESTS, VALLEYS AND ISLANDS · A MIX OF CULTURES
THE OLD AND THE NEW IN CONFLICT

The cultural complexity of Southeast Asia reflects both its fragmented island geography and its central location on ancient trading routes. Since earliest times, the sea has linked even the most distant groups, serving to transmit cultural ideas and behavior within the region – for example, women nearly everywhere play an important role in trade and ritual. Water and forest – the twin characteristics of the environment of Southeast Asia – have strongly influenced human activity: rice cultivation is almost universal and a number of common cultural features have arisen around it. But there is enormous diversity, too. Over the millennia immigrants and traders from outside the region (including most recently European colonialists) have introduced a succession of languages and religious cultures.

COUNTRIES IN THE REGION
Brunei, Burma, Cambodia, Indonesia, Laos, Malaysia, Philippines, Singapore, Thailand, Vietnam

POPULATION
Over 175 million Indonesia
Over 50 million Thailand, Philippines, Vietnam
10 million–40 million Burma, Malaysia
2 million–10 million Cambodia, Laos, Singapore
Under 1 million Brunei

LANGUAGE
Countries with one official language (Bahasa Indonesia) Indonesia; (Bahasa Malaysia) Malaysia; (Burmese) Burma; (Khmer) Cambodia; (Lao) Laos; (Thai) Thailand; (Vietnamese) Vietnam
Countries with two official languages (English, Malay) Brunei; (English, Pilipino) Philippines
Country with four official languages (Bahasa Malaysia, Chinese, English, Tamil) Singapore

RELIGION
Country with one major religion (B) Cambodia
Countries with two major religions (B,I) Laos; (B,M) Thailand
Countries with three or more major religions (B,C,T) Vietnam; (B,C,H,M) Indonesia; (B,C,I,M) Burma; Brunei; (B,M,P,RC) Philippines; (B,C,H,M,T) Singapore, Malaysia
Key: B–Buddhist, C–Various Christian, H–Hindu, I–Indigenous religions, M–Muslim, P–Protestant, RC–Roman Catholic, T–Taoist

PEOPLES OF THE FORESTS, VALLEYS AND ISLANDS

Until the 18th century Southeast Asia was a region of forests, and in comparison to India and China, its immediate neighbors, it was only relatively lightly populated. Except in a few areas, such as the island of Java and the Red river delta of northern Vietnam, land was abundant, and the earliest groups of people inhabiting the region were able to develop ways of life that took advantage of this fact. They worked with the forest, periodically clearing it for shifting cultivation or living a hunting–gathering existence within it. In the forested uplands of the region, small and dispersed groups of people survive in this way to the present day.

The water of life

Water has been no less important to the patterns of human settlement and culture in the region. An abundant supply of water was necessary for the wet cultivation of rice, the staple crop of Southeast Asia, and it was in the well-watered lowland valleys of the mainland and on the islands of Indonesia that the region's major agrarian kingdoms and civilizations developed. The control of water, both technologically, through sophisticated irrigation schemes, and culturally, through ritual and magic, are common threads that bind together the varied peoples of Southeast Asia. Another habit that links them all together is the use of the *ani ani*, or finger knife, in rice harvesting, which is to be found nearly everywhere in the region.

Water also provided the means by which people could exchange goods and ideas with each other, and with the wider world. The rivers and seaways that permeate Southeast Asia, act as corridors through dense forest and link the distant islands. The region lies on the maritime crossroads between India and China, and many of its early empires – including the Sumatra-based Srivijaya empire (from the 7th to the 14th century), the Majapanit empire centered on Java (from the 13th to the 15th century) and the empire established by the port of Malacca on peninsular Malaysia (from the 15th to the 16th century) – owed their power and prosperity to their ability to control the trade between east and west.

Cultural contacts

Over the last 2,000 years Southeast Asia has witnessed a series of cultural invasions from both east and west. The Chinese influence is strongest in Vietnam, where it can be traced back at least to the 2nd century BC. The Mandarin system of government, the philosophical, political and ethical doctrines of Confucianism – which stress the obligation of respect to elders, social conformity and the importance of ancestors and lineage – as well as the Chinese "religion" of Taoism, which is often regarded as standing in opposition to Confucianism – have all taken deep root here. Over the centuries there has also been considerable migration of Chinese people into the Malay Peninsula and into some of the islands of the region.

In the other direction, traders and migrants from the Indian subcontinent introduced a succession of cultural and religious influences. The earliest of these

Songkran celebrations (*above*) in the Thai city of Chiang Mai. The Buddhist New Year is greeted by a wave of festivities. Worshippers visit shrines and monasteries, taking food and gifts to the monks.

A cremation in Bali (*right*) Belief in reincarnation is very strong among the people of this small Indonesian island, the last stronghold of Hinduism in the region. Cremations, which liberate the souls of the dead for the onward journey, are performed with elaborate rites.

A floating market in Thailand (*left*) Waterways are a vital element of life in Southeast Asia, often providing the main links between and within communities.

was Hinduism, which is now adhered to only by the people of the Indonesian island of Bali. The missionary nature of Buddhism encouraged its spread into the region from the 3rd century BC onward: it remains the predominant religion of the mainland. From the 13th century AD Arab traders began to extend their activities into Southeast Asia and helped to establish Islam in the Malay Peninsula, on the islands of Indonesia, and on the southern Philippines.

They were followed not much later by European traders – first the Portuguese, and then the British and Dutch. Over the next three or four centuries European influence was transformed into colonial rule – the Portuguese, Dutch and British in Indonesia, Malaysia and Burma, the Spanish and United States in the Philippines, and the French in Indochina. They introduced, along with Christianity, Western-style systems of government and economic structures, and technology.

A MIX OF CULTURES

A number of clear patterns can be distinguished within Southeast Asia's cultural complexity. The traditions of the peoples of the islands are substantially different from those of the mainland, and there are similar distinctions between those of the uplands and the lowlands. In addition, within the region popular animist beliefs exist beside, and mingle with, the world's major religious faiths, and introduced cultural elements confront indigenous ways of life.

The island–mainland division is discernible in the distribution of languages within the region. On most of the islands a range of related Austronesian languages are spoken, which had their common source in a parent language originating in the Pacific region about 5,000 years ago. The mainland languages are much more varied, and their origins – whether from the Tibetan and Chinese families of languages, or from those of the Pacific and Indian subcontinent – are the subject of dispute. In more modern times, immigrant Chinese have introduced various Chinese languages, while indentured laborers from southern India have added Tamil. European languages, particularly English and French, are spoken among the urban elites.

Religious patterns

The distribution of religions shows the same broad divisions. Across the mainland, the practice of Theravada Buddhism is predominant, though in Vietnam the Mahayana Buddhism of eastern Asia is more common, as well as Taoism and Confucianism, reflecting the country's close cultural links with China. Among the islands, Islam prevails across Indonesia, Brunei, Malaysia and also the southern Philippine island of Mindanao. The Indonesian island of Bali remains the last stronghold of a once much stronger Hindu tradition, and most people in the Philippines are Roman Catholic – a legacy of Spanish colonial rule.

All these religions were introduced to the region from outside, and this has led some experts to regard Southeast Asia as a mere receptacle for foreign cultural influences. But, in reality, the religions are syncretic: they have been molded to meet the particular needs of their Southeast Asian adherents, and each incorporates a range of traditional animist beliefs in spirits and ancestors who are held to inhabit the natural world. Filipino Christians, for example, wear magic amulets, believe in spirits and perform supposedly Christian rites that have their origins in an animist past.

The Karen hill people of northern Thailand and Burma believe in the ancestral spirit (*bgha*) and life principle (*k'la*), and a multitude of other gods and spirits, but a significant proportion also accept Christian ideas. Living in isolated groups among the dense upland forests, these people had been little affected by the ideas of Buddhism, but Christian missionaries who first began to arrive in the area at the end of the 19th century succeeded in converting the Karen by drawing upon their myths to make comparisons between the Karen religion and Biblical Christianity. The Karen proved receptive to this approach and by 1919 nearly 20 percent of the Burmese Karens had become Christian. Many Christian missionaries continue to proselytize in the marginal areas of Southeast Asia, for example among the Dayaks of eastern Malaysia, the Papuans of Irian Jaya and among the hill peoples of Thailand.

Traditional beliefs

Many of the traditional beliefs and practices of rural people persist, reflecting their dependence upon the natural world for survival. Forests, trees and rivers contain spirits, while water is revered and honored. Festivals, magic and ritual ensure the arrival of the monsoon rains,

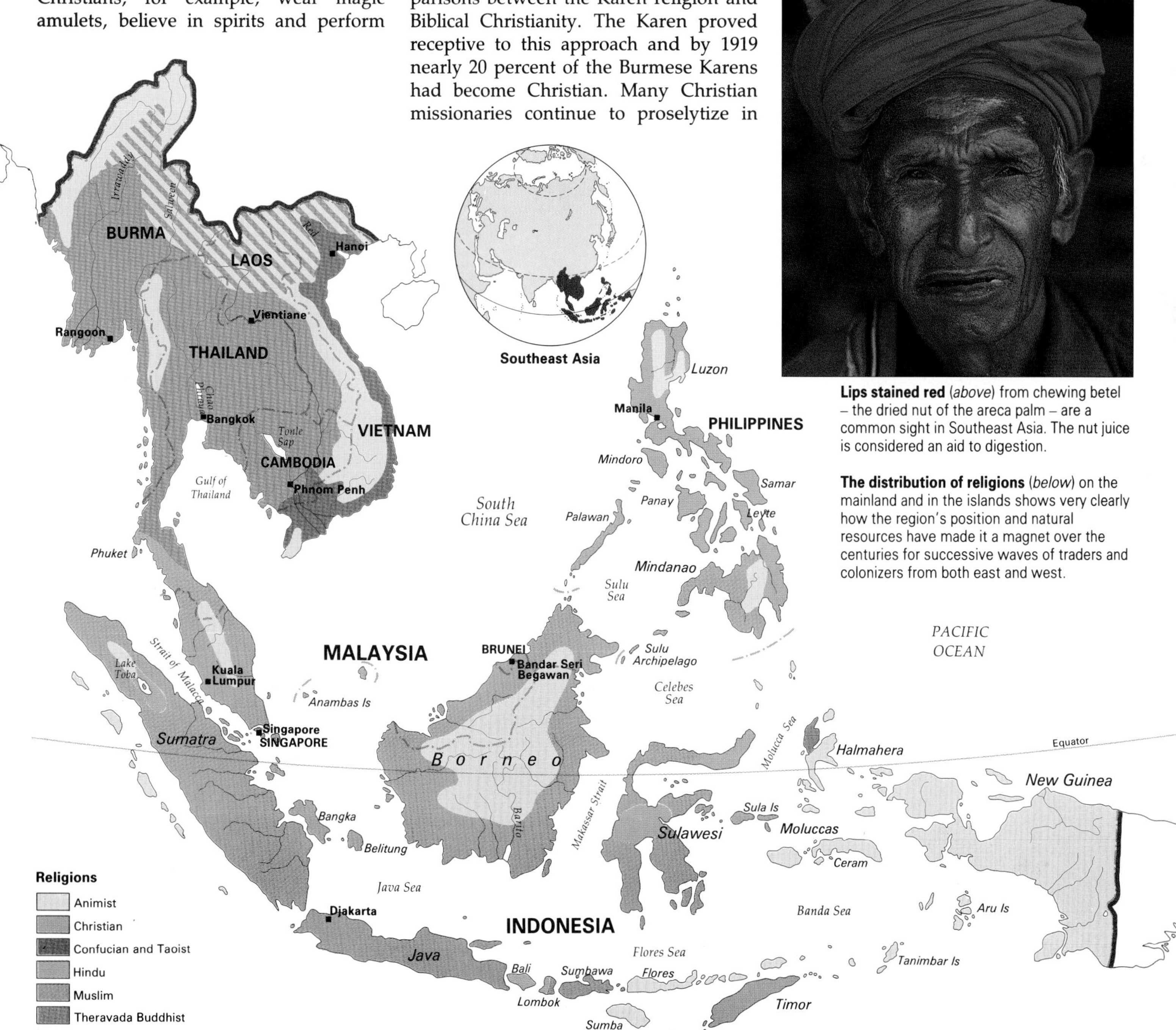

Lips stained red (*above*) from chewing betel – the dried nut of the areca palm – are a common sight in Southeast Asia. The nut juice is considered an aid to digestion.

The distribution of religions (*below*) on the mainland and in the islands shows very clearly how the region's position and natural resources have made it a magnet over the centuries for successive waves of traders and colonizers from both east and west.

The public face of Islam Two young Malaysian Muslim girls walk home together past a mosque in a busy street in Kuala Lumpur. The *dakwah*, or Islamic revival, has persuaded increasing numbers of women to resume wearing traditional headgear in public.

which mark the beginning of the rice-growing cycle. In rural parts of Thailand, during periods of drought, female cats, held to personify dryness, are carried in procession and drenched while villagers entreat for rain. The rice season is marked by rituals to propitiate the spirits of the soil, and of the rice itself, thus ensuring a good harvest.

Commercialization, consumerism, new technologies and Western cultural values have all impinged upon traditional life in modern Southeast Asia. Even the remote, forested areas have become accessible to the outside world. As a result the cultural diversity existing among small, widely dispersed groups is being eroded. The customs and exotic dress of the hill peoples of northern Thailand and of Borneo may serve as attractions for tourists, but their ways of life – the bases of their particular cultures – are rapidly giving way to change as they are assimilated into a wider national, and even international, culture.

It is in the cities, however, that Western culture, economic activity and political ideology have had the most dramatic impact. It is there, born in the struggle for independence from colonial rule, that nationalist revolutionary movements took root, and the postcolonial transformation of the traditional economy has wrought enormous social change. Large communities of immigrants, such as the Chinese, are concentrated in these crowded urban areas, and slum settlements contain millions of people whose way of life seems far removed from their traditional roots in the countryside – in Bangkok alone there are an estimated 1.2 million such displaced slum dwellers.

Yet, despite the harsh physical alteration of their living conditions, the cultural outlook and values of these people may be substantially unchanged, continuing to influence their lives as they did in the past. Young Buddhist men still strive to enter the monkhood and, in contrast to the urbanized societies of South America, ties between city and countryside remain strong. Rural migrants send their earnings home to their families, and most intend to return to the village, buy some land, and settle down as farmers.

THE WAYANG PUPPET THEATER OF INDONESIA

Wayang is a Javanese word meaning shadows. The earliest record of *Wayang* theater is found in a poem of the 11th century, by which time it was already well established in Java. Since then it has spread to all the other islands of the Indonesian archipelago.

The performers enact dramas by using puppets of leather, cloth and wood, or painted paper, and sometimes even live actors, to cast shadows on a white screen. Performances are usually staged to mark important moments in an individual's life – for example, birthdays, circumcision, marriage – and provide far more than just entertainment: the characters stand as role models and the stories contain ethical lessons and frequently comment on current affairs. Originally the puppets represented ancestral spirits, and the puppeteer was probably a shaman, a traditional healer and priest. Later, under Indian influence, the epic poems that tell of the Hindu gods and heroes, the *Ramayana* and *Mahabharata*, were incorporated into the repertoire.

The oldest and commonest form of *Wayang* uses two-dimensional, finely carved and painted leather puppets, jointed at the elbows and shoulders and manipulated by horn rods. To enact the entire repertoire of 177 plays, some 200 puppets are used; a single performance may need as many as 60 and last up to nine hours. The puppets are highly stylized and immediately recognizable to the audience – key characters in the drama may be represented by a number of puppets to depict different moods. A coconut oil lamp was used to cast the shadows, but today electric light is employed, which diminishes some of the mystery of the occasion.

THE OLD AND THE NEW IN CONFLICT

Rapid economic and social change of the kind that the countries of Southeast Asia have experienced since 1945 inevitably leads to friction as the lifestyles, cultural values, religious ideals and economic aspirations of different groups of the population come into conflict with one another. The need to build stable nation states in the years after decolonization led many governments to attempt to integrate, by persuasion and sometimes by force, those groups living on the periphery, both in a geographical and in a cultural sense.

Consequently, in Indonesia the forest peoples of Kalimantan (Borneo) and Irian Jaya (New Guinea) have been encouraged to abandon their traditional systems of shifting agriculture and to settle, grow rice, learn the national language, and convert to the national religion of Islam: in short, to become wholly "Indonesian". In Kalimantan, for example, traditional practices such as marriage ceremonies are reported to have been banned by the authorities. In resisting cultural imperialism of this kind, groups in Irian Jaya have taken up arms to protect their cultural as well as their political independence. The hill peoples of Burma have also resisted the attempts of the lowland-dominated government to assimilate them culturally.

The attraction of the West

There is a wider, but corresponding, conflict between traditional and Western ways of life. Many Southeast Asians feel that in the rush to modernize their economic and social structures they have jeopardized their cultural identity. In material terms, young people in particular are ignoring their own traditions in favor of Western-style houses, food, clothes and art. Throughout the region pizza and burger restaurants are becoming increasingly popular; jeans, T-shirts and suits are replacing the *sarong*, the loose skirt worn by both men and women; and Western fashions are becoming the hallmark of success.

Many people, especially the older generation, are fearful that a similar but more sinister change is overtaking nonmaterial cultural values. Respect for elders is declining, crime and violence are on the increase, and traditional religious practices are ignored – all changes that are, rightly or wrongly, seen to be characteristic of Western culture. So far had this process gone that the Singapore government, which had done more than any other to promote a Western world view, saw the need, some 25 years after independence, to cultivate a distinctive Singaporean identity among its predominantly Chinese citizens. Accordingly, Confucianist ethics and virtues were stressed, Western individualism frowned upon, and the greater use of Mandarin Chinese encouraged.

Modernization has led many people throughout the region to become more secular in their outlook, provoking a resurgence of various forms of religious fundamentalism in reaction. In Malaysia, the *dakwah*, or Islamic revival, has promoted Islam as a way of life, leading to demands for an Islamic state and Islamic law (*syariah*). In schools and universities girls are encouraged to wear Islamic headgear and are dissuaded from mixing with male classmates.

Winners and losers

In rural areas, the favoring of commercialized farming methods over traditional subsistence systems has had widespread impact on people's lives. Some farmers have had their land repossessed by moneylenders as market forces turned against them, compelling them to become landless wage laborers. Responding to this threat of dispossession, there has been a demand for a return to the traditional village ideals of self-help and self-reliance. The *Wattanatham Chumchon* ideology ("community culture") in Thailand is one such movement that builds on folk culture and local concepts.

Cultural conflict (*above*) Forest dwellers wait outside a store in Irian Jaya. Many traditional peoples have found themselves marginalized as governments try to impose some form of national identity on a previously diverse society.

Two-wheeled traffic (*below*) dominates the streets of Ho Chi Minh City. The bicycle is the favored form of transportation in many Asian countries: the cost of motorized vehicles is prohibitive for most people, though motorbikes are increasing in popularity.

In the cities, competition for economic resources is at the root of conflict between different ethnic groups. Large numbers of Chinese, and to a lesser degree Indian, immigrants not only introduced new cultural traditions but also became prosperous as traders and manufacturers. In Malaysia, resentment by Malays of the economic success of the Chinese community led to antagonism between the two groups, which in 1969 resulted in sectarian riots. The Malay-dominated government introduced positive discrimination to redress the economic balance in favor of the Malays, but in spite of some narrowing of inequalities, considerable enmity remains.

A PEOPLE DIVIDED BY WAR

Since World War II, ideological division, rooted in the nationalist struggle for independence, has brought terrible conflict, physical devastation and cultural turmoil to the countries of Indochina. On the people of Vietnam, divided by war from 1947 to 1979, the effects have been enormous. The presence of large numbers of United States' forces had great impact on the south: Saigon, the capital, had 56,000 registered prostitutes and countless drug addicts. Since "reunification" in 1975 the government of the Socialist Republic of Vietnam has tried to reform and rehabilitate these people using innovative methods such as acupuncture, self-criticism and re-education: success rates are reported to be high.

In former North Vietnam the effects of the war were equally dramatic. To protect industries and people from the American bombing campaign, perhaps as much as three-quarters of the population of the capital, Hanoi, were evacuated to the countryside: some 750,000 people. The process of socialist development replaced traditional patterns of land ownership and labor, and the war caused major disruption to village life throughout Vietnam.

Years after reunification, the differences between the northern and southern halves of the country are marked. In the south Ho Chi Minh City (formerly Saigon) bustles with life. Young people are relatively Westernized and consumer goods are widely available. In the north, entrepreneurs are few, the bureaucracy more stifling and life is harder. Vietnam still consists of two very different halves – almost two countries.

Theravada Buddhism

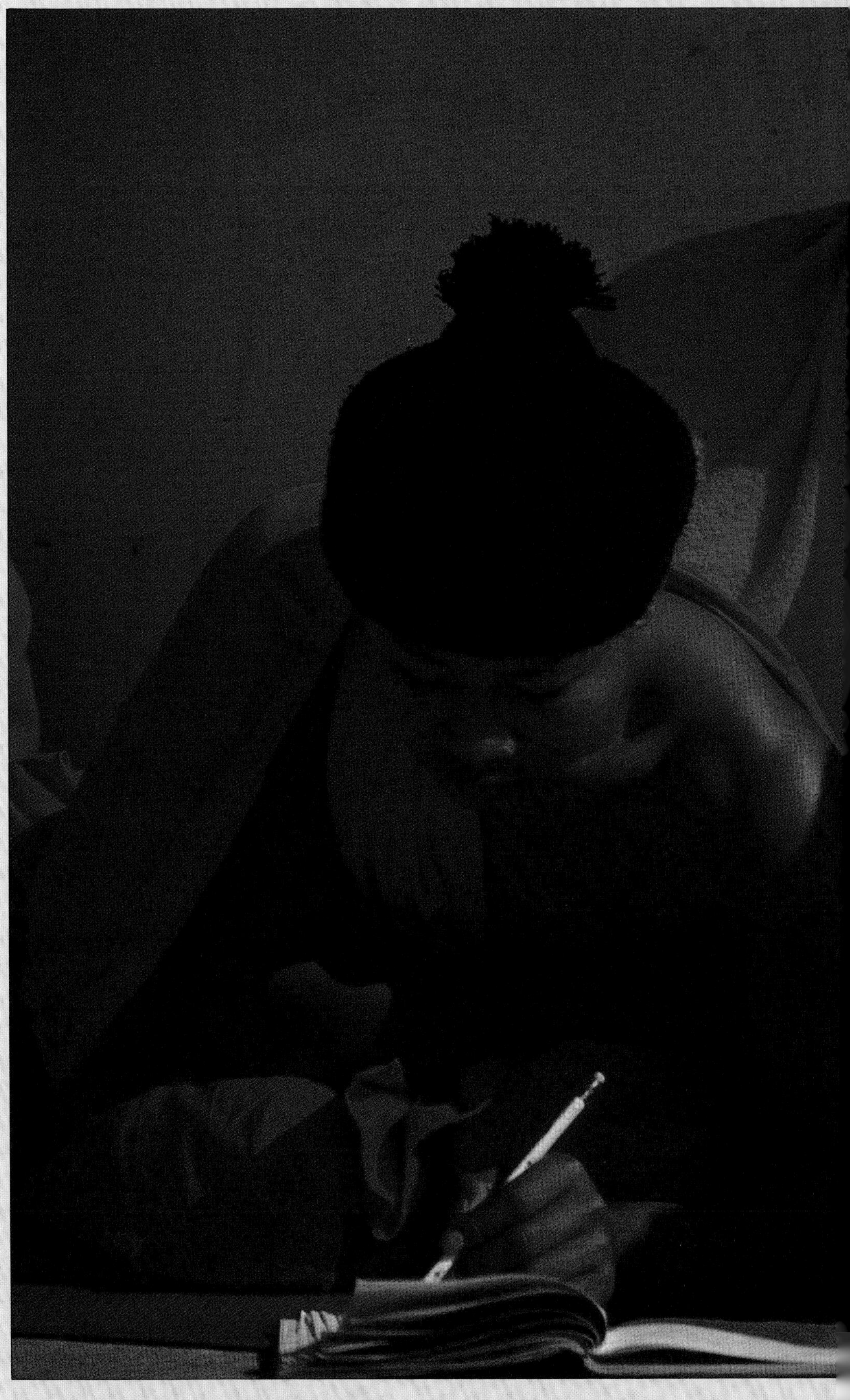

Although the practice of Buddhism, the religion and philosophy founded by Siddhartha Gautama – the Buddha, or Enlightened One (c. 563–483 BC) – in India in the 5th century BC, had spread to Southeast Asia as early as the 3rd century AD, the school of Buddhism embraced by most Southeast Asians today, Theravada Buddhism, was not introduced until the 12th century by monks from Sri Lanka. It is both a universal and, importantly, a popular religion: it emphasizes the composite nature of all things.

The performance of meditative techniques and rituals as the path to *nirvana*, the state of enlightenment – becoming one with God – that Buddhists aspire to, often appears to conflict with the noisy ceremonies designed to help the dead on their way, and the copious consumption of alcohol that is part of the ordination of the Buddhist monk. However, *nirvana* is a remote goal for Theravada Buddhists. Instead, they mainly concentrate their efforts on living within the law of *Karma*, the reduction of suffering.

An individual's position in society and the path his or her life takes is explained in terms of this law and of the accumulation of merit acquired in the course of innumerable deaths and rebirths. Theravada Buddhists try to perform good deeds and avoid being wicked in this lifetime, not only to live a holy life now, but also to avoid providing a cause for unhappiness in a future life or lives. *Karma* is not fate, as so commonly thought in the West. It is a direct result of, and may be changed by, an individual's actions throughout his life – it involves free will.

A religion for everyday

Theravada Buddhism in Southeast Asia blends quite comfortably a number of elements taken from different Indian religions – Brahmanism and Hinduism – and from animism and ancestor worship. Amulets are worn for protection from harm; spirits and ghosts are appeased; various shrines and natural objects are believed to have supernatural powers, and astrologers are avidly consulted and their instructions followed. These aspects of the religion provide worldly assurance for its adherents and are perceived to be complementary to Karmic law.

In Burma, Thailand, Cambodia and Laos, the countries of Southeast Asia where Buddhism is the dominant faith, almost every village contains a monastery and most young men are expected to become monks at least once in their lives, usually during the three-month Buddhist Rains Retreat, from July to October, but sometimes for considerably longer. This period of monkhood is invaluable in a number of respects. It enables a man to study Buddhist teachings, to further his general education, to gain merit for his family, particularly his mother, and to prepare for a responsible moral life – in short to become complete and mature. Theravada Buddhist monks live solely on

what is given to them, and they begin each day by walking through the village or town to accept offerings of food – usually from women.

Theravada Buddhism has had to struggle to come to terms with a rapidly changing secular world. Monks have become used to dealing with money and many are interested in politics. Some have embraced Marxist ideas, others have become anticommunist. In 1976, the Thai monk Kittivuddho announced that it was not a wicked act to kill communists – indeed, he stated that it would gain an individual some merit. More commonly, however, monks have become concerned with questions of how to help the poor and disadvantaged, both in a spiritual and in an economic sense.

Novice Theravada Buddhist monks attend class in a monastery school in Tha Ton in northern Thailand. Nearly every Buddhist male spends time as a monk in his early life – considered part of the necessary training for manhood.

Peoples of the highlands

When the first Europeans, as late as the 1930s, penetrated the exceptionally mountainous and remote interior of Irian Jaya, the western end of the island of New Guinea, now belonging to Indonesia, they found hundreds of small Neolithic ethnic groups, speaking over 1,000 separate languages. Often isolated in small mountain valleys, they had had little contact with anyone but their immediate neighbors, though there must have been links in the past, as pigs were introduced to the region perhaps 10,000 years ago, sweet potatoes 300 years ago, and cowrie shells, used for decoration, would have been obtained by trade with coastal peoples.

These highland peoples are shifting cultivators. Each clan or sub-clan possesses a core area of intensive cultivation, where pigs are raised on sweet potatoes for both ritual slaughter and feasting and for exchange with nearby groups. Around the core, land is cleared and cultivated for two or three years before being abandoned and a new plot started.

Settlements are generally based on patrilineal groups, and men take wives from other clans in exchange for pigs, shells or plumes from birds of paradise. At the center of each settlement stands the men's house, larger than the other dwellings, which is lived in by the married men and initiated youths. It is here that the men's sacred objects are kept and major decisions taken. Men decide when to cultivate new areas, start wars with neighboring groups or conclude them by peace treaty. Women and uninitiated youths live in smaller dwellings, sometimes scattered so as to be close to the garden plots that the women tend. If a settlement's population should become too large for the available land, then a quarrel is taken as a pretext for one or more families to move off and start their own garden.

From the dawn of time Mist rolls gently back from the valley as day breaks over a village in the mountains of Irian Jaya.

HIDDEN SOCIETIES

IMITATION AND ISOLATION · NEIGHBORS AT A DISTANCE
LIFE IN THE FAST LANE

Until comparatively recent times, the peoples of Japan and Korea were scarcely known to the outside world. For over 200 years, until the mid 19th century, foreigners were forbidden on pain of death to enter Japan, and any Japanese who went overseas were unable to return to their country. During the same period, Korea also had minimal contact with foreigners; even today, North Korea is almost unknown by outsiders. Once isolation ended, for Japan earlier than for South Korea, both countries embarked on rapid social and economic change; more recently they have moved within a generation from the despair of wartime defeat and occupation to overwhelming industrial success. Today both countries are living with the social strains that their postwar economic miracles have wrought.

COUNTRIES IN THE REGION

Japan, North Korea, South Korea

POPULATION

Japan	122.78 million
North Korea	22.42 million
South Korea	42.5 million

LANGUAGE

Countries with one official language (Japanese) Japan; (Korean) North Korea, South Korea

RELIGION

Japan Most Japanese are adherents both of Shinto (93.1%) and Buddhism (73.9%); Christian (1.4%)

North Korea Nonreligious or atheist (67.9%), traditional beliefs (15.6%), Ch'ondogyo (13.9%), Buddhist (1.7%), Christian (0.9%)

South Korean Nonreligious or atheist (57.4%), Christian (20.7%), Buddhist (19.9%), Confucian (1.2%), Ch'ondogyo (0.1%))

IMITATION AND ISOLATION

If Japan and Korea's physical geography once made long periods of self-isolation possible, their proximity to each other and to the hearth of Chinese civilization has encouraged cultural exchange. Many elements of Japanese culture, including the Chinese script, Confucianism and Mahayana Buddhism, first entered the country from China, usually transmitted via Korea.

Early cultural exchange

In neither country is there a complete archaeological record, and the origins of their peoples are somewhat uncertain. The Korean peninsula was settled from the north, mainly by various Chinese groups but also by Mongols and forest peoples from Manchuria on the northeastern margins of China, and from the south. It was the Manchurians who introduced the religious influences of Confucianism and Buddhism in the first few centuries AD.

By the 8th century the peoples of the Korean peninsula had been unified, first under the Silla and then under the Koryŏ kingdoms. In 1259 the Mongols invaded, but a century later power was seized back by the Yi dynasty, which ruled Korea as a tributory state of China from 1392 to 1910. Confucianism was made the dominant ethical system. In the late 16th century a Japanese invasion of Korea was defeated with Chinese assistance, but not before many Korean craftsmen and scholars had been taken back to Japan. In 1640 Korea cut itself off from the world, banning foreign travel for 250 years and maintaining relations only with China.

The Japanese islands were probably first settled by groups of people from the Asian mainland, including the ancestors of the Ainu, an aboriginal Caucasian people today found only in the northernmost island of Hokkaido. At the end of the Neolithic period (about 250 BC) a new wave of people moved into Kyushu island and began to spread rapidly eastward and northward.

The first unified Japanese state emerged with the Yamato court, headed by a hereditary emperor, around the Inland Sea between Kyushu and Honshu in the 5th century. For a time it controlled part of Korea. A military–feudal society, based on *daimios*, or warlords, and a warrior class of the *samurai*, under an overall general known as the *shogun*, developed. In time the emperor became subordinate to the powerful families who had control of this system.

Japanese court life took many elements from Chinese culture, including its laws and its administrative structure, written script, styles of dress, tea-drinking, lacquer work, poetry and drama. Shinto, the indigenous Japanese religion with a belief in the divinity (*kami*) of nature, was modified by Buddhism, as it had earlier been by Confucianism, and the Japanese altered other cultural borrowings to create their own particular culture. By the 11th century this had developed to produce *The Tale of the Genji*, a classic work of literature that many claim as the world's first novel, and by the 1300s the unique *No* style of drama had been elaborated.

A long religious tradition *(above)* Monks celebrate the Buddha's birthday at a temple in South Korea. Buddhism entered Korea from China and subsequently spread to Japan.

New influences *(left)* Japanese children competing in a school baseball league. Baseball – imported during the American occupation after World War II – rivals Sumo wrestling as Japan's most popular sport.

Outsiders

Portuguese traders arrived in the islands in 1542, to be followed by missionaries and by traders from the Netherlands and Britain. The Europeans supplied firearms that enabled the protracted feudal wars of the period to be resolved in favor of the Tokugawa shogunate (1603–1867). Immediately, a policy of isolation was put into effect: Christian missions and Buddhist monasteries were closed, adherents persecuted or forced to recant and all foreign trade closely controlled. Over the next two centuries Japanese society, cut off from outside contact, became highly centralized and rigidly stratified.

Following the ending of *sakoku* (the period of exclusion) in 1853, Japan embarked on a rapid program of economic and social modernization, freely adopting Western ideas and knowledge such as medicine from Germany, ship-building technology from Britain and forms of industrial organization from the United States. Under severe demographic pressure it began to expand its territories, first into Hokkaido, resulting in the cultural assimilation of the last of the Ainu, and then into Korea.

During Japan's occupation of Korea, from 1910 to 1945, use of the Korean language was forbidden and up to 2 million people were taken as forced labor to Japan. Many aspects of Korean life – its education, policing, banking and judicial systems – still bear a Japanese imprint. With the collapse of Japan's imperial ventures in Asia war divided the peninsula, and North and South Korea are now hostile strangers, mindful of Japan's harsh occupation.

NEIGHBORS AT A DISTANCE

On the surface at least, Japan and Korea are both culturally uniform societies enjoying the fruits of an economic prosperity of their own making. Only North Korea stands apart from this, but even the deep ideological gulf that divides North from South Korea is narrowed by the fact that 10 million southerners, a quarter of the population, have relatives on the other side of the border. However, contact is virtually impossible: there are no mail or telephone links between the two halves and no crossings from one to the other.

Both Japan and Korea are fiercely proud of their traditions, and this can make it very difficult for outsiders – or even for Japanese and Koreans who have lived some time overseas – to become part of society. Their respective languages are unique to each country and difficult for others to learn; they are not mutually intelligible. Both place value on their social homogeneity, which is regarded as a source of strength since the lack of linguistic, ethnic, class or even urban and rural differences means that individuals wish to improve the economic state of the country as a whole rather than just one interested section.

In part, this outlook derives from Japan and Korea's common Confucian tradition. Confucianism is a moral code rather than a religion and concentrates on teaching the proper behavior between individuals. In particular, it holds that social hierarchies are necessary for societies to function smoothly; it teaches that juniors should show respect and loyalty to their seniors, while seniors should repay them with benevolent and wise leadership. Confucianism thus lies behind the respect for authority and the decorum that is daily displayed in social interaction in both countries. The first thing Japanese and Koreans do when they meet is exchange namecards to indicate precisely what position they hold and how they should be addressed. Confucianism even has a strong effect on the Japanese language, since the way in which people address each other is determined by their relative positions. For example, there are at least five different ways of saying "to go" in Japanese, depending on who is speaking to whom, where, and when.

Ancient remedies A street trader in Seoul, South Korea, dispenses from a large range of herbal medicines. Even in the cities, traditional beliefs and customs have withstood the changes brought by rapid modernization: many people, for example, still turn to natural medicines rather than modern pharmaceutical products to cure their illness.

The influence of Confucian ethics is stronger in Korea than anywhere else in the world and lies behind the South Korean government's periodic purification drives, with moral exhortations to curb crime and conspicuous expenditure. Displays of wealth and privilege are discouraged; private yachts and private school tutors are banned (in North Korea, even bicycles are the property of the state). Income distribution in South Korea is relatively even by international standards, and there are few wealthy individuals or large landowners.

This egalitarian strain produces social tensions that are evident, for example, in the education system, which turns out far more qualified graduates than there are professional positions available. Strains are also found in the area of press freedom, which – while great by contrast with North Korea – is still curbed by the government; there are no independent television stations, for example.

The Japanese also have a strong ideology of both solidarity and hierarchy, which rapid postwar economic growth helped maintain; 90 percent describe themselves as middle class, the old distinction between *samurai* and peasant has faded, and there are no longer any aristocrats. There are no class accents, and status in business comes from age and length of service. But in recent years land price inflation has begun to create a divide between families that can afford their own house and those that cannot.

In order to maintain a sense of social homogeneity, it is official policy to play down the existence of minority groups, the largest of which are the *burakumin*, descendants of the premodern *eta* or "untouchable" castes who once served as leather workers and animal-slaughterers; they number between 2 and 3 million. Only 16,000 Ainu are left in Hokkaido, where some have taken to opening their villages to the tourist trade as part of an attempt to recover their ethnic heritage. Despite a clear emphasis on group activity and loyalty, the Japanese also express their strong personal ambition and determination in their literature and many artistic skills.

Religious attitudes

The peoples of Japan and Korea display strong differences in religion. In Japan,

A wedding couple in the formal splendor of traditional Japanese dress pose for photographs after the Shinto ceremony at a local temple. Although arranged marriages – once the norm – are no longer as common as they were, parents or older friends very often play an informal role in initiating a suitable match.

A UNIQUE WAY OF WRITING

South Korea is possibly unique in having a public holiday to celebrate its alphabet. The Korean script – known as *han'gul* – is a source of national pride and, in its singular appearance, a sign of national identity, even for overseas Korean communities. It came into being during the reign of Sejong the Great (1397–1450), the fourth of the rulers of the Yi dynasty, who headed a cultural and intellectual revival. He founded a royal academy for study and research, and it was here that a phonetic system for writing the Korean language was first developed in 1443.

Before that date, the Chinese system of ideograms, which directly express Chinese thoughts, had been used, but now the Koreans were able to discard them and express their own views of the world for the first time. The first work written down in the new script was the *Songs of the Flying Dragons* cycle, a celebration of the founding of the Yi dynasty that contains the main concepts of Korean Confucianism.

Han'gul replaced 50,000 Chinese characters with just 28 letters – a triumph of scholarship and ingenuity. For many centuries both writing systems existed in tandem, Chinese script being retained by scholars and upper-class Koreans. However, the use of *han'gul* was strengthened after Korea's years of isolation were ended. It was officially banned during the Japanese occupation, when its illicit use became the symbol of Korean nationalism. But the ease with which it can accommodate new words makes it vulnerable to imported foreign words.

individuals are not expected to stand out too much from their fellows, and religion – for most people a blending of Shinto and Buddhist ideas – is expressed in communal visits to local shrines and temples and the observance of annual festivals. In South Korea, people tend to be more individualistic, and there is consequently greater diversity of religous practice. Buddhism and Confucianism are both important, and many Koreans turn to *madang* or shamans to contact ancestors' spirits for them. Most *madang* are women, who do not meet with the approval of the state.

The contrast between the two is most apparent in the influence of Christianity in each society. Despite constant missionary activity in Japan during the 16th century and since the mid-19th century, less than 1 percent of people are Christian. In South Korea the figure is one-fifth, mostly belonging to Protestant sects, and numbers are increasing each year. One well-known religious sect, with followers worldwide, originated in South Korea in the mid 1950s under its leader, the Reverend Sun Myung Moon. Called the Unification Church, it has often been accused of brainwashing its followers.

The secret North

Since 1948, under the brand of communism of Kim Il Sung, religion in North Korea has been officially forbidden, and traditional culture ruthlessly repressed and replaced by new socialist ideals. Churches and temples were razed, inheritance of property banned and families dissolved by taking children into state nurseries. Marriage must be sanctioned by the state and traditional funeral rites have been simplified. North Korea remains one of the most hidden societies in the world. Until the late 1980s – when the government began to set up joint tourist ventures with Japan and France – it was virtually impossible for foreigners to travel anywhere in the country.

LIFE IN THE FAST LANE

The postwar material advances made by both South Korea and Japan cannot be overestimated; by contrast, North Korea – the more industrialized half of the country at the time of the Korean War – has become markedly poorer since. Such rapid economic growth has brought profound changes to people's lives. In 1970, for example, only 6 percent of South Korean households had television sets; today they are virtually universal. Thirty years ago 80 percent of people were landed peasants; today the figure has fallen to 15 percent.

Throughout Japan, the idea of progress and scientific advancement is viewed with approval; the public has accepted the installation of electronic "policemen" in city streets to answer queries and give directions, and vending machines carrying an extensive range of goods can be found in even the most remote areas. Although Western goods and styles have been widely adopted, Japanese culture has not necessarily been weakened. For example, more and more couples are choosing to follow their civil wedding, at which Western dress is usually worn, with a Shinto ceremony, for which they assume traditional costume.

Changes to family life

The costs, especially to family life, of both countries' economic miracles are far-reaching. Men face long hours at work and long journeys commuting to and from home, while in Japan socializing with workmates after hours is the norm. In South Korea the number of workhours per week in the manufacturing sector is close to 55, compared with 40 in the United States; overall, they work 25 percent more hours than the Japanese. Both peoples have developed a reputation for not taking holidays; Japan has 18 public holidays a year, but many people ignore a third of them. The government has insisted that banks and stockmarkets close on Saturdays, and civil servants remain at home two Saturdays a month.

Time with the family is therefore restricted, and for those with small dwellings and large families time alone with a spouse is rare. In general, families are growing smaller; both populations are youthful, which will result in a large proportion of elderly people early next century. In Japan, only 5 percent of the population were over 65 in 1960; the figure will be 20 percent by 2040.

Talking business (*above*) The work ethic is deeply engrained in Japanese culture; after office hours, male colleagues expect to socialize in cafés or bars rather than return to the family home.

Urban rebels (*left*) These youths have adopted the uniform of Western revolt to defy convention. Urbanization and changes to family life can weaken the bonds of duty and respect that the young traditionally owe their elders.

Devil incarnate (*right*) A theatrical representation of *Oni* – or demon – at a festival in Kyoto, central Honshu. A local religious celebration can be found taking place at virtually any time of the year in Japan.

The old are generally respected in both societies, but how to support the elderly generation in the future will be a major problem. At present 40 percent of Japan's elderly people live with their children, but smaller two-generation families are increasingly the norm. In Korea, retirement homes and day care centers have begun to appear.

In both Japan and Korea women have been expected to leave full-time employment in their mid 20s in order to marry and devote themselves to raising children and looking after their husbands. In Japan women make up only a fifth of the university population, and a woman who stays in a serious career after marriage is

still considered an oddity, though many work part-time. However, a drop of almost 25 percent in the number of people entering the labor force in Japan during the early 1990s means that employers are having to turn to women to meet future labor needs. Laws passed in the 1980s to ensure equal employment opportunities are opening up the professions, and in Japan two-fifths of the waged work force are now women, though their average pay is still half that of men. They are confined to clerical positions with little hope of rising to managerial status, except with foreign firms.

Japan's rate of marriage is the highest in the world and its divorce rate among the very lowest; divorce is rising but still frowned upon, and can mean penury for the woman. However, a trend toward later marriages and fewer children means that Japanese women are achieving more time to themselves. Recent evidence suggests that they are no longer prepared to be solely regarded as homekeepers, and have embraced the changing mores of marriage faster than men.

As in Korea, arranged marriages are declining, but unmarried men and women are allowed few opportunities to meet, and up to a third of households in urban areas may be single person. Korea is facing a desperate shortage of women as a result of demographic changes: in 1979 there were 79 men for every 100 women; by 2000, however, it is estimated that there will be 120.

New workers?

Japan and Korea have long opposed large-scale immigration as a threat to social homogeneity. There are several thousand guest workers in Japan, from Southeast Asian countries such as Malaysia and the Philippines, and from Korea, but their economic position is precarious; up to 300,000 illegal immigrants exist, mainly employed by small firms in the fast-food and laundry sectors. About 750,000 Koreans, many of them descendants of wartime forced laborers, live in Japan. Until 1990 they had to carry alien registration cards and be fingerprinted, while many public positions were closed to them. The extension of full civil rights to the third generation of Koreans is seen as evidence of an improvement in relations between the two countries.

Over the past 30 years, almost half a million middle-class South Koreans have migrated overseas, generally to the United States. Many were either the wives of American servicemen stationed in Korea, or were wealthy people fearful of another war. Now the majority are professionals looking for employment, or students searching for postgraduate education. The government has been trying to halt the brain drain, but the existence of large and increasingly rich Korean communities in cities such as Los Angeles and New York in the United States, and even Tokyo, provide attractive alternatives to the difficulties of advancement at home.

STATE AND RELIGION IN JAPAN

The Shinto religion is unique to Japan. It concentrates on the relationship between man and nature: sacred power in the form of gods (*kami*) are found in a large number of natural sites such as rivers, trees, rocks and mountains: Mount Fuji, the best-known mountain in Japan and one of its national symbols, is regarded as a sacred object and there is a shrine at its summit. Shinto has neither a founder nor any sacred scripts (though there is a large mythology based on orally transmitted stories of the *kami*), and comprises a rather loosely structured set of practices and rituals. There is no regular worship, but it offers a strong communal focus through annual festivals at local shrines and temples. Most Japanese people combine Buddhist and Shinto beliefs and ceremonies in their daily lives without any conflict of loyalties: most will choose to have a Shinto wedding, for example, but a Buddhist funeral is usually preferred.

Between 1870 and 1945 Shinto was made a state religion and was used by ultranationalists to promote the idea of the Japanese people's continuity and uniqueness. In 1947 the new constitution formally separated religious and state organizations, symbolized by the emperor's renunciation of divinity. However, the accession of Emperor Akihito in 1990 to the chrysanthemum throne, the 125th of a single line of emperors, led many to question the role of Shinto ritual in the enthronement ceremony, and care was taken to stress the traditional cultural aspects of the occasion rather than its religious significance. Today, for most people, Shinto has reverted to its original role as a community religion based around the local shrine.

The future generation

Most people in Japan and South Korea's predominantly young societies have grown up within the years of the economic miracle: unlike their parents, they do not remember the privations of wartime and occupation, and are accustomed to a life of affluence, supported by hard work. This is particularly true of today's generation of schoolchildren. One survey found that half of Japan's four- to nine-year-olds had their own television sets, radios and tape recorders, while a third had telephones. Living in smaller families, they are more likely to have their own rooms, when once they would rarely have left their mothers' side. Changes in diet to a much greater emphasis on meat make the younger Japanese visibly taller and fatter than their parents.

In both Japan and South Korea, success in education is almost the only way of ensuring a good job and high status for life. This motivation, and the fear of failure, explains why children work so hard, in Japan doing at least two hours of homework a day, about four times as much as children in the United States. Eight out of ten Japanese children attend preschool classes, almost all stay in education until 18, and a third go to university. More would go if there were more places available. The figures are nearly the same in South Korea; however, here graduate unemployment is twice the national average. Consequently, in both countries, there is intense competition to get into one of the top universities to ensure the best chance of future status. During their teens Japanese and South Korean children have very little free time; the vast majority have to attend cram schools as well as their normal studies.

In Japan and South Korea's meritocratic societies there is much emphasis on equality of chance in the education system. Pupils are not picked out for preferential treatment, classes are large by international standards, and all students use the same textbooks, take the same examination (usually in the form of multiple-choice answers) and, in Japan, wear the same uniforms. Uniforms were banned in South Korea in 1983, and here the selection of middle schools is made by lottery rather than by examination.

In order to pass the examinations to ensure them places in the most sought-after schools and universities, Japanese children have to learn enormous numbers of facts, so that they become a test of hard work and perseverance rather than ability or creativity. As a result, children tend to be obedient and conformist. Although they consistently score well in international comparative tests, especially in science and mathematics, anxieties about their education systems have been expressed in both Japan and South Korea. Parents in both countries complain that their children are suffering from excessive competition and leading unhappy lives – described as a "gray youth" – as a result of examination pressure. However, rates of child suicide are no higher than in many Western countries.

Faces in a crowd (*above*) These uniformed children, enjoying a school outing, have robust physiques and a healthy appearance, a reflection of the changes in the Japanese diet: more meat and dairy products are now being consumed than ever before.

Riding high (*right*) Students uproariously celebrate success in the entrance exam to the prestigious University of Tokyo. For these lucky ones, perseverance at school work has been well rewarded – a university education is seen as an invaluable ticket to career success in the future.

Other critics argue there is too much emphasis on quantity rather than quality and that the individual creativity required by industry is missing – some employers feel that more emphasis should be given to training individuals who would then be able to invent their own new products for their export-driven economies rather than take ideas that originated in the West, improve them, and then sell them back. For the present, however, most people are unwilling to change a system that has clearly been so beneficial for building successful economies, though there are increasing signs of a more questioning attitude among younger people.

A new outlook?

The emergence of a new generation – the *Shinjinrui*, or new breed – in Japan, open to Western influences but increasingly interested in their own traditions, perhaps heralds a new era. They are prepared to work hard, but not to allow their bosses to interfere in their private lives and are increasingly reluctant to be sent abroad for long periods by their companies. They are less group-oriented than the older generation, having grown up with the individualistic pursuits of television and comic books.

The older generation shows anxiety about the new breed. Several commentators complained about the lack of competitiveness of Japanese athletes in the Seoul Olympics in 1988, especially when compared with the performance of their age-old rivals, the Koreans. But Korea is also feeling such strains, as the newly urbanized young reject the village-based Confucian ethics of filial duty and subordination. The sacrifices their parents made to send them to good universities have resulted in exposure to different ideals and Western ways of life. The student population is at the forefront of challenges to the government; their often violent confrontations with the police have been accompanied by a number of suicide protests. How both countries will meet the future is an open question.

Sacred gateways

In earlier times, the innumerable *kami* or spirits of the Shinto religion were worshiped at sites such as rocks and waterfalls. These places were chosen by the *kami* themselves, though they did not necessarily reside there permanently. Each dynastic family, and the district it contolled, would have its own *kami* site. From medieval times onward shrines were built at these places, always using natural materials for construction, and always marked by a *torii* or gateway that defines the boundary of sacred space and announces the existence of the shrine.

There are over 80,000 such shrines in Japan, not including household shrines. As many as half of the public shrines, and many household ones, are dedicated to Inari, the deity traditionally associated with success in agriculture, though subsequently aligned also with good luck in business. The Fushimi Inari shrine in Kyoto, Japan's former capital, is the central Inari shrine, and one of the country's most popular. It was transferred there from Mount Inari in the 9th century and was originally the shrine of the Hata family. In Kyoto, Inari was also identified with the Buddhist guardian deity, Dakini. On its annual festival, 9th April, it is visited by crowds of people from all over Japan.

Torii are found on natural features such as rocks or summits without an associated shrine building. When a *torii* close to Tokyo International Airport was moved to make way for a carpark, a series of accidents at the airport persuaded the authorities to replace it. It still stands as a symbol of security for air travelers.

Avenue of *torii* The Fushimi shrine is noted for its long avenue of *torii*, painted vermilion and placed so close together as to almost shut out the sun. It winds for 5 km (3 mi) across the hillside. Such avenues are characteristic of Inari shrines.

奉

THE QUEST FOR IDENTITY

PEOPLES OF THE SOUTH · OLD TIES AND NEW INFLUENCES · CONTEMPORARY QUANDARIES

Great diversity of peoples and lifestyles is to be expected within Australasia and Oceania, spread as the region is across thousands of kilometers of ocean. Yet elements of a common origin exist among the indigenous peoples of the region, and there are similarities in their responses and adaptations to Western influences. Although the area was first settled by Asiatic peoples many thousands of years ago, for the past 200 years its cultures have been dominated by European colonists. The indigenous peoples sought, often in vain, to preserve their cultures and lifestyles while the new settlers in their turn attempted to forge a national identity that was independent of that of their homelands. Today these national identities are having to adapt to new balances in ethnic composition as a result of changing immigration patterns.

COUNTRIES IN THE REGION

Australia, Fiji, Kiribati, Nauru, New Zealand, Papua New Guinea, Solomon Islands, Tonga, Tuvalu, Vanuatu, Western Samoa

POPULATION

Over 16 million Australia

3 million–4 million New Zealand, Papua New Guinea

Over 700,000 Fiji

100,000–300,000 Solomon Islands, Vanuatu, Western Samoa

Under 100,000 Kiribati, Nauru, Tonga, Tuvalu

LANGUAGE

Countries with one official language (E) Australia, Fiji, Kiribati, New Zealand, Papua New Guinea, Solomon Islands; (N) Nauru

Countries with two official languages (E,To) Tonga; (E,Sa) Western Samoa

Country with three or more official languages (B,E,F) Vanuatu

Country with no official language Tuvalu

Other significant languages in the region include French, and a great variety of indigenous Melanesian and Polynesian languages

Key: B–Bislama, E–English, N–Nauruan, Sa–Samoan, To–Tongan

RELIGION

Countries with one major religion (P) Tuvalu

Countries with two major religions (P,RC) Kiribati, Nauru, Tonga, Western Samoa; (C,I) Vanuatu

Countries with three or more major religions (C,H,M) Fiji; (C,I,RC) Papua New Guinea; (A,I,P,RC) Solomon Islands; (A,N,P,RC) New Zealand; (A,J,M,N,O,P,RC) Australia

Key: A–Anglican, C–Various Christian, H–Hindu, I–Indigenous religions, J–Jewish, M–Muslim, N–Nonreligious, O–Orthodox, P–Protestant, RC–Roman Catholic

PEOPLES OF THE SOUTH

Human settlement of the Pacific probably dates from about 50,000 years ago, when nomadic hunter–gatherers from Southeast Asia began to move into the region. At least 40,000 years ago they reached Australia through Papua New Guinea, where they dispersed inland, successfully adapting their hunting–gathering way of life to the harsh climate. The highland people of Papua New Guinea were among the world's first agriculturalists – there is evidence to date cultivation there to about 9,000 years ago.

Pacific migrations

About 5,000 years ago the development of seagoing outrigger canoes enabled a second movement of people from Southeast Asia into the island groups of the Pacific – the ancestors of the present-day peoples of Micronesia and Polynesia in the mid and southeastern Pacific. On these scattered volcanic islands and atolls land was scarce and highly valued, and consequently status came to be associated with control over land: the cultures of both Micronesia and Polynesia show a high degree of social stratification that is lacking among the peoples of the larger, more mountainous islands of Melanesia (Papua New Guinea, New Caledonia, Vanuatu, the Solomon Islands and Fiji).

Considerable interaction took place between the peoples of all the Pacific island groups. In the Fijian islands, for example, large double-hulled canoes were used for trading between islands, and regular exchange festivals brought hundreds of people together. Elaborate seaborne trading systems were also characteristic of coastal areas of Papua New Guinea. Knowledge of stars, currents, bird migrations and signs of distant land was such that controlled journeys over long distances were possible – one of the better-known migrations was the movement of Polynesian peoples to what is now New Zealand in about 750 AD.

Today known as Maoris, the way of life of these settlers reflected both their Polynesian roots and their new physical environment. The cultivation of sweet potatoes was adapted to the colder conditions of the New Zealand climate by winter storage, while hunting, fishing and food-gathering replaced cultivation in the southern part of South Island. Maori society consisted of groups of various sizes, membership of which was based on descent from a common ancestor. Their culture was marked by deep attachment to the land, which by custom was held in communal possession.

A Great South Land

Western contact with the South Pacific began in the 17th century, when explorers came searching for the Great South Land, or Terra Australis, that had filled the imagination of Europeans for centuries. In 1788, the British annexed Port Jackson (Sydney) on the southeast coast of Australia, in order to use it as a penal colony. This was the start of a forced migration

that lasted until the mid 19th century as thousands of British convicts were transported to Australia in order to provide labor in the new colony. Settlement gradually advanced into the interior, following the search for grazing land and minerals (especially gold). Throughout the 19th century Australia was the goal of many emigrants seeking an escape from rural poverty and economic depression in the British Isles.

From 1800 onward, New Zealand was also colonized by British settlers, including significant numbers from Scotland, drawn there by its timber, mineral resources and agricultural potential. British sovereignty over New Zealand was formally established in 1840 with the signing of the Treaty of Waitangi by local Maori chiefs. The Pacific islands also attracted the attention of a number of European powers. The most intense period of Western contact was between 1840 and 1900.

For most of the indigenous peoples of the region, the European invasion was a disaster. The settlers seized their lands, disrupted their economies and spread diseases to which they had no resistance. In Australia, the Aborigines (as they were called by the Europeans) were rapidly displaced from agricultural and pastoral lands near the coast. Inland the harsh conditions kept Europeans at bay, thus allowing some Aboriginal people to continue their hunting–gathering way of life for a longer period.

In New Zealand, too, the Maori population declined sharply after colonization, and tribal structures were changed by introduced economic and cultural values. Fierce attachment to their ancestral lands fueled Maori resistance to further encroachment by European settlers after 1840, but failed to prevent appropriation of most of their best land.

In a Melanesian longhouse women of the Mendi ethnic group in Papua New Guinea receive their male visitors. The women and children of one line of descent live together, keeping their pigs in cages along the walls; the menfolk live in another house.

OLD TIES AND NEW INFLUENCES

Both Australia and New Zealand have seen considerable changes to their ethnic composition this century, as a result of largescale immigration programs set up after World War II. In Australia, these led to an influx of migrants from southern Europe and later from Asia, particularly Indochina, the Philippines and Malaysia: New Zealand has sizable Chinese and Indian minorities as well as a growing community of Pacific islanders from Western Samoa and the Cook Islands. Some 20 percent of Australia's current population of 17 million was born overseas, and some 15 percent of New Zealand's 3.3 million. The ethnic diversity of these migrant populations and the still-steady stream of arrivals means that national cultures are gradually changing.

A distinctive Australian outlook began to emerge early in colonial times. The struggle to tame the unyielding bush became a central part of the Australian character and helped develop the typical Australian male traits of improvization, loyalty to companions or "mates", disrespect for class divisions and a wry sense of humor. Such traits are typified by the country's unofficial national anthem of "Waltzing Matilda", with its evocation of the life of the "outback", and by the figure of Ned Kelly (1855–80), the most famous of Australia's rural outlaws of the 19th century. Regarded by many as the champion of the oppressed rural worker against the large landowner, he perpetrated a series of daring robberies and ambushes against the police that have become enshrined in national mythology.

Links with Britain remained strong in New Zealand, which until recently was a major agricultural producer for the former, and cultural attitudes reflected typical rural conservatism and religious (particularly Protestant Christian) conformity. In both countries, World War I provided a great stimulus to the forging of national identity. The cost to Australia and New Zealand of their involvement in the war in Europe in terms of the numbers of dead and wounded was enormous, and the courage of their soldiers – particularly those who took part in the disastrous campaign at Gallipoli (Turkey) in 1915 – became a source of enduring national pride to both. This legend was deliberately fostered by the governments of both countries in the setting up of many war memorials and the establishment of official Anzac Day commemorations on 25 April each year.

Vibrant streetlife Shoppers stroll past colorful graffiti painted on a wall in a Brisbane market. New immigrants from Southeast Asia have added to Australia's cultural mix to create a lively and cosmopolitan urban society.

Indigenous populations

These European-based national identities almost completely ignored the presence of the countries' indigenous peoples, though some cultural borrowings took place – for example, a Maori war chant is used by the New Zealand rugby team as part of its prematch ritual of preparation, along with the singing of the national anthem. The Maori people and their culture, however, proved comparatively resilient to the European invasion, as is evident from the landscape in the survival of more than 800 *marae* (meeting centers), many of them ornately carved, which form the focal point of Maori social and family connections. There has recently been a resurgence of Maori culture, with a subsequent revival of traditional dances and ceremonies.

Maori meeting house The local meeting house is the center for Maori social and religious life, reflecting strong communal traditions. Intricate carving characterizes Maori architecture; skilled craftsmen keep the art alive on many new buildings.

In Australia the Aboriginal people have not fared so well. Dispossessed of their lands by European settlers, they were generally assumed to be a dying race that could readily be assimilated into white Australia. However, they refused to be assimilated or to die out, and so eventually were granted their own reserves in which to follow their traditional lifestyles. By 1988, about 10 percent of Australia was held by Aboriginal people, almost all in the center and northern coastal regions.

Island identities.

Substantial movement has long been a characteristic of the Pacific island peoples, resulting in a highly blended population. There are regional variations, however. The Australian territories of Christmas Island and the Cocos Islands contain populations of Chinese, Malays and Europeans. One-third of the population of Norfolk Island, 1,650 km (1,000 mi)

A nation of cricket-lovers Thousands of Australians gather to cheer their team on in a Test Match between Australia and England. Every Australian supporter fervently hopes that the "Poms" (English) will be given a hiding by their national heroes.

SPORTING NATIONS

In both New Zealand and Australia sport has played a major part in creating a national image and in establishing both countries on the world stage. The New Zealand landscape lends itself to outdoor activities, and the successful sportsperson is seen as someone who successfully battles against the odds, the embodiment of the New Zealand way of life. Rugby Union football is regarded as the national sport, but in recent years New Zealanders have also achieved international prowess in a variety of other sports. These successes are a source of intense national pride, and competition with Australia is particularly intense.

More than 6.5 million of Australia's citizens regularly participate in outdoor sports. In organized sports, the two main passions are football (either played according to Australian Rules or those of the even rougher Rugby League), and cricket – few Australian males have not played cricket at some stage in their lives, though many now prefer to watch games on television than attend them. The premier event in the cricket calendar is the England–Australia "Ashes" series of Test Matches, which have been played regularly since 1877.

With the good surf that is found along much of the coast, and the proximity of most of the population to it, it is not surprising that the more individual sport of surfing has also become a popular national pastime in recent years.

southeast of Sydney, claim descent from the mutineers of the British warship HMS *Bounty* who were resettled there from Pitcairn Island in 1856. On Fiji the descendants of Indian sugar estate workers brought there as indentured laborers in the 19th century now outnumber the Fijian population. There is a large European community, of mainly French descent, in New Caledonia.

A legacy of the colonial period is the widespread use of English throughout Oceania, but even where it (or French as in New Caledonia) is the official language, other languages are widely used, and linguistic diversity is a feature of many island groups. More than 100 languages are recorded in the islands that make up Vanuatu, for example. In Papua New Guinea, between 700 and 800 languages have been identified and, to cope with this, widespread use of Pidgin (a mix of English and the indigenous language) has developed.

As a result of the proselytizing activity of Western missionaries, Christianity is widespread throughout the islands: on Fiji there is also a strong Hindu affiliation among the Indians in the population. Evangelizing Christian groups from the United States and Europe are still zealous in winning converts among remote groups of people, particularly in Papua New Guinea. Yet, though most people in that country claim to be Christians, traditional practices are still prevalent and religious belief is strongly influenced by magic and pre-Christian ritual.

In the Melanesian islands religious observance traditionally aimed at enlisting supernatural help to ward off physical harm and achieve material success. This belief was carried over into Christianity, with converts tending to regard the new faith as a way to achieve Western material goods in the same way that traditional religion had provided items of food, fiber and trade. They believed that these Western-style goods ("cargo" in Pidgin) would arrive soon after conversion via a ship, airplane or supernatural source. These cargo cults, as they are known, later became the vehicle for anti-European sentiment and revolt when the expected goods failed to arrive.

CONTEMPORARY QUANDARIES

Throughout the region the creation of more confident and independent national identities, free of former loyalties, has not been achieved without a certain amount of soul-searching. In Australia, for example, cultural diversity is increasingly replacing the former national stereotypes, particularly in the large cities, where most immigrants have settled. Greek, Italian and Asian restaurants now offer a variety of food choices instead of the ubiquitous roast dinner or barbecue, and the annual Australia Day celebrations have taken on an exciting multicultural character, incorporating music, dance and foodstuffs from many parts of Europe, Asia and the Middle East. This has given an entirely new character to the occasion, which previously centered on the ceremonial hoisting of the British flag.

Debating the issues

In Australia, as elsewhere in the region, the old ties with colonial Europe no longer set the cultural tone. A wide range of languages are now offered in Australian schools and community radio stations in the major cities currently cater specifically for their non-English-speaking audiences. Questions about what Australia is, what it means to be an Australian, and what directions should be set for the future, were brought into focus during the 1988 bicentennial celebrations of Australia's annexation by Britain. Opinion polls gave voice to public concern about Australia's changing ethnic and racial composition, and widespread dissatisfaction was expressed at the idea of mass migration continuing to Australia from anywhere in the world.

Other issues have also been fiercely debated in recent years. The tension between the Aboriginal community and white Australians has spilled over into riots and violent confrontation with the police. The feminist challenge to male domination has been fought against a mainstream culture that traditionally attaches great value to male macho characteristics. There has been a steady but slow increase in the number of women rising to higher executive positions in government and administration, the legal profession and commerce.

Conservation has provided another arena for cultural dissent. A developed "sense of place" and better knowledge and understanding of the ancient Australian continent have encouraged an increasing number of people to defend their environment from despoliation. Similar concerns are felt throughout the region, which was used in the past as a nuclear testing ground by the United States and European military powers: New Zealand took an independent lead in the 1980s in banning nuclear warships from its ports.

Caring for the environment Conservation workers at a garbage dump in Canberra, Australia's capital, recover waste for recycling. Concern over the preservation of Australia's unique natural environment has boosted support for ecological movements.

Although the proportion of immigrants in the New Zealand population is smaller than in Australia, it too is a multicultural society as a result of post World War II immigration. Nevertheless, the numbers involved have not been great enough to have the same sort of cultural impact as in Australia, and the debate on contemporary cultural identity has consequently been more closely restricted to the issue of indigenous rights.

Maori concerns about their cultural survival, especially in the cities where they form a large urban "underclass", were given renewed expression during the 1990 commemoration of the 150th signing of the Treaty of Waitangi, which ceded sovereignty to the British. When Maoris seek to rectify an injustice or to settle a dispute over land or fishing rights, they appeal to the Treaty's guarantee of undisturbed possession of their lands. Despite the considerable efforts made by the New Zealand government to present the Treaty today as a "pact of friendship" and a "national symbol of unity", it still arouses bitter controversy, and there is considerable dispute about its current significance.

In all the Pacific islands, political independence, economic development and growing urbanization contrast strongly with the traditional communal and village-based lifestyles that are still lived by many peoples, such as the Wola in Papua New Guinea. As a result, the search for national cultural identities remains elusive. Linguistic and cultural divisions among the diverse peoples of Papua New Guinea, the Solomon Islands and Vanuatu present great problems of national cohesion, while racial tension between Asians and Fijians, and between Europeans and indigenous groups of peoples, is so high in Fiji and New Caledonia respectively that it has erupted into violence.

A new cultural influence

As European influence throughout the region has waned during the 20th century, that of the United States has grown. It has been at its most imperialistic in the scattered islands of the mid-Pacific where the United States has political control. Government administration in

Competing for status Success in a ceremonial pig hunt in Papua New Guinea confers influence and respect on a man and his family. Competitive exchange rituals have replaced tribal warfare and are the accepted means of establishing a social hierarchy.

the Trust Territory of the Pacific is supported almost entirely by funds from the United States – withdrawal of these funds would seriously threaten the collapse of the economy. The islanders have been swamped with American values and lifestyles. Guam has one of the biggest McDonalds fast-food restaurants in the world, and the traditional kissing of elder people's hands to show respect has been replaced by a handshake.

The challenge now facing most island states, particularly those that are independent, is how to achieve the best mixture of their traditional cultures with features that were introduced by colonial powers, and those now being transmitted through the media, foreign travel, trade and military alliances. The resilience of many cultures so far to the pressures of technology, different beliefs and social organization holds out some optimism for this process.

The impact of American culture extends beyond the Pacific islands. Australia is now a consumer culture strongly influenced by North American models, as communicated through films, television and a whole range of consumer products. American influences are reflected in the growing popularity of baseball and the proliferation of fast-food chains and clothing fashions. Americanized spelling is common and Disneyland-type theme parks are increasing – part of the growing commercialization of leisure.

Young Australians, keen to imitate their American idols, have dropped much of the vivid Australian vernacular language, with its distinctive use of slang. Conversely, however, the Australian film and television industry, which gained an international reputation in the 1970s and 1980s, found many of its themes in Australia's cultural past and transmitted them to a growing worldwide audience in such films as *Gallipoli* and *Crocodile Dundee*. Authors such as Patrick White (1912–90) winner of the Nobel Prize for Literature in 1973 and, from a younger generation, Peter Carey, have contributed to the rising reputation of Australian arts and literature, while international business success has also increased worldwide awareness of Australia.

EXCHANGE CEREMONIES OF THE WOLA

Trade between coastal and inland peoples, and between those of different islands, has always been a feature of Pacific island culture, and among the peoples of Melanesia elaborate exchange rituals, in which surpluses of pigs (the islanders' main meat source) and root crops are exchanged for ceremonial valuables, have developed. For the Wola of Papua New Guinea – whose society is not organized into political groups under acknowledged leaders – such ceremonial exchange systems are especially important since they give order to social life and establish the means by which men maintain smooth relations with one another. Excellence in exchange has become a way for an individual to achieve high status and respect in society.

The Wola follow a long cycle of exchange ceremonies, known as the *Sa* cycle. During the feasts that accompany the exchange of goods dances and mass pig kills are held. There is a competitive element to these ceremonies, and the destruction of property may even be involved. The kind of goods that are typically exchanged include shells, salt, stone axes, cassowaries (large flightless birds), marsupials and pigs.

Mass kills take place every five years. The cassowaries are either obtained by trade or trapped as chicks and then reared by the Wola women. Cycles of pig production are orchestrated so that there are sufficient animals for the ceremonies, which extend over two days. On the first day the birds and pigs are killed as part of a dance ceremony – the number of animals an individual slaughters is taken as an indication of his ability to manage wealth. The following day the kill is butchered and cooked, and the meat distributed to relatives and friends.

Australia's Aboriginal people

The 1988 bicentennial celebrations in Australia commemorated the foundation of a country that, for the majority of its inhabitants, is only 200 years old. However, archaeological evidence, based on the carbon dating of charcoal in hearth sites, shows that there has been human settlement in Australia for at least 40,000 years. The first raising of the British flag on Australian soil on 26 January 1788 marked the beginning of an unrelenting process of dispossession of Australia's indigenous population from its ancestral lands, based – until very recently – on an attitude of racial superiority that considered the Aboriginal culture to be primitive and unsophisticated.

Central to the way of life and network of beliefs of the Aboriginal people is the idea that they are inseparable from the land they occupy. Natural sites and landmarks were consequently endowed with mythological significance and became the focus of religion and ceremony. These places are associated with the "Dreaming" – a sacred, heroic time long ago when people and nature came to be as they are, but a time that also continues into the present in complex ways connecting people and their ancestors. Every person living and dead has a place within a physically and spiritually united world.

In the highly variable Australian environment, where lengthy periods of drought are frequent, Aboriginal people moved both regionally and seasonally in order to find food. Various social and religious practices were associated with these migrations. Aboriginal nomadism was not a random wandering but was based on knowledge of what the environment offered in different places and at different times.

Particular Aboriginal groups have a special bond with certain places and landmarks. For example, the Pitjantjatjara and Yankunjatjara are the traditional guardians of the Uluru lands on which Ayers Rock – the gigantic monolith of sedimentary rock that rises from the surrounding sand plains of Australia's "red center" – is situated. Shallow caves on the base of the rock contain Aboriginal carvings and paintings.

Cultural survival

The "Dreaming" stories were given full artistic and religious expression in possibly the oldest living continuous art tradition in the world. Although the past

Dispossessed (*above*), a couple wait for social security payments. Deprived of their spiritual and cultural link with the land, many Aboriginal people have been forced into an unsatisfactory urban existence.

At one with nature (*left*) Living a nomadic life of hunting and gathering, Aboriginal people have gained a detailed knowledge of the seasonal availability of food – both plant and animal – and water.

200 years has seen much erosion of the traditional Aboriginal lifestyle, a rich body of art has survived, especially among those on large reserves in the remote areas of central Australia, where it has been a powerful force in maintaining cultural identity. Carvings and paintings are now sometimes used to confront white society with the plight of the Aboriginal people, particularly those living in urban areas.

Attempts by displaced Aboriginals to regain their cultural heritage have centered on the issue of land rights. However, it was not until 1977 that Aboriginal ownership was properly recognized, in the form of land rights in reserves under government control in the Northern Territory. Aboriginal cultural revitalization has taken other directions. In music, a number of Aboriginal bands are blending traditional and contemporary indigenous music to produce a striking mix of Aboriginal themes and modern rock.

In some Australian cities, enclaves of Aboriginals have begun to assert their Aboriginal identity and to attack both the history of oppression and continuing injustices. In 1988, 15,000 Aboriginals marched in Sydney to protest against the celebratory nature of the bicentennial activities. The powerful image presented by their simple black and orange flag with its central full sun was a striking symbol of the Aboriginal cultural resurgence, and of the ancient continent of Australia itself.

GLOSSARY

Aborigine
INDIGENOUS PEOPLES of a region, commonly used to describe the original inhabitants of Australia.
African-American
A person of African ancestry born in the United States of America.
AIDS
Acquired Immune Deficiency Syndrome, a fatal disease caused by the Human Immunodeficiency Virus (HIV) which damages a human's immune system and exposes it to rare infections; first recognised in 1981.
Altaic
A sub-family of related languages including TURKIC, Mongolian and Tungusic groups, spoken in parts of eastern Europe, the Soviet Union and Asia.
alternative culture
A culture or lifestyle characterized by the rejection of mainstream or conventional cultural values, for example hippies and GAYS; also called counter-culture.
Amerindian
INDIGENOUS PEOPLES of the Americas, particularly Central and South America.
ancestor worship
The religious practice of worshiping the spirits of dead ancestors who are believed to be able to intercede in human affairs.
Anglican
A PROTESTANT CHURCH founded in England in the 16th century, which includes the Church of England and other churches throughout the world in communion with it.
animism
RELIGIONS that believe that spirits inhabit natural phenomena such as the sky, sun, water and animals.
anticlericalism
Opposition to the CHURCH or the clergy, usually objecting to their influence over civic life or government.
antisemitism
Hostility to JEWS, or more rarely, to other semitic peoples such as Arabs.
apartheid
An organization of society that keeps the races apart. It is generally used in the case of South Africa, where it was introduced by the National Party after 1948 as a means of ensuring continued white political dominance.
Aryan
A people from Central Asia who spread to Mesopotamia, Asia Minor and India in the second millennium BC; INDO-EUROPEAN languages originate from them.
Ashkenazy
A JEW from eastern and central Europe; the Ashkenazim originally possessed distinct traditions and rituals that distinguished them from SEPHARDIC Jews, and they developed the YIDDISH language.
assimilation
A process by which a minority group or people reduces or completely loses its identifying cultural characteristics and blends into the dominant group.
atheist
A person who does not believe in the existence of GOD or gods.
Austro-Asiatic
A language family spoken in southeast Asia, including Vietnamese and Khmer, among others.
Austronesian
A language family spoken in Southeast Asia and the countries of the Pacific Basin; it includes Hawaian, Balinese and Tagalog, the main language of the Philippines.
Aymara
An INDIGENOUS PEOPLE of Peru and Bolivia, and their language.

Bantu
A sub-family of languages spoken by peoples of central and southern Africa; the term is also used to refer to the peoples themselves; they originated in present-day Cameroon and spread across sub-Saharan Africa. In South Africa it was a blanket description for all Black Africans.
baptism
A religious initiation ceremony involving ritual purification by immersing a person in water or sprinkling water on them; it marks admission to the CHURCH, and among CHRISTIAN churches may be performed on infants or, in some SECTS, adults.
bar mitzvah
Among JEWS, an initiation ceremony marking the passage from childhood to adulthood undergone by boys at 13 years of age.
Berber
An INDIGENOUS PEOPLE of North Africa and their language; many live in Algeria and Morocco.
Bible
The book of scriptures of CHRISTIANITY and JUDAISM. The Jewish Bible contains many books in common with the Christian version describing historical events and prophetic teachings, but the latter also includes accounts of the life and teachings of Jesus Christ.
Bishop
A clergyman within the CHRISTIAN CHURCHES, ranking above a PRIEST.
bourgeois
Name given to the 19th-century middle classes to distinguish them from hereditary landowners on the one hand and waged laborers on the other.
Brahmin
A member of the highest or priestly CASTE of HINDUISM.
bridewealth
A cultural practice in which the husband's kin makes a payment to the wife's kin to compensate them for the loss of her labor after marriage.
Buddhism
A RELIGION founded in the 6th and 5th centuries BC and based on the teachings of Siddartha Gautama; it is widely observed in south and Southeast Asia and contains a number of distinct traditions such as MAHAYANA and THERAVADA Buddhism.
Byzantine empire
The Christian empire that succeeded the ROMAN EMPIRE in the eastern Mediterranean, centered on Byzantium, now Constantinople. It existed between 395 and 1453.

capitalism
A system of economic organization, based upon private ownership of property, the production of commodities for sale in unregulated markets and the use of waged labor in their production, to achieve maximum profits.
cargo cult
A movement found in 20th-century Melanesia. Followers believe that their ritual observance will result in the arrival of Western consumer goods, or cargo, the overturning of the colonial social order, and the end of the world.
carnival
A traditional CHRISTIAN festival that takes place before the period of abstinence known as Lent; also any period of public revelry involving street parades and masquerades, particularly as held in Caribbean communities throughout the world.
caste
A hierarchically ranked social order based on closed social categories or castes. An individual is born into the caste of his or her parents, cannot leave it and must marry within it. In HINDUISM, caste or *jati* is based on ideas of purity and pollution.
Caucasian
A racial classification based on white or light skin color; inhabitants of the Caucasus region and their language, part of the INDO-EUROPEAN family.
Celt
Member of an INDO-EUROPEAN iron-working group of peoples originating in southwest Germany and dominating large parts of Europe in the 5th and 6th centuries BC; speakers of Celtic languages such as Breton, Gaelic and Welsh are sometimes described as Celts in the present day.
Chicano
A person of Mexican ancestry born in the United States of America.
Christianity
A RELIGION based on the teachings of Jesus Christ and originating in the 1st century AD from JUDAISM. Its main beliefs are found in the BIBLE and it is now the world's largest religion, divided into a number of CHURCHES and SECTS, including ROMAN CATHOLICISM, PROTESTANTISM and ORTHODOX churches.
church
A public place of worship in the CHRISTIAN RELIGION; an organized body of Christianity and other religions sharing doctrine, administration and clergy.
circumcision
A cultural and religious practice of cutting the male or female genitalia; the circumcision ceremony is a key rite of passage to adulthood in many societies.
clan
A LINEAGE in which the original ancestors or the links with them are either unknown or, sometimes, non-human, for example animals; a basic form of political organization.
class
A group of people sharing a common economic position, for example large landowners, wage-laborers or small business-owners.
colonialism
The control of a foreign territory by a state or people for purposes of settlement and economic expansion.
commune
A group of individuals sharing property, goods and dwellings, often motivated by religious or UTOPIAN ideals and governing themselves.
communion
In CHRISTIANITY, a religious rite in which believers join together to take bread and wine in celebration of Jesus Christ; also refers to the fellowship of the Christian community.
communism
A social system based on the common ownership of all property, individual freedom and the absence of a state.
Confucianism
A RELIGION or moral code based on the teachings of the Chinese philosopher Confucius (c551–479 BC) that formed the foundations of Chinese imperial administration and ethical behavior; also followed in Korea and other east Asian countries.
convert
A person who changes from one belief, usually religious, to another.
creole
A PIDGIN language that has become more sophisticated and serves as a MOTHER TONGUE; also the cultural mixing of Africans and Europeans in the Caribbean and bordering regions.
culture
The beliefs, customs and social relations of a people; the system of meanings and interpretations of the world made by a people.
Cultural Revolution
A radical attempt to continue the revolution in China from 1966 to 1976, involving the purging of the Communist party and instigated by its Chairman, Mao Zedong.

democracy
A form of decision-making in which policy is made by all the people affected by the decisions; usually refers to elected and accountable governments.
denomination
A religious SECT or CHURCH.
dialect
A distinct form of a language, often with unique elements of vocabulary and pronunciation, usually based on a region or class.
Dravidian
A family of languages spoken in south India and Sri Lanka, including Tamil and Telugu; also the peoples speaking them.
Druze
A SECT of SUNNI ISLAM founded in 1021 by the followers of Caliph al-Hakim bi-Amr Allah and practiced in parts of Lebanon and Syria.
dowry
A cultural practice in which the wife's kin makes a payment to the husband's kin as part of her inheritance upon marriage.
dynasty
A hereditary line of rulers.

Easter
A CHRISTIAN festival marking the resurrection of Jesus Christ and observed on a Sunday in March or April each year.

ecological movement
A broad-based political movement and IDEOLOGY concerned with halting environmental despoilation and creating a society that sustains itself without excessive damage to the natural environment; it includes political parties and pressure groups.
endogamy
Marriage in which both partners come from the same culturally defined group.
Episcopalian
A PROTESTANT CHURCH, part of the ANGLICAN communion, found mainly in the United States of America and Scotland; also refers to any Protestant church recognising the authority of bishops, as distinct from PRESBYTERIANISM for example.
ethics
Ideas concerning issues of morality and proper conduct; the study of such ideas.
ethnicity
A cultural and social identity based upon shared language, religion, customs and/or common descent or kinship; defines an ethnic group.
evangelism
A movement within a CHURCH (usually CHRISTIAN) that preaches personal conversion and attempts to gain CONVERTS.
exchange ceremony
A culturally significant occasion in which a transaction takes place, for example a marriage ceremony or the exchange of gifts between two social groups.
extended family
A social unit consisting of two or more parent-child groupings and two or more generations.

federalism
A form of constitutional government in which power is shared between two levels – a central or federal government and a tier of provincial or state governments.
feminism
A political and cultural movement and IDEOLOGY advocating women's rights and equality between women and men.
fengshui
A Chinese ANIMIST belief that powerful spirits of ancestors, dragons and tigers occupy natural phenomena and should not be disturbed. Hence special geomancers are required to understand the landscape and choose appropriate sites for such things as burial mounds.
fetish
An inanimate object thought to possess spiritual powers and used to influence the behavior of others.
feudal
A hierarchical form of society based upon landowners collecting dues from agricultural producers or PEASANTS in return for military protection.
folk culture
The CULTURE of nonurban, nonindustrial peoples, commonly PEASANTS, often used by NATIONALISTS as representing the authentic expression of a people.
folklore
A body of customs, stories, legends and superstitions of a FOLK CULTURE, usually an oral rather than a written tradition.
Franks
A group of GERMANIC peoples inhabiting the Rhine region in the 3rd century who overthrew the ROMAN EMPIRE and in 800, under Charlemagne, founded the Holy Roman Empire.
fundamentalism
The religious belief in the infallibility or inerrancy of scriptures, usually stressing traditional interpretations and opposed to modern ideas or LIBERALISM.

gay
Popular and preferred term for homosexuals, usually male, and their cultural lifestyle.
Germanic
A sub-family of related languages, part of the INDO-EUROPEAN family, including German, Dutch, YIDDISH and Afikaans, among others.
ghetto
A part of a city, commonly a slum, in which all or most of the residents belong to a disadvantaged ethnic or racial minority; originally ghettoes were areas set aside for JEWS in medieval European cities, and they may still be the result of enforced or unwanted segregation.
god
A superhuman being or deity worshiped and believed to have powers over both nature and humanity.
Greco-Roman
Relating to ancient Greek and Roman civilizations; often used to describe their distinctive architectural or artistic legacies.
Gypsy
A member of a nomadic ethnic group, thought to originate in India and now found worldwide, though concentrated in Spain and eastern Europe; their estimated number is 5 million.

hajj
The pilgrimage to Mecca in Saudi Arabia by MUSLIMS that takes place once a year during the 12th month of the lunar calendar.
Han
A Chinese ethnic group, founders of the Han DYNASTY that ruled imperial China between 202 BC and 220 AD.
Hebrew
Another term for JEWS, and also the official language of Israel.
Hinduism
A RELIGION or SYNCRETISM of religions originating in India in the second millennium BC and still the dominant religion of south Asia. Its beliefs and practices are based on the VEDAS and other scriptures.
Holi
A spring festival celebrated by Hindus.
homophobia
Hostility toward GAYS.
hunter–gatherer
An individual whose way of life is based on gathering naturally available plants and occasional hunting.

ideology
A coherent system of ideas; the ideas of a CLASS or other social group with distinct interests, often claimed to be universal or natural.
imam
In SUNNI ISLAM the leader of prayers in a MOSQUE. In SHI'A ISLAM the hereditary successors of 'Ali and thus the legitimate leaders of Islam; SECTS differ as to whether there were 7 or 12 imams.
imperialism
The process whereby one country forces its rule on another country, usually with the aim of economic exploitation; it usually refers to a period of European domination from the late 19th century to the mid 20th century.
Inca
An empire of South America based in Peru, lasting from 1200 to 1537, that developed sophisticated irrigation and transportation systems; it was overthrown by Spanish conquerors.
indentured laborer
A worker hired according to fixed contractual conditions, usually for a given period and on such terms that it is impossible to leave employment easily; in practice it is virtually slavery.
indigenous peoples
The original inhabitants of a region, usually referring to precapitalist or traditional peoples.
Indo-European languages
The largest family of languages, including most of those spoken in Europe and south Asia.
Inuit
INDIGENOUS PEOPLES of the Arctic and tundra regions of North America, Greenland and Siberia.
Islam
A RELIGION based on the revelations of God to the prophet Muhammed in the 7th century AD, which are contained in the QU'RAN; practiced throughout North Africa, the Middle East and parts of Southeast Asia.

Jainism
A RELIGION established in the 6th century BC in opposition to HINDUISM; it is still observed by over 2 million people in India.
Jew
An adherent of JUDAISM, and/or a descendant of the ancient HEBREWS; there are 13 million Jews worldwide, mostly in the United States of America and Israel.
Judaism
A RELIGION founded in 2000 BC among the HEBREWS and practiced by JEWS; it is MONOTHEISTIC and its main beliefs as revealed to prophets such as Moses are contained in the BIBLE.

kinship
A system of rights and obligations among a group culturally defined as related to one another by birth or marriage; also the social network defined by these relations.

lama
A monk in BUDDHISM.
Latin
A language of the Italic sub-family of the INDO-EUROPEAN family, from which all ROMANCE languages are derived; no longer spoken as a MOTHER TONGUE, but the main language of medieval European scholarship and the official language of ROMAN CATHOLICISM until the mid 20th century.
Latino
A person of Central or South American ancestry born or living in the United States of America, usually, but not necessarily, also speaking Spanish and practicing ROMAN CATHOLICISM.
liberalism
A political IDEOLOGY that believes in the greatest practical autonomy of the individual and tolerance of the beliefs and lifestyles of others.
liberation theology
A movement within the CHRISTIAN CHURCH holding that Christ's teachings can and should be used to champion the causes of social justice and end poverty; usually refers to a movement within the ROMAN CATHOLIC church of Latin America.
lineage
A social group that defines itself by one or more known and shared ancestors, and which forms a basic unit of social and economic organization.
lingua franca
A language used, often in trading situations, as a means of communication between people whose MOTHER TONGUES are mutually intelligible, for example, SWAHILI.
Longhouse religion
A NATIVE-AMERICAN RELIGION founded in 1800 by a young warrior named Handsome Lake; it combines traditional and CHRISTIAN ideas and is still practiced among the Iroquois peoples.
Lutheranism
A PROTESTANT CHURCH founded on the principles of the German priest Martin Luther (1483–1546), with adherents mainly in Germany, Scandinavia and the United States of America.

Magyar
A NOMADIC people originating in the Caucasus region around 460 AD, from where they conquered Hungary and adjacent regions in the 9th century; also refers to their language.
Mahayana Buddhism
A SECT of BUDDHISM common in Nepal, Tibet, China, Korea and Japan; it holds that humans can seek the aid of spiritual beings.
Mandarin
A language of the Sino-Tibetan sub-family, spoken in China; it is the world's most spoken language.
Maori
INDIGENOUS PEOPLES of New Zealand, of Polynesian descent.
Marxism
The system of thought derived from the 19th-century political theorist Karl Marx, in which politics is interpreted as a struggle between economic CLASSES. It promotes communal ownership of property when it is practiced, so is commonly termed COMMUNISM.
matriarchy
A social organization in which women head families and, by implication, make important political decisions; common in parts of Africa and some NATIVE-AMERICAN peoples.
matrilineal descent
The cultural practice of tracing descent on the mother's side of the family through the female line; common in small-scale agricultural societies, for example in Africa.

matrilocal
The cultural practice of a newly married couple taking up residence with the bride's kin.
mestizo
Peoples of mixed AMERINDIAN and Spanish or Portuguese descent in South America.
Methodism
A PROTESTANT CHURCH based on the EVANGELICAL teachings of the English theologian, John Wesley (1703–91); it has 20 million adherents, mainly in the British Isles and the United States of America.
Métis
Peoples of mixed NATIVE-AMERICAN and European descent in Canada.
migration
The movement of people from one place to another with the intention of staying in the region of destination, usually to find work.
missionary
An individual who travels to another region or country with the aim of converting the inhabitants to his or her RELIGION; particularly of the CHRISTIAN church.
Mogul
A dynastic empire in India, 1526–1857, founded by invaders from Afghanistan with MONGOL ancestry; during the Mogul empire Indo-Islamic art and architecture flourished.
Mongol
A NOMADIC peoples from central Asia who conquered huge areas of the continent in the 13th century, from China to Europe; today they number 3 million and inhabit parts of China, the Soviet Republics and Mongolia.
monogamy
The cultural practice of having only one marriage partner at a time.
monotheism
The belief in and worship of a single GOD or deity.
Moor
A NOMADIC people from North Africa who conquered Spain in the 8th century, and created an ISLAMIC civilization based in Cordoba before their defeat and expulsion in 1492.
mosque
A public place of worship for MUSLIMS.
mother tongue
The original or native language of a person or people.
multiculturalism
The policy of and belief in the equal treatment of cultural groups in society, involving the right to retain distinct cultural or ethnic identities, for example language. It is usually contrasted with the ideal of assimilation, or the creation of a single, shared national culture.
Muslim
An adherent of ISLAM.

nation
A group or community of persons who believe they consist of a single people, based upon historical and cultural criteria and sharing a common territory. Sometimes used interchangeably with state.
nationalism
An IDEOLOGY that assumes all NATIONS should have their own state, a nation-state, in their own territory, the national homeland.
Native-American
INDIGENOUS PEOPLES of North America.
Nazi
Member of the National Socialist German Workers' Party, founded by Adolf Hitler in the 1920s; also popularly used to describe a member of any far-right political party, particularly those that preach RACISM and ANTISEMITISM.
neolithic
A period of human cultural development marked by the use of stone tools, domestication of animals, cultivation of grains and the setting up of permanent villages, which emerged within the last 10,000 years.
neolocal
The cultural practice of a newly married couple setting up their own residence separately from either the man's or the woman's kin.
New Age
A general term for a loose collection of beliefs, quasi-religions and spiritual practices that involve mind and body enhancement and forms of therapy, common in Western countries in the late 20th century.
Nilo-Saharan
A family of languages spoken in east Africa that include the Nilotic languages comprising Dinka, Nuer and Masai among others.
nomad
A person whose way of life, usually PASTORAL, is based on constant movement in order to exploit geographical variations in the natural environment.
nuclear family
A family based on two parents and their children.

official languages
Languages used by governments, schools, courts and other official institutions in countries where the population has no single common MOTHER TONGUE.
Orthodox Church
A group of CHRISTIAN CHURCHES that took the Patriarch of Constantinople as their head, and broke away from the ROMAN CATHOLIC church in the 9th century: includes the national churches of Greece, Russia, Armenia, Romania and Ethiopia among others.
Ottoman empire
Empire founded in the 13th century by Ottoman Turks. It controlled the area where Asia, Europe and Africa meet; it fragmented at the end of World War I. Turkey is its modern remnant.

Paleo-Siberian
A group of languages spoken in parts of Russia.
pastoralism
A way of life based on tending of herding animals such as sheep, cattle, goats or camels; often NOMADIC, it involves moving herds according to the natural availability of pasture and water.
patois
Another term for a distinct DIALECT, commonly used to describe dialects of formerly colonized countries, for example in the Caribbean.
patriarchy
A social organization in which males head families and make the important political decisions; the system of ideas and practices by which men dominate women.
patrilineal descent
The cultural practice of tracing descent on the father's side of the family through the male line; widespread among pastoral NOMADS, for example.
patrilocal
The cultural practice of a newly married couple taking up residence with the male's kin.
peasant
An individual whose way of life is based on small-scale agriculture, usually involving both cultivation and livestock rearing.
Phoenician empire
Trading empire of the 2nd millennium BC located in present-day Lebanon, but also including settlements throughout the Mediterranean region such as Carthage (present-day Tunis).
philosophy
Any system of ideas or beliefs concerned with the discovery of fundamental truths about such things as human existence and nature; the systematic study of such ideas.
pidgin
A simplified language, often a LINGUA FRANCA, allowing for basic communication between persons whose MOTHER TONGUES are mutually unintelligible; usually based on a European language and developed in former colonized countries.
plural society
A society created under colonial conditions and consisting of several distinct cultural groups whose religious, linguistic or other differences are such that the society may be thought to lack cohesion or national solidarity.
polygamy
The culturally sanctioned practice of having more than one marriage partner at a time.
polytheism
The belief in and worship of many GODS or many manifestations of the same god.
Pope
The usual title of the BISHOP of Rome, who is the spiritual head of the ROMAN CATHOLIC CHURCH.
Presbyterianism
A PROTESTANT CHURCH in which lay members of the congregation are involved in the church's administration.
priest
An official within a CHURCH or RELIGION who conducts religious ceremonies, possesses special knowledge and/or intercedes with GOD or gods on behalf of adherents.
Protestant
Term describing a number of CHURCHES that share a common rejection of the authority of the POPE and of ROMAN CATHOLICISM; originated in the 16th century European Reformation.
purdah
The cultural practice of secluding women from all men other than their closest relatives, for example by confining them to certain areas of a dwelling or requiring them to wear veils in public; common among adherents of ISLAM and HINDUISM and commonly practiced in Pakistan, Saudi Arabia and Iraq.

Quechua
A language spoken in South America and the people who speak it; common in Peru and Bolivia. It was the language of the INCA rulers.
Qu'ran
The sacred text of ISLAM that contains the revelations made by God to the prophet Muhammed; it contains the legislative, theological and ethical principles of Islam.

racism
The IDEOLOGY or belief that persons can be classified into races, that racial type explains social or cultural behavior, and that races can be ranked in terms of superiority and inferiority; also used generally to describe any hostility toward a person or group of a different physical appearance.
Ramadan
A period of abstinence observed by MUSLIMS taking place in the ninth month of the lunar year and corresponding to the time in which Muhammad first received God's revelations.
Rastafarianism
A RELIGION founded in Jamaica in the mid 20th century, based on a reinterpretation of the BIBLE that claims that Black Africans are the lost tribe of Israel, that Ethiopia is their spiritual home and that the Ethiopian Emperor Haile Selassie was their spiritual leader. Adherents are found in Caribbean communities throughout the world.
religion
A system of beliefs, practices and related institutions founded on the belief in and worship of one or more GODS or spirits; it may also include prescriptions and practices addressing ETHICS, government and personal conduct, as well as accounts of the origin and nature of the world itself.
reservation/reserve
An area created by a colonial power for INDIGENOUS PEOPLES as a means of removing them from their lands and subjugating them.
rite
A ceremony or occasion of special cultural significance, for example religious worship or a person's transition between stages of a life cycle; it is usually conducted according to prescribed rules of procedure.
Roman Catholic
A CHRISTIAN CHURCH headed by the POPE that traces its origin and authority to St Peter, one of the disciples of Jesus Christ and the first BISHOP of Rome; it is the largest of the Christian churches.
Romance language
A sub-family of related languages, part of the INDO-EUROPEAN family, all of which are derived from LATIN; includes Italian, French, Portuguese and Spanish.
Roman empire
An empire founded in the first century BC from the Roman Republic, which began around 500 BC in present-day Italy; at its height it controlled the Mediterranean, large parts of western Europe and the Middle East. It was conquered in the 5th century and divided, the eastern half becoming the BYZANTINE EMPIRE.

Romany
Another name for GYPSY; also refers to their language, belonging to the INDO-EUROPEAN family.

Sanskrit
A language of the INDO-EUROPEAN family, now extinct, originating in ancient India; the VEDAS were originally written in Sanskrit.
sect
A religious body sharing doctrine and/or organization, usually separate or dissenting from an established CHURCH.
sectarianism
The division of society into hostile and competing factions or SECTS.
secularization
The process of the declining significance of RELIGION in society, as revealed for example, by falling numbers of adherents or a weakening of the CHURCH'S influence over government or civic life.
separatism
A political movement in a state that supports the secession of a particular minority group, within a defined territory, from that state.
Sephardic
A JEW of Spanish or Portuguese descent or origin, whose religious traditions and prayer rites differ from the ASHKENAZIM.
shaman
A PRIEST, generally one who uses hallucinogenic or drug-induced states to communicate with the GODS or spirits.
Shari'a
The code of behavior of SUNNI ISLAM as derived from the QU'RAN and the recorded teachings of Muhammad. It is divided into four schools of law, and interpreted by the *ulama* or learned scholars.
Shi'a
The smaller of the two main divisions of ISLAM. Followers recognize Muhammad's son-in-law, 'Ali, and his descendants, the IMAMS, as his true successor and legitimate leaders of Islam. It has a number of SECTS, including the Isma'ilis, DRUZE and so-called Twelver Shi'ites and is prominent in Iraq and Iran.
shifting cultivation
A way of life based upon cultivating plants on land, usually forest, that is periodically cleared and then returned to fallow; also known as swidden.
Shintoism
A POLYTHEISTIC RELIGION, SYNCRETIC with BUDDHISM and practiced in Japan; it was the official state religion from the 19th century to World War II.
Sikhism
A RELIGION established by Guru Nanak in the 15th century in the Punjab, in the northwest Indian subcontinent, that opposed HINDUISM; it is now observed by the worldwide community of Sikhs, many of whom are fighting for their own autonomous homeland in the Punjab.
Sino-Tibetan
A language family spoken in central and Southeast Asia, including Chinese, Thai, Tibetan and Burmese.

Slav
A European group of peoples originally from the area of the Carpathians in Eastern Europe. In the present day Slavs include a number of distinct ethnic groups; West Slavs (Poles, Czechs and Slovaks); East Slavs (Russians, Ukrainians and Belorussians); and South Slavs (Serbs, Croats, Slovenes, Macedonians and Bulgars).
socialism
An economic system and a political IDEOLOGY based upon the principles of equality between people, the redistribution of wealth and property and equal access to such things as health care and education.
subculture
A distinct culture within a society, usually youth-based.
Sunni
Larger of the two main divisions of ISLAM. Its followers recognize the Caliphs as the successors to Muhammad and follow the *sunna* or way of the prophet as recorded in the *hadith*, the teachings of Muhammad.
Swahili
A LINGUA FRANCA widely spoken in east Africa made up of BANTU and Arabic linguistic elements.
synagogue
A public place of worship for JUDAISM.
syncretism
The mixing of beliefs and practices from two or more distinct RELIGIONS, often involving the transfer of identity of one set of GODS or spirits to another.

taboo
A cultural sanction or culturally forbidden practice or belief.
Tantric Buddhism
A tradition of BUDDHISM that places emphasis on meditation.
Taoism
A Chinese philosophy and RELIGION usually ascribed to the teachings of Lao Zi (7th century BC) and later SYNCRETIC with MAHAYANA BUDDHISM; banned by the Chinese communist government but widely practiced in Taiwan.
Theravada Buddhism
The earliest SECT of BUDDHISM, originating in Sri Lanka. It believes that humans must look to themselves and not to spirits for their salvation, and is widely practiced in Southeast Asia.
Torah
Name given to the first five books of the Jewish BIBLE that contain the written law as revealed by God to the prophets, notably Moses; also refers to all the moral, theological and ritual principles of JUDAISM.
tradition
Any custom, practice or belief passed on from one generation to another of a distinct people or CULTURE over a long or unknown and uninterrupted period of time; it has the implication of being an authentic expression of a people,though many traditions are in fact modern reinventions of past practices and customs.
tribalism
The identification with a TRIBE and the pursuit of its interests and aims, usually in contrast to the wider aims and interests of a NATION.
tribe
A group of people united by a common language, religion, customs and/or descent and KINSHIP; often used to describe the social groups of peoples who have no developed state or government and whose social organization is based on CLANS.
Turkic
A group of related languages, part of the ALTAIC family, including Uzbek, Azeri and Kazakh, and spoken in central Asia.

untouchable
A member of the lowest CASTE of HINDUISM, usually considered as being so polluted as to be below and outside the caste hierarchy; also called *harijans*.
Uralic
A sub-family of related languages, including Finnish, Hungarian and Lapp, spoken in parts of Europe and Russia.
utopia
An ideal society, especially in its laws, government and social conditions, that serves as a model for a group sharing the same political and religious views.

Vedas
The oldest scriptures of HINDUISM. They consist of the Samhita (concerning prayer, hymns and spells), the Brahmana (the principles of sacrifice), the Aranyakas (dealing with meditation), and the Upanisads (writings on mysticism).
Viking
A sea-faring warrior people of Scandinavia who, between the 9th and 11th centuries, conquered much of the Atlantic coast of Europe and settled Greenland.
voodoo
A SYNCRETIC RELIGION of West African beliefs and ROMAN CATHOLICISM practiced in parts of the Caribbean, notably Haiti.

welfare state
A social and economic system based on state provision of, and responsibility for, such things as health care, pensions and unemployment benefit and financed by general contributions from the working population; access to these services was intended to be equally available to all and free from charge. It originated in Britain at the start of the 20th century and became widespread in Europe after World War II.

Yiddish
A language originating among the ASHKENAZY JEWS of Central and Eastern Europe that contains words from German, Hebrew and other languages.

Zoroasterianism
A RELIGION founded by the Persian prophet Zoroaster (c628–c551 BC), based on the idea of a struggle between good and evil spirits. Although almost completely swept away by ISLAM, it is still practiced by the Parsis of India, and some Iranians.

Further reading

Anderson, Benedict, *Imagined Communities: Reflections on the Origin and Spread of Nationalism* 2nd edition (Verso, London, 1991)
Hobsbawm, Eric, and Ranger, Terence (editors), *The Invention of Tradition* (Cambridge University Press, Cambridge, UK, 1983)
Jackson, Peter, *Maps of Meaning: an Introduction to Cultural Geography* (Unwin Hyman, London, 1989)
Jordan, Terry G., and Rowntree, Lester, *The Human Mosaic: a Thematic Introduction to Cultural Geography* 4th edition (Harper & Row, Cambridge, Mass., 1986)
Keesing, Roger M., *Cultural Anthropology in Perspective* 2nd edition (Holt, Rinehart & Wilson, Fort Worth, Texas, 1981)
Lewis, I. M., *Social Anthropology in Perspective* 2nd edition (Cambridge University Press, Cambridge, UK, 1985)
Moore, Henrietta L., *Feminism and Anthropology* (Polity Press, Cambridge, UK, 1988)
The New Encyclopedia Britannica, 15th edition (Encyclopedia Britannica, Chicago, Illinois, 1989)
Reader, John, *Man on Earth* (Penguin Books, Harmondsworth, Middlesex, 1988)
Smith, Anthony D., *The Ethnic Origin of Nations* (Basil Blackwell, Oxford, 1986)
Wolf, Eric R., *Europe and the People without History* (University of California Press, Berkeley, California, 1982)

Acknowledgments

Picture credits
Key to abbreviations: AAA Ancient Art & Architecture Collection, Middlesex, UK; **AGE** AGE Fotostock, Barcelona, Spain; **APA** Andes Press Agency, London, UK; **Asp** Allsport, London, UK; **BAL** Bridgeman Art Library, London, UK; **BFI** British Film Institute Stills, Posters and Designs, London, UK; **C** Collections, London, UK; **CPL** Cephas Picture Library, Surrey, UK; **E** Explorer, Paris, France; **EU** Eye Ubiquitous, East Sussex, UK; **FSP** Frank Spooner Pictures, London, UK; **HDC** Hulton Deutsch Collection, London, UK; **HL** Hutchison Library, London, UK; **IB** The Image Bank, London, UK; **IF** Image Finders, Vancouver, Canada; **IP** Impact Photos, London, UK; **JSLI** JS Library International, London, UK; **M** Magnum Photos Limited, London, UK; **N** Network Photos, London, UK; **PNWA** Peter Newark's Western Americana, Bath, UK; **PP** Panos Pictures, London, UK; **Psp** Picturesport Associates Limited, London, UK; **Rf** Redferns, London, UK; **RHPL** Robert Harding Picture Library, London, UK; **SAP** South American Pictures, Suffolk, UK; **SGA** Susan Griggs Agency, London, UK; **SPP** Scandia Photopress, Sweden; **SRG** Sally & Richard Greenhill, London, UK; **Z** Zefa Picture Library, London, UK.

t=top, b=botton, l=left, r=right.

6–7 HL **8–9** E/Jacques Brun **10** HL **10–11** SGA/Gra Momotiuk/John Eastcott **11** HL/Sarah Errington **12–13** RHPL **13** HDC/Bettmann Archive **14** AAA/Ronald Sheridan **14–15** E/Nicolas Thibault **16–17** FSP/Patrick Piel **17t** HDC/Bettmann Archive **17b** FSP/Roger Bull **18–19t** IB/Stockphotos Inc/Marks Production **18–19b** IB/Guido Alberto Rossi **20t** E/Guy Philippart de Foy **20b** Christine Osborne **21** M/Ian Berry **22** E/François Varin **22–23** M/Ian Berry **23** Christine Osborne **24** M/Abbas **25t** Christine Osborne **25b** E/Gérard Boutin **28** AGE **28–29** AGE/Richard Nowitz **30** Christine Osborne **30–31** Christine Osborne **32** M/Stuart Franklin **32–33** M/Bruce Davidson **33** RHPL/Walter Rawlings **34–35** HL/Lesley McIntyre **34** C/Brian Shuel **35** M/Bruno Barbey **36–37** M/Burt Glinn **37** M/Steve McCurry **38–39** RHPL/Robin Hanbury-Tenison **39** M/Paul Fusco **40** T M J Fisher **40–41** FSP/Gamma/Pool Mondiale **45tl** AGE **45bl** E/Marc Moisnard **45r** M/Gilles Peress **46–47** AGE **47** Asp **49** Psp **50–51** RHPL **51t** FSP/Pono Press **51b** Brian & Cherry Alexander **52–53** IF/Chris Speedie **55t** IP/Michael Mirecki **55b** SGA/Nathan Benn **56** AGE **56–57** M/Steve McCurry **58** M/Ernst Haas **58–59** M/Eve Arnold **59** Asp/Nathan Bilow **60tPNWA** **60b** PNWA **60–61** SGA/Ted Spiegel **62–63** M/Paul Fusco **64** M/Harry Gruyaert **64–65** M/Rene Burri **66** M/Peter Marlow **66–67** N/Pillitz **67** M/Alex Webb **68** M/David Hurn **68–69** M/Dennis Stock **70** SAP/Hilary Bradt **70–71** RHPL **72–73** M/Abbas **73** M/Burt Glinn **74** EU/David Cumming **74–75** EU/David Cumming **76t** SGA/Adam Woolfitt **76b** SGA/Adam Woolfitt **77** SGA/Adam Woolfitt **78–79** SGA/Adam Woolfitt **80–81** SAP/Kimball Morrison **82–83** SAP/Tony Morrison **83** Asp/Billy Stockland **84t** SAP/Tony Morrison **84b** Ken Heyman **85** RHPL **86–87** SAP/Tony Morrison **89** Z/Everts **90** SGA/Tor Eigeland **91** SPP/Ronny Johannesson **92–93** RHPL **93t** IP/Daniel Reed **93b** SGA/Michael St Maur Sheil **94** AGE **94–95** E/Bernard Maltaverne **87t** SRG **97b** EU/David Cumming **98–99** M/Harry Gruyaert **99t** T M J Fisher **99b** RHPL **100** IP/John Arthur **101** APA/Carlos Reyes **102** CPL/Mick Rock **102–103** RHPL **104–105** M/Jean Gaumy **105** RHPL **106–107** AGE **107t** M/Richard Kalvar **107b** E/B & J Dupont **108** M/G Pinkhassov **110** E/Pierre Tetrel **111** E/Hug **112–113** E/Yann Arthus-Bertrand **114–115** E/Jose Dupont **116l** E/Jacques Joffre **116r** RHPL/Paul Van Riel **117** Anne-Marie Bazalik **118** IP/Jon Lister **118–119** IP/Penny Tweedie **120** M/Stuart Franklin **121** BAL **122–123** FSP/Gamma/Photo News **124–125** E/Charles Lenars **125t** Christine Osborne **126** M/Harry Gruyaert **126–127** EU/David Cumming **128t** Ken Heyman **128b** M/Fred Mayer **130t** M/Thomas Hoepker **130b** SGA/Robert Frerck **130–131** SGA/Robert Frerck **132–133** M/Fred Mayer **134t** M/Fred Mayer **134b** M/F Scianna **134–135** Life File/Terence Waeland **136t** EU/Julia Waterlow **136t** Life File/Patrick Roberts **137** BFI **138** M/Dennis Stock **138–139b** N/Barry Lewis **138–139t** M/Leonard Freed **141l** M/Thomas Hoepker **141r** M/Erich Hartmann **142–143** CP/D Burnett **143t** IP/Rolf Hayo **143b** BAL/National Museum, Stockholm **144–145** M/G Mendel **145** M/Bruno Barbey **146–147** FSP/Gamma/Uli Weyland **147** Life File/Patrick Roberts **148–149** M/Erich Hartmann **150–151** RHPL **152–153** RHPL **153t** M/Steve McCurry **153b** FSP/Art Zamur/Gamma **154** RHPL/Andrew Mills **155** RHPL/Paul Van Riel **156** RHPL/Israel Talby **156–157** RHPL/Israel Talby **158–159** M/Bruno Barbey **161** RHPL **162–163** M/Abbas **163l** FSP/Gamma/Blanche **163r** RHPL/Gascoigne **164–165** AGE **165t** EU/Nick Wiseman **165b** IP/Paul Forster **166t** Private Collection **166b** from "Vladimir Tatlin and the Russian Avant-Garde", Yale University Press **166–167** Life File/Alan Gordon **168–169** M/Abbas **170–171** Christine Osborne **171t** M/Abbas **171b** Christine Osborne **172–173** FSP/Gamma/Jean Michel Turpin **173t** M/Ara Guler **173b** AGE **174** Christine Osborne **175** JSLI **176** Z/Nagold Leidman **177** HL/Sarah Errington **178–179** Edward Parker **178** EU/Thelma Sanders **179** PP/Martin Adler **180** HL/Melanie Friend **180–181** HL/Liba Taylor **182** HL/E Nairn **182–183** HL/Dave Brinnicombe **184–185** HL **186–187** HL/Nick Haslam **187** HL/Sarah Errington **188** AGE **189** HL **190** EU/Thelma Sanders **190–191** AGE **192t** Edward Parker **192b** Rf/Tim Hall **192–193** HL/Bernard Régent **195t** M/Ian Berry **195b** HL/T Beddon **196** IP/David Reed **196–197** IP/Penny Tweedie **198** IP/P Cavendish **198–199** M/G Mendel **200t** HL **200b** IP/P Cavendish **200–201** IP/Penny Tweedie **203l** AGE **203r** E/Gérard Boutin **205t** **RHPL** **205b** CPL/Nigel Blythe **206–207** M/Bruno Barbey **207** RHPL **208** Christine Osborne **208–209** AGE **210–211** RHPL **212** EU/Julia Waterlow **213** PP/Trevor Page **214** EU/Julia Waterlow **214–215** EU/Julia Waterlow **216** PP/Alain le Garsmeur **216–217** EU/Julia Waterlow **218l** PP/Alain le Garsmeur **218r** EU/Julia Waterlow **218–219** Life File/Joan Batten **220–221** Christine Osborne **221t** RHPL **221b** E/René Mattes **222** Edward Parker **223** E/C Lenars **224–225** Adrian Arbib **225** M/Philip Jones-Griffiths **226–227** EU/David Cumming **228–229** Brian & Cherry Alexander **230–231** AGE **231** EU/Frank Leather **232** AGE **233** EU/Frank Leather **234** M/Bruno Barbey **234–235** CPL/Nigel Blythe **235** EU/Frank Leather **236** Martin Barlow **237** CPL/N J Kealey **238–239** M/Burt Glinn **240–241** Ken Heyman **242t** Helene Rogers/TRIP **242b** Life File/Emma Lee **243** Life File/Nicole Sutton **244** Helene Rogers/TRIP **245** HL/C Nairn **246–247** AGE **247** Christine Osborne

Editorial, research and administrative assistance
Nick Allen, Helen Burridge, Carol Busia, Joanna Chisholm, Reina Foster-de Wit, Pamela Mayo, Hilary McGlynn, Eelin Thomas, Steve Vertovec

Cartography
Maps draughted by Euromap, Pangbourne

Index
Barbara James

Production
Clive Sparling

Typesetting
Brian Blackmore, Niki Whale

Color origination
Scantrans pte Ltd, Singapore

INDEX

Page numbers in **bold** refer to extended treatment of topic; in *italic* to illustration or map caption